AF581662

Biocatalytic Applications in Biotechnology

Biocatalytic Applications in Biotechnology

Editors

Emmanuel M. Papamichael
Panagiota-Yiolanda Stergiou

MDPI • Basel • Beijing • Wuhan • Barcelona • Belgrade • Manchester • Tokyo • Cluj • Tianjin

Editors

Emmanuel M. Papamichael
University of Ioannina
Greece

Panagiota-Yiolanda Stergiou
University of Ioannina
Greece

Editorial Office
MDPI
St. Alban-Anlage 66
4052 Basel, Switzerland

This is a reprint of articles from the Special Issue published online in the open access journal *Catalysts* (ISSN 2073-4344) (available at: https://www.mdpi.com/journal/catalysts/special_issues/Biocatalytic_Biotechnology).

For citation purposes, cite each article independently as indicated on the article page online and as indicated below:

LastName, A.A.; LastName, B.B.; LastName, C.C. Article Title. *Journal Name* **Year**, *Volume Number*, Page Range.

ISBN 978-3-0365-6265-0 (Hbk)
ISBN 978-3-0365-6266-7 (PDF)

Contents

About the Editors

Emmanuel M. Papamichael

Prof. Emmanuel M. Papamichael, graduated (Diploma of Chemistry) from Athens' Kapodistrian University in 1972, and received his Doctoral degree (PhD) from University of Ioannina (1981). Currently, he is Professor Emeritus in the Department of Chemistry (University of Ioannina). His research interests comprise: the kinetics of enzymatic and heterogeneous reactions, Investigation of their mechanisms and the development of optimal mathematical models, the isolation and purification of hydrolytic enzymes, enzyme biotechnology, parametric and non-parametric statistical treatment of experimental data, experimental design, error analysis, computer programming. Prof. Papamichael has published 131 original papers (after peer-review) in international journals, international books and international conferences, has received < 1300 citations, and participates as an Editorial Board Member in many ISI Journals. He is also a reviewer in more than 100 international journals, and he was supervisor on < 30 PhD/Master/Diplomas. In the last 5 years he was/is scientific director in three competitive research programs funded by the EU.

Panagiota-Yiolanda Stergiou

Dr. Panagiota-Yiolanda Stergiou, graduated (Diploma of Chemistry) from the University of Ioannina in 2008, and received her Master of Science Diploma (2011) and her Doctoral degree (2017) from University of Ioannina. At present, she is a Postdoctoral Researcher at the Department of Chemistry, University of Ioannina, Greece (Enzyme Biotechnology and Genetic engineering group). Her MSc is in Chemistry and her PhD is in Enzyme Biotechnology, entitled "Analytical and kinetic methods, and Structural studies for the investigation of specific biotechnological applications of free and immobilized proteolytic enzymes". Her research interests are in the areas of enzyme biotechnology (food biotechnology, biofuel), the kinetics and mechanisms of enzymatic and heterogeneous reactions and the experimental design of bioprocesses. She has presented more than 30 papers in international and national conferences, and she has published 20 papers in peer-reviewed scientific journals.

Article

Regioselective Hydroxylation of Rhododendrol by CYP102A1 and Tyrosinase

Chan Mi Park [1], Hyun Seo Park [1], Gun Su Cha [2], Ki Deok Park [3] and Chul-Ho Yun [1,*]

[1] School of Biological Sciences and Technology, Chonnam National University, 77 Yongbongro, Gwangju 61186, Korea; cmpark0710@gmail.com (C.M.P.); hs6471@naver.com (H.S.P.)

[2] Namhae Garlic Research Institute, 2465-8 Namhaedaero, Namhae, Gyeongsangnamdo 52430, Korea; gscha450@gmail.com

[3] Gwangju Center, Korea Basic Science Center, 77 Yongbongro, Gwangju 61186, Korea; kdpark@kbsi.re.kr

* Correspondence: chyun@chonnam.ac.kr

Received: 12 August 2020; Accepted: 24 September 2020; Published: 25 September 2020

Abstract: Rhododendrol (RD) is a naturally occurring phenolic compound found in many plants. Tyrosinase (Ty) converts RD to RD-catechol and subsequently RD-quinone via two-step oxidation reactions, after which RD-melanin forms spontaneously from RD-quinone. RD is cytotoxic in melanocytes and lung cancer cells, but not in keratinocytes and fibroblasts. However, the function of RD metabolites has not been possible to investigate due to the lack of available high purity metabolites. In this study, an enzymatic strategy for RD-catechol production was devised using engineered cytochrome P450 102A1 (CYP102A1) and Ty, and the product was analyzed using high-performance liquid chromatography (HPLC), LC-MS, and NMR spectroscopy. Engineered CYP102A1 regioselectively produced RD-catechol via hydroxylation at the ortho position of RD. Although RD-quinone was subsequently formed by two step oxidation in Ty catalyzed reactions, L-ascorbic acid (LAA) inhibited RD-quinone formation and contributed to regioselective production of RD-catechol. When LAA was present, the productivity of RD-catechol by Ty was 5.3-fold higher than that by engineered CYP102A1. These results indicate that engineered CYP102A1 and Ty can be used as effective biocatalysts to produce hydroxylated products, and Ty is a more cost-effective biocatalyst for industrial applications than engineered CYP102A1.

Keywords: rhododendrol; cytochrome P450; tyrosinase; biocatalyst; hydroxylation; bioconversion

1. Introduction

Biocatalysts are biological systems, such as purified enzymes and whole cells, that can catalyze diverse chemical reactions. Enzymes can accept a wide range of molecules as substrates, and they exhibit high catalytic activities towards simple and complex chemicals. Consequently, they are used as biocatalysts in fields of chemical synthesis, biosensor production, and bioremediation. The biocatalysts can also be used to support several cascade reactions during organic compound synthesis. In particular, they are effective tools in the fine chemical and pharmaceutical industries because they exhibit chemo-, stereo-, and regioselectivity for chemical production in eco-friendly ways [1–5].

Generally, cytochrome P450 (P450 or CYP) enzymes catalyze the monooxygenase reaction (i.e., the insertion of one atom of molecular oxygen into a substrate and the reduction of the other oxygen atom to water) [6]. Required electrons for the reaction are provided from nicotinamide adenine dinucleotide cofactors (NADPH or NADH) via cytochrome P450 reductase (CPR). In addition, P450s catalyze diverse reactions such as C-hydroxylation; N-oxidation; N-, O-, and S-dealkylation; C–C bond cleavage; alcohol oxidation using diverse natural and non-natural compounds as substrates. Accordingly, CYPs have considerable potential as biocatalysts for novel drug development [7–9]. CYP102A1 from *Bacillus megaterium* (BM3) consists of the catalytic heme domain and the reductase domain in a single polypeptide [10]. CYP102A1 is more effective as a biocatalyst because of its high activity and solubility. The engineering of CYP102A1 through site-directed and random mutagenesis is reported to improve their activities toward various marketed drugs, steroids, and non-natural substrates. However, bioconversions via CYP102A1 have drawbacks on an industrial scale because they still require the use of a high-cost cofactor, NADPH [11–15].

Tyrosinase (Ty) is a copper-containing dioxygenase responsible for melanin formation in many organisms. Two copper ions, CuA and CuB, coordinate six conserved histidine residues in the active sites [16,17]. Ty catalyzes the ortho-hydroxylation of monophenols to ortho-diphenols (catechols) via monophenolase activity, as well as the subsequent oxidation of ortho-diphenols to ortho-quinones via diphenolase activity. The highly reactive ortho-quinones convert to melanin spontaneously, and are involved in skin pigmentation and fruit and vegetable browning [18,19]. Ty has a considerable potential to synthesize bioactive catechol derivatives because of regioselective ortho-hydroxylation via monophenolase activity. However, Ty-associated melanin formation should be inhibited for industrial applications of the bioactive catechol derivatives [20,21]. Protein engineering using site-directed and random mutagenesis has been attempted to improve Ty's activity and substrate specificity [22,23]. Furthermore, the inhibition of Ty's diphenolase activity using catechol, NADH, and ascorbic acid was found to enhance the production of catechol derivatives [24–26]. For the industrial use of Ty to produce catechol compounds, the issue of further oxidation of catechol needs to be solved, but Ty seems to be a more cost-effective biocatalyst compared with P450 because it does not require the redox partners and NADPH.

Rhododendrol (RD, 4-(4-hydroxyphenyl)-2-butanol) is a naturally-occurring phenolic compound found in many plants such as *Acer nikoense* and *Betula pubescens* [27–29]. In melanocytes, RD is metabolized to RD-quinone via RD-catechol by a Ty-catalyzed reaction. Subsequent spontaneous reactions convert RD-quinone to melanin [30]. RD metabolites and RD-derived melanin generated by Ty cause oxidative stress and exhibit cytotoxicity in melanocytes [31]. RD is also cytotoxic in human lung cancer cells [32], but not in keratocytes and fibroblasts [33]. However, the bioactive or adverse effects of various RD metabolites are not well established because metabolites with high purity cannot be obtained on a large scale.

The aim of this study was to produce a highly pure RD-catechol product using engineered CYP102A1 and Ty as biocatalysts. We found that engineered CYP102A1 and Ty can catalyze the regioselective hydroxylation at the ortho position of RD to produce an RD-catechol with high selectivity (Figure 1). In addition, we compared the yield, final product concentration, and productivity for RD-catechol production by engineered CYP102A1 and Ty. To the best of our knowledge, this study is the first report of the production of a highly pure RD-catechol using engineered CYP102A1 and Ty.

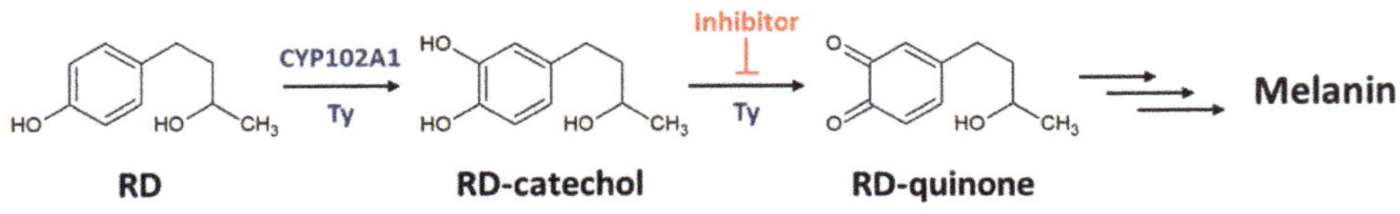

Figure 1. Scheme for rhododendrol (RD)-catechol production using cytochrome (CYP)102A1 and tyrosinase (Ty).

2. Results and Discussion

2.1. Hydroxylation of RD by Engineered CYP102A1

To find a CYP102A1 enzyme with a high activity towards RD, the catalytic activities of the CYP102A1 wild type (WT) and its engineered enzymes toward RD were investigated at a fixed substrate concentration of 200 μM for 30 min of incubation. An NADPH regenerating system (NGS) was added to provide NADPH to the CYP102A1. The formation of RD-derived products by CYP102A1 was analyzed using high-performance liquid chromatography (HPLC). One major product was generated by engineered CYP102A1. The retention times of RD and its major product were 13.4 and 8.4 min, respectively (Figure 2). Among this study's tested enzymes, WT CYP102A1 apparently could not oxidize RD, whereas ten engineered CYP102A1 showed high activity toward RD with more than 200 nmol product/nmol P450. Engineered CYP102A1 #16 (M16) was selected for further experiments because it showed the highest activity with 645 nmol product/nmol P450 among the tested 45 engineered CYP102A1 (Figure S1).

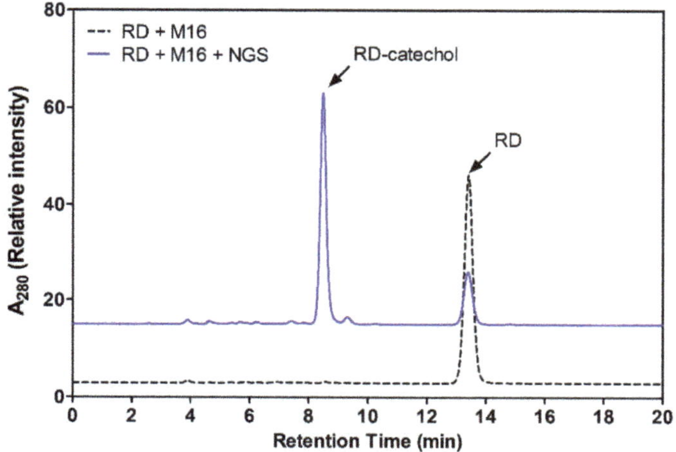

Figure 2. High-performance liquid chromatography (HPLC) chromatogram of RD and RD-catechol produced by engineered CYP102A1. The reaction mixtures contained 0.2 μM M16 or 200 μM RD with or without NADPH regenerating system (NGS). RD and RD-catechol were eluted at 13.4 and 8.4 min, respectively.

M16 might produce RD-catechol or RD-resorcinol via hydroxylation at the ortho or meta positions of the RD's phenol ring. The major RD-derived product was identified using liquid chromatography-mass spectrometry (LC-MS) and nuclear magnetic resonance (NMR) spectroscopy. Although the molecular mass of RD is 166.22 g·mol^{-1}, its mass to charge (*m/z*) value was calculated as 149 via LC-MS [34]. The *m/z* value of the major product was 165, which was 16 higher compared to RD (Figure S2). The ^{1}H and ^{13}C NMR spectra of RD and the product of M16 are shown in Figures S3–S6. Chemical structure and chemical shift assignments for RD and the product of M16 are shown in Figure S7 and Table S1. These results indicate that engineered CYP102A1 regioselectively produces an RD-catechol via hydroxylation at the ortho position.

2.2. RD-Catechol Production by Engineered CYP102A1

To increase the product concentration, M16 catalytic reactions were performed with various RD concentrations. RD-catechol production increased with RD concentration, and the productivity increased up to 200 μM RD (Figure 3). In this condition, product yield, final product concentration, and productivity were 50.4%, 18.4 mg·L^{-1}, and 36.7 mg·L^{-1}·h^{-1}, respectively. The maximum RD-catechol was produced when RD concentration was 200 μM. The RD-catechol yield increased until 100 μM RD, and then decreased. When RD concentration was 100 μM, the yield was highest at 53.8%. These results indicate that M16 catalytic reactions below 200 μM RD to produce RD-catechol is efficient Further purification steps might be needed to obtain highly pure RD-catechol, as the yield of RD-catechol was lower than 90%. Although M16 can catalyze the regioselective hydroxylation of RD, M16 has limitations for industrial-scale applications due to its low yield and demand for

NADPH. Therefore, it is necessary to develop alternative enzymes that can reduce cost and time of RD-catechol production.

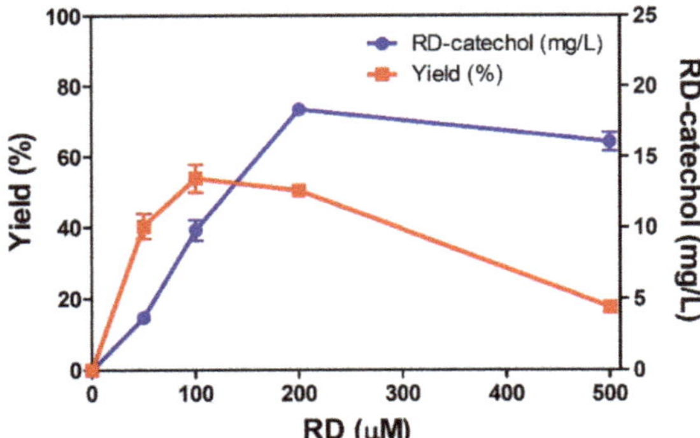

Figure 3. RD-catechol production by engineered CYP102A1 with increasing RD concentration. The reaction mixtures contained 0.2 μM M16 with indicated RD concentration up to 500 μM, and NGS. RD-catechol produced by M16 was analyzed using HPLC.

2.3. *Inhibition of RD-Quinone Formation by Ascorbic Acid Analogs*

To develop an alternative enzyme to overcome the limitations of the CYP102A1 catalysis requiring NADPH, we attempted to produce RD-catechol using Ty as a biocatalyst. Ty produces RD-catechol and RD-quinone from RD by catalyzing two-step oxidations of ortho-hydroxylation and the subsequent oxidation. RD-quinone is spontaneously converted to RD-melanin via a non-enzymatic pathway [30]. The absorbance of RD at 280 nm changed after the Ty-catalyzed reactions. Within 3 min, absorbance near 400 nm increased, and absorbance at 280 and 340 nm also increased over time (Figure S8). This indicate that Ty rapidly produces RD-quinone and secondary quinone from RD [31]. Therefore, secondary oxidation by Ty should be inhibited to selectively produce RD-catechol. It has been reported that L-ascorbic acid (LAA) increases catechol production via the reduction of ortho-quinone to ortho-catechol [26]. In addition, LAA and D-ascorbic acid (DAA), an isomer of LAA, act as a suicide substrate against Ty in aerobic conditions [35]. In this study, we investigated the inhibitory effects of LAA, D-ascorbic acid (DAA), and dehydroascorbic acid (DHA), an oxidized form of LAA, on Ty-induced RD-quinone formation by monitoring absorbance at 400 nm. The inhibitory effects of LAA and its analogs increased with increasing their concentrations. When Ty reacted with RD at 37 °C for 30 min, 1 mM LAA, DAA, and DHA inhibited RD-quinone formation by 86%, 95%, and 8%, respectively (Figure 4). The inhibitory effects of LAA and its analogs at 25 °C were similar to those at 37 °C (Figure S9). These results indicate that LAA analogs act as inhibitors of Ty, not as antioxidants for the reduction of ortho-quinone to ortho-catechol. In addition, it can be suggested that Ty has potential as an alternative biocatalyst for RD-catechol production because RD-quinone formation is inhibited in the presence of LAA analogs.

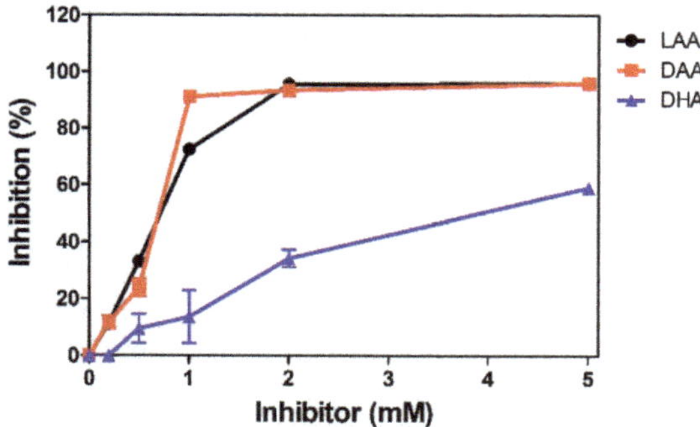

Figure 4. Inhibition of RD-quinone formation by L-ascorbic acid (LAA), D-ascorbic acid (DAA), and dehydroascorbic acid (DHA) at 37 °C.

2.4. Regioselective Production of RD-Catechol by Ty

For selective production of RD-catechol by Ty, LAA was selected as an inhibitor of its diphenolase activity. Ty produced one major product at a temperature range of 25–45 °C when the reaction mixture contained Ty, RD, and LAA. RD's conversion to the product by Ty was the highest at 37 °C (Figure S10). In addition, production of the major product by Ty was investigated when 5–100 mM LAA contained in reaction mixture at 37 °C. RD's conversion by Ty increased over time and was saturated within 30 min. Ty produced the highest product of 84.1 mg·L^{-1} when the reaction mixture containing 10 mM LAA was incubated at 37 °C for 30 min (Figure 5).

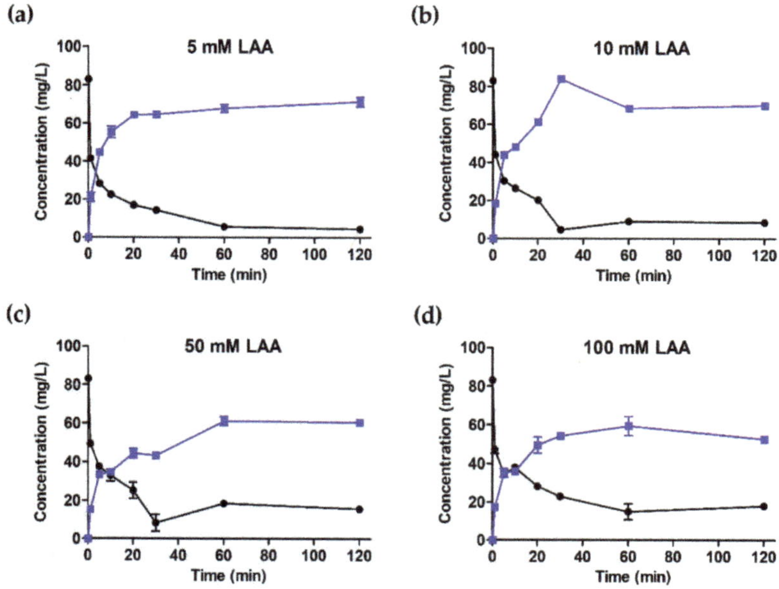

Figure 5. Effect of LAA concentration on the production of RD-catechol. The reaction mixtures contained five units of Ty and 500 µM RD in 250 µL of a 100 mM potassium phosphate buffer (pH 6.5) with 5 (**a**), 10 (**b**), 50 (**c**), or 100 mM LAA (**d**). The reaction mixtures were incubated at 37 °C for 0–120 min. The RD-catechol produced by Ty was analyzed using HPLC. The circle (•) and square (■) lines indicate concentration of RD and RD-catechol, respectively.

The major product of Ty eluted at 8.4 min on HPLC (Figure S11), which was the same retention time with the major product of M16 (Figure 2). To identify Ty's product, LC-MS and NMR analyses were performed. The *m/z* value of this product was 165 on LC-MS, which was 16 higher compared to the RD (Figure S12). The ^{1}H and ^{13}C NMR spectra of Ty's product were the same as those of M16's product (Figures S13 and S14). These results indicate that Ty catalyzes the ortho-hydroxylation to regioselectively produce RD-catechol when LAA acts as an inhibitor of Ty's diphenolase activity.

To increase RD-catechol production, Ty catalytic reactions were performed at high RD concentrations. Ty's RD-catechol production increased with RD concentration, and it became saturated from 2 mM RD. In this condition, yield, product concentration, and productivity were 69.3%, 252 mg·L^{-1}, and 505 mg·L^{-1}·h^{-1}, respectively. The RD-catechol yield increased until 500 μM and then decreased. When the RD concentration was 500 μM, the yield was highest at 93.9%. The final product concentration and productivity were 85.6 mg·L^{-1} and 171 mg·L^{-1}·h^{-1}, respectively (Figure 6). These results suggest that Ty can be used as an alternative biocatalyst for the selective production of RD-catechol with high productivity.

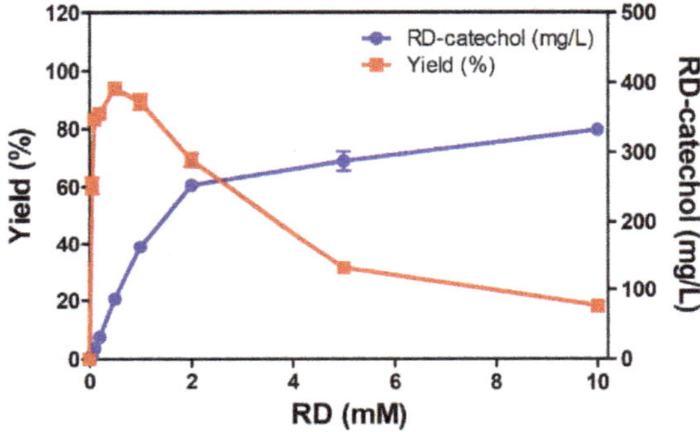

Figure 6. RD-catechol production by Ty with increasing RD concentration. The reaction mixtures contained five units of Ty, indicated concentrations of RD, and 10 mM LAA. RD-catechol produced by Ty was analyzed using HPLC.

The production capability of enzymes is evaluated in terms of productivity (mg·L^{-1}·h^{-1}) and final product concentration (mg·L^{-1}). For pharmaceutical production, the estimated minimum productivity and final product concentration are 1 mg·L^{-1}·h^{-1} and 100 mg·L^{-1}, respectively [36–39]. The production abilities of M16 and Ty were compared when the RD concentration was 500 μM (Table 1). In this condition, both biocatalysts produced mainly RD-catechol, and their selectivity values exceeded 90%. Ty yielded 93.9% RD-catechol from RD, whereas M16 produced 17.6% RD-catechol from RD. The productivities of M16 and Ty were 32.1 mg·L^{-1}·h^{-1} and 171 mg·L^{-1}·h^{-1}, respectively. These results indicate that Ty produced 5.3 times more RD-catechol compared with M16. In addition, one-pot synthesis of RD-catechol by Ty is possible without additional purification steps when the RD concentration is 500 μM.

Table 1. Comparison of production abilities of engineered CYP102A1 and Ty.

Biocatalyst	Selectivity (%)	Yield (%)	Product Concentration (mg·L^{-1})	Productivity (mg·L^{-1}·h^{-1})
CYP102A1 [1]	93.5 ± 1.4	17.6 ± 0.7	16.1 ± 0.7	32.1
Ty [2]	90.4 ± 0.4	93.9 ± 1.8	85.6 ± 1.6	171

[1] The reaction mixture contained 0.2 μM M16, 500 μM RD, and NGS in 250 μL of 100 mM potassium phosphate buffer (pH 7.4) and was incubated at 37 °C for 30 min (Figure 3). [2] The reaction mixture contained five units of Ty, 500 μM RD, and 10 mM LAA in 250 μL of 100 mM potassium phosphate buffer (pH 6.5) and was incubated at 37 °C for 30 min (Figure 6).

In conclusion, we found that engineered CYP102A1 and Ty can catalyze the regioselective hydroxylation of RD and produce RD-catechol. Additionally, LAA and its analogs inhibit Ty's diphenolase activity and contribute to selective RD-catechol production by inhibiting RD-quinone and melanin formation. Compared with engineered CYP102A1, Ty can reduce the cost and time of RD-catechol production because its productivity and yield were quite high at 171 mg·L^{-1}·h^{-1} and 93.9%, and it does not require high-cost NADPH. Taken together, these findings indicate that engineered CYP102A1 and Ty can be used as efficient biocatalysts and immobilized systems to make hydroxylated products. Ty is a cost-effective biocatalyst and is more appropriate for industrial applications compared with engineered CYP102A1.

3. Materials and Methods

3.1. Materials

Glucose-6-phosphate dehydrogenase (G6PDH, from yeast), glucose-6-phosphate (G6P), $NADP^+$, Ty (from *Agaricus bisporus*), LAA, DAA, and DHA were obtained from Sigma-Aldrich (St. Louis, MO, USA). RD was obtained from Tokyo Chemical Industry (Tokyo, Japan). All the other chemicals were of analytical grade.

3.2. Preparation of Engineered CYP102A1

The engineered CYP102A1 enzymes used in this study are known to have high activities toward non-natural substrates, and they were constructed by site-directed and random mutagenesis [14,40–42]. The changed amino acid residues of engineered CYP102A1 are shown in Table S2. CYP102A1 WT and engineered enzymes were expressed in *Escherichia coli* and were purified as previously described [40]. The plasmid of CYP102A1 enzymes (in pCW vector) was transformed to *E. coli* DH5α F′-IQ strain. The transformed cells were grown overnight in 5 mL of Luria–Bertani (LB) broth with ampicillin (100 μg·mL^{-1}) at 37 °C. Precultured cells were inoculated to 250 mL of terrific broth (TB) containing ampicillin (100 μg·mL^{-1}) and were grown at 37 °C with shaking at 250 rpm until reaching 0.8 of A_{600}. The expression of CYP102A1 enzymes was induced by adding 0.5 mM isopropyl β-D-1-thiogalactopyranoside (IPTG) and 1 mM δ-aminolevulinic acid. The culture was grown at 30 °C with shaking at 190 rpm for 24 h. The cells were harvested by centrifugation at 5000× *g* for 15 min. The cell pellet was resuspended in a TES buffer (100 mM Tris-HCl, pH 7.6, 0.5 mM EDTA, and 500 mM sucrose) and lysed by sonication. After the lysate was centrifuged at 100,000× *g* for 90 min at 4 °C, the soluble cytosolic fraction was collected and used for the activity assay. The CYP102A1 enzymes' concentration in the cytosolic fraction was determined from the CO-difference spectra using $\varepsilon = 91$ mM^{-1}·cm^{-1} [43].

3.3. Hydroxylation of RD by Engineered CYP102A1

The catalytic activity of RD hydroxylation by CYP102A1 was measured as described previously [44]. To measure the activity of the CYP102A1 WT and engineered enzymes, the reaction mixtures contained 0.2 μM enzyme, 200 μM RD, and NGS in 250 μL of 100 mM potassium phosphate buffer (pH 7.4). To investigate the effects of RD concentration on RD-catechol production by M16, the reaction mixtures contained 0.2 μM M16 and NGS in 250 μL of 100 mM potassium phosphate buffer with 0, 50, 100, 200, or 500 μM RD. The reactions were initiated by adding NGS (10 mM G6P, 2 mM $NADP^+$, and 0.5 μg yeast G6PDH in 250 μL) and were stopped with 25 μL of 1 N HCl containing 2 M NaCl and 600 μL of ice-cold ethyl acetate after incubating for 30 min at 37 °C. After centrifugation of the samples at 3000 rpm for 10 min, an organic layer was transferred to a new tube and evaporated under nitrogen gas. The samples were dissolved using 180 μL of mobile phase. The reaction products were analyzed using an SPD-20A UV-Visible detector (Shimadzu, Kyoto, Japan) on an LC-20AD HPLC (Shimadzu, Kyoto, Japan). The products were separated by a Gemini C18 column (150 × 4.6 mm, 5 μm; Phenomenex, Torrance, CA, USA) with a mobile phase of 10 mM formic acid in water/methanol (60:40, *v*/*v*). The sample was injected with 30 µL using a SIL-20A auto sampler (Shimadzu, Kyoto, Japan), and the flow rate was 0.5 mL/min. The RD and products were detected at 280 nm.

3.4. Inhibition of RD-Quinone Formation

RD-quinone formation by Ty was monitored using a UV-160 IPC UV–Visible spectrophotometer (Shimadzu, Kyoto, Japan). The reaction mixtures contained 20 units of Ty, 500 µM RD, and 1 mM LAA, DAA, or DHA in 1 mL of 100 mM of potassium phosphate buffer (pH 6.5), and were incubated at 25 or 37 °C for 30 min. The inhibitory effects were expressed as percentage of the absorbance decreased relative to the absorbance at 400 nm of the sample without the inhibitor.

3.5. Production of RD-Catechol by Tyrosinase

The Ty reaction mixtures contained five units of Ty and 500 µM RD in 250 µL of 100 mM potassium phosphate buffer (pH 6.5) with 5, 10, 50, or 100 mM LAA, and were incubated at 37 °C for 0, 1, 5, 10, 20, 30, 60, or 120 min. To increase RD-catechol production, five units of Ty; 0, 0.05, 0.1, 0.2, 0.5, 1, 2, 5, or 10 mM RD; and 10 mM LAA were incubated in 250 µL of 100 mM potassium phosphate buffer (pH 6.5) at 37 °C for 30 min. The reaction mixtures were initiated by adding enzyme and were stopped with 25 µL of 1 N HCl containing 2 M NaCl and 600 µL of ice-cold ethyl acetate. After centrifugation at 1000× *g* for 10 min, an organic layer was transferred to a new tube and evaporated under nitrogen gas. The samples were dissolved using 180 µL of mobile phase, and the products were analyzed via HPLC as described above.

3.6. Identification of RD-Catechol by LC-MS

To investigate the mass of the major product, we used a CYP102A1 reaction mixture containing 0.2 µM M16, 500 µM RD, and NGS in 250 µL of 100 mM potassium phosphate buffer (pH 7.4). The Ty reaction mixture contained five units of Ty, 500 µM RD, and 10 mM LAA in 250 µL of 100 mM potassium phosphate buffer (pH 6.5). These mixtures were incubated at 37 °C for 30 min. Injection samples were created as described above. The mass values of RD and its products were analyzed using a TSQ Quantum™ Access MAX Triple Quadrupole Mass Spectrometer on an Accela 1250 HPLC system (Thermo Fisher Scientific, Waltham, MA, USA). The samples were separated on a ZORBAX SB-C18 column (250 mm × 4.6 mm i.d. 5 µm; Agilent, Santa Clara, CA, USA) at a flow rate of 1 mL/min. The mobile phases were (A) 10 mM formic acid in water and (B) methanol. The isocratic flow of mobile phase was (A) 60%: (B) 40% on HPLC. The injection volume was 10 µL. The mass spectra were recorded by electrospray ionization in positive mode to identify the RD metabolites. The collision energy and scan rate were 10 V and 0.5 spectra/s, respectively.

3.7. Identification of RD-Catechol by NMR Spectroscopy

To identify the products of engineered CYP102A1 and Ty, reaction mixtures and injection samples were prepared as described above. The samples were separated using HPLC, and the major product eluted at 8.4 min was collected. The collected samples were dried using a freezing drier (Operon, Gimpo, Korea). NMR experiments were performed at 25 °C on a Varian VNMRS 600 MHz NMR spectrometer (Varian Inc., Palo Alto, CA, USA) equipped with a carbon-enhanced cryogenic probe. Methanol-d_4 was used as the solvent, and the chemical shifts for the proton NMR spectra were measured in parts per million (ppm) relative to tetramethylsilane. All of the NMR experiments were carried out with standard pulse sequences in the VNMRJ (v. 3.2) library and processed with the same software (Agilent, Santa Clara, CA, USA).

3.8. Statistical Analysis

All experiments were performed three times. The values are presented as means with a standard error of mean (SEM).

Supplementary Materials: The following are available online at http://www.mdpi.com/2073-4344/10/10/1114/s1: Figure S1: Total turnover numbers of engineered CYP102A1 enzymes. Figure S2: Total ion chromatography and mass scan of RD and product of M16. Figure S3: ^{1}H NMR spectra of RD (a) and the major product of M16 (b), Figure S4: Comparison of aromatic region in the ^{1}H NMR spectra of RD (a) and the major product of M16 (b), Figure S5: ^{13}C NMR spectra of RD (a) and the major product of M16 (b), Figure S6: Comparison of quaternary peaks in the ^{13}C NMR spectra of RD (a) and the major product of M16 (b), Figure S7: Chemical structures of RD and RD-catechol, Figure S8: RD-quinone formation by Ty was analyzed by UV–Vis spectrophotometer, Figure S9: Inhibition of RD-quinone formation by L-ascorbic acid (LAA), D-ascorbic acid (DAA), and dehydroascorbic acid (DHA) at 25 °C, Figure S10: Effects of temperature on conversion rate of CYP102A1 and Ty, Figure S11: HPLC chromatogram of RD and RD-catechol produced by Ty, Figure S12: Total ion chromatography and mass scan of RD and product of Ty, Figure S13: ^{1}H NMR spectra of products of M16 (a) and Ty (b), Figure S14: ^{13}C NMR spectra of products of M16 (a) and Ty (b), Table S1: ^{1}H and ^{13}C chemical shifts of RD and RD-catechol, Table S2: Changed amino acid residues of the engineered CYP102A1 used in this study.

Author Contributions: Conceptualization, C.M.P., G.S.C., and C.-H.Y.; validation, C.M.P., H.S.P., and G.S.C.; formal analysis, C.M.P.; investigation, C.M.P., H.S.P., G.S.C., and K.D.P.; writing—original draft preparation, C.M.P.; writing—review and editing, C.-H.Y.; visualization, C.M.P.; supervision, C.-H.Y.; funding acquisition, C.M.P. and C.-H.Y. All authors have read and agreed to the published version of the manuscript.

Funding: This research was supported by Basic Science Research Program through the National Research Foundation of Korea (NRF) funded by the Ministry of Education (2020R1I1A1A01055345) and the Basic Research Lab Program (NRF-2018R1A4A1023882), National Research Foundation of Korea, Republic of Korea.

Conflicts of Interest: The authors declare no conflict of interest.

References

1. Schmid, A.; Dordick, J.S.; Hauer, B.; Kiener, A.; Wubbolts, M.; Witholt, B. Industrial biocatalysis today and tomorrow. *Nature* **2001**, *409*, 258–268. [CrossRef] [PubMed]
2. Torres, J.A.; Nogueirab, F.G.E.; Silvac, M.C.; Lopes, J.H.; Tavaresa, T.S.; Ramalho, T.C.; Corrêaa, A.D. Novel eco-friendly biocatalyst: Soybean peroxidase immobilized onto activated carbon obtained from agricultural waste. *RSC Adv.* **2017**, *7*, 16460–16466. [CrossRef]
3. Choi, J.M.; Han, S.S.; Kim, H.S. Industrial applications of enzyme biocatalysis: Current status and future aspects. *Biotechnol. Adv.* **2015**, *33*, 1443–1454. [CrossRef] [PubMed]
4. Nitti, A.; Bianchi, G.; Po, R.; Swager, T.M.; Pasini, D. Domino Direct Arylation and Cross-Aldol for Rapid Construction of Extended Polycyclic pi-Scaffolds. *J. Am. Chem. Soc.* **2017**, *139*, 8788–8791. [CrossRef] [PubMed]
5. Osburn, P.L.; Bergbreiter, D.E. Molecular engineering of organic reagents and catalysts using soluble polymers. *Prog. Polym. Sci.* **2001**, *26*, 2015–2081. [CrossRef]
6. Nelson, D.R. The cytochrome p450 homepage. *Hum. Genom.* **2009**, *4*, 59–65. [CrossRef]
7. Guengerich, F.P. Common and uncommon cytochrome P450 reactions related to metabolism and chemical toxicity. *Chem. Res. Toxicol.* **2001**, *14*, 611–650. [CrossRef] [PubMed]
8. Guengerich, F.P.; Munro, A.W. Unusual cytochrome p450 enzymes and reactions. *J. Biol. Chem.* **2013**, *288*, 17065–17073. [CrossRef]
9. Urlacher, V.B.; Girhard, M. Cytochrome P450 Monooxygenases in Biotechnology and Synthetic Biology. *Trends Biotechnol.* **2019**, *37*, 882–897. [CrossRef]
10. Munro, A.W.; Daff, S.; Coggins, J.R.; Lindsay, J.G.; Chapman, S.K. Probing electron transfer in flavocytochrome P-450 BM3 and its component domains. *Eur. J. Biochem.* **1996**, *239*, 403–409. [CrossRef]
11. Lussenburg, B.M.A.; Babel, L.C.; Vermeulen, N.P.E.; Commandeur, J.N.M. Evaluation of alkoxyresorufins as fluorescent substrates for cytochrome P450 BM3 and site-directed mutants. *Anal. Biochem.* **2005**, *341*, 148–155. [CrossRef] [PubMed]
12. Yun, C.H.; Kim, K.H.; Kim, D.H.; Jung, H.C.; Pan, J.G. The bacterial P450 BM3: A prototype for a biocatalyst with human P450 activities. *Trends Biotechnol.* **2007**, *25*, 289–298. [CrossRef] [PubMed]
13. Whitehouse, C.J.; Bell, S.G.; Wong, L.L. P450 (BM3) (CYP102A1): Connecting the dots. *Chem. Soc. Rev.* **2012**, *41*, 1218–1260. [CrossRef] [PubMed]
14. Kang, J.Y.; Ryu, S.H.; Park, S.H.; Cha, G.S.; Kim, D.H.; Kim, K.H.; Hong, A.W.; Ahn, T.; Pan, J.G.; Joung, Y.H.; et al. Chimeric cytochromes P450 engineered by domain swapping and random mutagenesis for producing human metabolites of drugs. *Biotechnol. Bioeng.* **2014**, *111*, 1313–1322. [CrossRef] [PubMed]

15. Urlacher, V.B.; Eiben, S. Cytochrome P450 monooxygenases: Perspectives for synthetic application. *Trends Biotechnol.* **2006**, *24*, 324–330. [CrossRef] [PubMed]
16. Ismaya, W.T.; Rozeboom, H.J.; Weijn, A.; Mes, J.J.; Fusetti, F.; Wichers, H.J.; Dijkstra, B.W. Crystal structure of *Agaricus bisporus* mushroom tyrosinase: Identity of the tetramer subunits and interaction with tropolone. *Biochemistry* **2011**, *50*, 5477–5486. [CrossRef] [PubMed]
17. Ramsden, C.A.; Riley, P.A. Tyrosinase: The four oxidation states of the active site and their relevance to enzymatic activation, oxidation and inactivation. *Bioorg. Med. Chem.* **2014**, *22*, 2388–2395. [CrossRef]
18. Pillaiyar, T.; Namasivayam, V.; Manickam, M.; Jung, S.H. Inhibitors of Melanogenesis: An Updated Review. *J. Med. Chem.* **2018**, *61*, 7395–7418. [CrossRef]
19. Goldfeder, M.; Kanteev, M.; Adir, N.; Fishman, A. Influencing the monophenolase/diphenolase activity ratio in tyrosinase. *Biochim. Biophys. Acta* **2013**, *1834*, 629–633. [CrossRef]
20. Halaouli, S.; Asther, M.; Sigoillot, J.C.; Hamdi, M.; Lomascolo, A. Fungal tyrosinases: New prospects in molecular characteristics, bioengineering and biotechnological applications. *J. Appl. Microbiol.* **2006**, *100*, 219–232. [CrossRef]
21. Hernandez-Romero, D.; Sanchez-Amat, A.; Solano, F. A tyrosinase with an abnormally high tyrosine hydroxylase/dopa oxidase ratio. *FEBS J.* **2006**, *273*, 257–270. [CrossRef] [PubMed]
22. Shuster Ben-Yosef, V.; Sendovski, M.; Fishman, A. Directed evolution of tyrosinase for enhanced monophenolase/diphenolase activity ratio. *Enzym. Microb. Technol.* **2010**, *47*, 372–376. [CrossRef]
23. Molloy, S.; Nikodinovic-Runic, J.; Martin, L.B.; Hartmann, H.; Solano, F.; Decker, H.; O'Connor, K.E. Engineering of a bacterial tyrosinase for improved catalytic efficiency towards D-tyrosine using random and site directed mutagenesis approaches. *Biotechnol. Bioeng.* **2013**, *110*, 1849–1857. [CrossRef] [PubMed]
24. Lee, N.; Kim, E.J.; Kim, B.G. Regioselective hydroxylation of trans-resveratrol via inhibition of tyrosinase from *Streptomyces avermitilis* MA4680. *ACS Chem. Biol.* **2012**, *7*, 1687–1692. [CrossRef]
25. Lee, N.; Lee, S.H.; Baek, K.; Kim, B.G. Heterologous expression of tyrosinase (MelC2) from *Streptomyces avermitilis* MA4680 in *E. coli* and its application for ortho-hydroxylation of resveratrol to produce piceatannol. *Appl. Microbiol. Biotechnol.* **2015**, *99*, 7915–7924. [CrossRef]
26. Lee, S.H.; Baek, K.; Lee, J.E.; Kim, B.G. Using tyrosinase as a monophenol monooxygenase: A combined strategy for effective inhibition of melanin formation. *Biotechnol. Bioeng.* **2016**, *113*, 735–743. [CrossRef]
27. Tallent, W.H. d-Betuligenol from Rhododendron maximum L. *J. Org. Chem.* **1964**, *29*, 988–989. [CrossRef]
28. Fujita, T.; Hatamoto, H.; Iwasaki, T.; Takafuji, S.I. Bioconversion of rhododendrol by *Acer nikoense*. *Phytochemistry* **1995**, *39*, 1085–1089. [CrossRef]
29. Pan, H.; Lundgren, L.N. Rhododendrol glycosides and phenyl glucoside esters from inner bark of *Betula pubescens*. *Phytochemistry* **1994**, *36*, 79–83. [CrossRef]
30. Ito, S.; Wakamatsu, K. Biochemical Mechanism of Rhododendrol-Induced Leukoderma. *Int. J. Mol. Sci.* **2018**, *19*, 552.
31. Ito, S.; Ojika, M.; Yamashita, T.; Wakamatsu, K. Tyrosinase-catalyzed oxidation of rhododendrol produces 2-methylchromane-6,7-dione, the putative ultimate toxic metabolite: Implications for melanocyte toxicity. *Pigment Cell Melanoma Res.* **2014**, *27*, 744–753. [CrossRef]
32. Yang, E.J.; An, J.H.; Son, Y.K.; Yeo, J.H.; Song, K.S. The Cytotoxic Constituents of *Betula platyphylla* and their Effects on Human Lung A549 Cancer Cells. *Nat. Prod. Sci.* **2018**, *24*, 219–224. [CrossRef]
33. Yang, L.; Yang, F.; Wataya-Kaneda, M.; Tanemura, A.; Tsuruta, D.; Katayama, I. 4-(4-Hydroroxyphenyl)-2-butanol (rhododendrol) activates the autophagy-lysosome pathway in melanocytes: Insights into the mechanisms of rhododendrol-induced leukoderma. *J. Dermatol. Sci.* **2015**, *77*, 182–185. [CrossRef]
34. Kasamatsu, S.; Hachiya, A.; Nakamura, S.; Yasuda, Y.; Fujimori, T.; Takano, K.; Moriwaki, S.; Hase, T.; Suzuki, T.; Matsunaga, K. Depigmentation caused by application of the active brightening material, rhododendrol, is related to tyrosinase activity at a certain threshold. *J. Dermatol. Sci.* **2014**, *76*, 16–24. [CrossRef] [PubMed]
35. Munoz-Munoz, J.L.; Garcia-Molina, F.; García-Ruiz, P.A.; Varon, R.; Tudela, J.; García-Cánovas, F.; Rodriguez-Lopez, J.N. Stereospecific inactivation of tyrosinase by l- and d-ascorbic acid. *Biochim. Biophys. Acta. Proteins Proteom.* **2009**, *1794*, 244–253. [CrossRef] [PubMed]
36. Porter, J.L.; Rusli, R.A.; Ollis, D.L. Directed Evolution of Enzymes for Industrial Biocatalysis. *ChemBioChem* **2016**, *17*, 197–203. [CrossRef] [PubMed]

37. Julsing, M.K.; Cornelissen, S.; Buhler, B.; Schmid, A. Heme-iron oxygenases: Powerful industrial biocatalysts? *Curr. Opin. Chem. Biol.* **2008**, *12*, 177–186. [CrossRef]
38. Pollard, D.J.; Woodley, J.M. Biocatalysis for pharmaceutical intermediates: The future is now. *Trends Biotechnol.* **2007**, *25*, 66–73. [CrossRef]
39. Straathof, A.J.J.; Panke, S.; Schmid, A. The production of fine chemicals by biotransformations. *Curr. Opin. Biotech.* **2002**, *13*, 548–556. [CrossRef]
40. Kim, D.H.; Kim, K.H.; Kim, D.H.; Liu, K.H.; Jung, H.C.; Pan, J.G.; Yun, C.H. Generation of human metabolites of 7-ethoxycoumarin by bacterial cytochrome P450 BM3. *Drug Metab. Dispos.* **2008**, *36*, 2166–2170. [CrossRef]
41. Park, S.H.; Kim, D.H.; Kim, D.; Kim, D.H.; Jung, H.C.; Pan, J.G.; Ahn, T.; Kim, D.; Yun, C.H. Engineering bacterial cytochrome P450 (P450) BM3 into a prototype with human P450 enzyme activity using indigo formation. *Drug Metab. Dispos.* **2010**, *38*, 732–739. [CrossRef] [PubMed]
42. Jang, H.H.; Ryu, S.H.; Le, T.K.; Doan, T.T.; Nguyen, T.H.; Park, K.D.; Yim, D.E.; Kim, D.H.; Kang, C.K.; Ahn, T.; et al. Regioselective C-H hydroxylation of omeprazole sulfide by *Bacillus megaterium* CYP102A1 to produce a human metabolite. *Biotechnol. Lett.* **2017**, *39*, 105–112. [CrossRef] [PubMed]
43. Omura, T.; Sato, R. The Carbon Monoxide-Binding Pigment of Liver Microsomes. II. Solubilization, Purification, and Properties. *J. Biol. Chem.* **1964**, *239*, 2379–2385. [PubMed]
44. Nguyen, N.A.; Jang, J.; Le, T.K.; Nguyen, T.H.H.; Woo, S.M.; Yoo, S.K.; Lee, Y.J.; Park, K.D.; Yeom, S.J.; Kim, G.J.; et al. Biocatalytic Production of a Potent Inhibitor of Adipocyte Differentiation from Phloretin Using Engineered CYP102A1. *J. Agric. Food Chem.* **2020**, *68*, 6683–6691. [CrossRef] [PubMed]

Article

Exploring Metagenomic Enzymes: A Novel Esterase Useful for Short-Chain Ester Synthesis

Thaís Carvalho Maester [1,2,†], Mariana Rangel Pereira [1,2,‡], Aliandra M. Gibertoni Malaman [1], Janaina Pires Borges [3], Pâmela Aparecida Maldaner Pereira [1] and Eliana G. M. Lemos [1,*]

1 Department of Technology, São Paulo State University (UNESP), Jaboticabal, SP 14884-900, Brazil; thaismaester@ecobiotech.com.br (T.C.M.); mr629@cam.ac.uk (M.R.P.); aligibertoni@gmail.com (A.M.G.M.); pamela.maldaner@unesp.br (P.A.M.P.)
2 Institute of Biomedical Sciences (ICB III), University of São Paulo (USP), São Paulo, SP 05508-900, Brazil
3 Institute of Biosciences, Languages and Exact Sciences, Department of Chemistry and Environmental Sciences, São Paulo State University (UNESP), São José do Rio Preto, SP 15054-000, Brazil; janaina-pires@hotmail.com
* Correspondence: egerle@fcav.unesp.br; Tel.: +55-16-3209-7409
† Current address: Supera Innovation and Technology Park, Ecobiotech Company, Ribeirão Preto, SP 14056-680, Brazil.
‡ Current address: Department of Biochemistry, University of Cambridge, Cambridge CB2 1TN, UK.

Received: 13 August 2020; Accepted: 24 August 2020; Published: 23 September 2020

Abstract: Enzyme-mediated esterification reactions can be a promising alternative to produce esters of commercial interest, replacing conventional chemical processes. The aim of this work was to verify the potential of an esterase for ester synthesis. For that, recombinant lipolytic enzyme EST5 was purified and presented higher activity at pH 7.5, 45 °C, with a T*m* of 47 °C. Also, the enzyme remained at least 50% active at low temperatures and exhibited broad substrate specificity toward *p*-nitrophenol esters with highest activity for *p*-nitrophenyl valerate with a K_{cat}/K_m of 1533 s^{-1} mM^{-1}. This esterase exerted great properties that make it useful for industrial applications, since EST5 remained stable in the presence of up to 10% methanol and 20% dimethyl sulfoxide. Also, preliminary studies in esterification reactions for the synthesis of methyl butyrate led to a specific activity of 127.04 $U{\cdot}mg^{-1}$. The enzyme showed higher esterification activity compared to other literature results, including commercial enzymes such as LIP4 and CL of *Candida rugosa* assayed with butyric acid and propanol which showed esterification activity of 86.5 and 15.83 $U{\cdot}mg^{-1}$, respectively. In conclusion, EST5 has potential for synthesis of flavor esters, providing a concept for its application in biotechnological processes.

Keywords: lipolytic enzymes; metagenome; family V; esterification; flavor esters

1. Introduction

With the increased environmental awareness, the utilization of enzymes may have special relevance. Industrial biocatalysts, in comparison to chemical catalysis, lead to cost savings in the application process, result in improved product quality, generate less waste, are more energy efficient, are able to reduce or eliminate the formation of by-products, and are safe and environmentally friendly [1–3]. The use of industrial enzymes has arisen as a significant solution for green and sustainable industrial products [4] and can be incorporated into the concept of the "circular economy" where nothing is wasted [5].

In this context, lipolytic enzymes are highly useful in many industrially significant processes due to their stability in the presence of solvents; exquisite chemo-, regio- and enantioselectivities; activity over a broad range of substrates; and no need for cofactors [6–10]. Such characteristics allow lipolytic enzymes to be used in the production of pharmaceuticals; in the food industry (with the global enzyme market being dominated by these two fields [11]); in the leather and paper industries; as additives in detergents; in the synthesis and degradation of plastics and biopolymers; for the production of biodiesel; and in bioremediation processes [12–16]. Thus, they are promising and have high-growth potential in the valuable world industrial enzymes market; the global lipase market is expected to reach $590.5 million in 2020 [17,18].

Lipolytic enzymes, comprising carboxylesterases (EC 3.1.1.1) and triacylglycerol lipases (EC 3.1.1.3), are members of the α/β hydrolase fold family and differ from each other in their biochemical properties. Carboxylesterases act on short-chain and usually water-soluble carboxylic esters, while lipases preferentially hydrolyze ester bonds of triglycerides with long-chain fatty acids which are often insoluble in water [19,20]. A classification of the bacterial lipases into eight families has been reported previously by Arpigny and Jaeger (1999), based on their conserved sequence motifs and some fundamental biological properties. Currently, many novel lipolytic enzymes have been discovered from different microorganisms, as yeasts [21,22], fungi [23–25] and bacteria [26–29].

Lipases and esterases catalyze hydrolytic reactions in aqueous media as well as the synthesis of esters from triglycerides/free fatty acids and alcohols in organic solvents or nonaqueous media [30]. Esterification reactions of carboxylic acids and alcohols generate important products as emulsifiers, biopolymers, and flavor esters. These products have been widely applied in cosmetics, plastics, medicine, food, etc. [12,16,31].

Conventionally, esterification reactions are performed using chemical catalysts, as acid solutions, which do not meet the concept of sustainable development. Therefore, enzyme-based biochemically produced carboxylic esters by lipolytic enzymes are interesting alternatives to their chemical counterparts [32,33], which lack specificity, resulting in the production of unwanted by-products; encounter difficulty in product recovery; consume more time and energy; have final products that may contain residues toxic to human health; and involve environmentally harmful production processes [33,34].

Therefore, characterization of novel lipases and esterases are potentially useful for industrial processes. Several proposals are feasible to search for new enzymes, such as metagenomics. Many lipolytic enzymes from metagenomic libraries have been discovered and characterized: the esterase Est906 of the family V from the metagenome of paper mill wastewater sediments that presented relevant characteristics for application in the detergent industry [35] and the lipase and its cognate foldase Lip-LifMF3 isolated from a metagenomic library from soil contaminated with fat that presented potential for application in biocatalysis [36]. In addition, we previously reported the characterization of the esterases ORF2, Est16 and EST3 from a metagenomic library isolated from a microbial consortium specialized in diesel oil degradation [37–39].

In a previous work, we sequenced the clone PL14.H10 and identified putative genes responsible for lipolytic activity. Cloning, expression and characterization of one of them was previous reported [39]. In this work, a novel enzyme from PL14.H10 was cloned and recombinantly expressed. Biochemical characterization of the enzyme was described, including substrate specificity, effect of additives on enzyme activity and effect of temperature and pH on enzyme conformation. In addition, the potential application of the enzyme as a catalyst for ester synthesis was highlighted.

2. Results and Discussion

2.1. Analysis of EST5

The sequence analysis of the insert DNA from clone PL14.H10 showed the presence of one 912 bp open reading frame (ORF), encoding a protein of 340 amino acids with predicted molecular mass of 37 kDa and theoretical pI of 7.3, without putative signal peptide. The putative esterase gene was designated *est5*. BlastP analysis based on the information in the GenBank database revealed highest identity (78%) of the corresponding protein EST5 with a lipase/esterase from uncultured bacterium derived from sea sediment sample (accession number: ADM63077.1), followed by 74%

identity with an alpha/beta hydrolase fold protein from *Parvibaculum lavamentivorans* (accession number: WP012110575.1), and 62% identity with an alpha/beta hydrolase from α-Proteobacterium Mf 1.05b.01 (accession number: WP029641682.1). The DNA sequence of *est5* gene has been deposited in GenBank with the accession number KY563703.

To analyze the phylogenetic relationship of EST5 with other lipolytic enzymes, a neighbor-joining tree was constructed based on amino acid sequences of lipolytic enzymes representing the eight families proposed by Arpigny and Jaeger (1999). As shown in Figure 1A, EST5 is most closely related to esterases Est16 and ORF2, and all three are derived from the same metagenomic library. They formed a separated group and showed identities of 50% and 40% to EST5, respectively. EST5 contains a catalytic triad that is typical of proteins with α/β hydrolase fold: Ser120-Asp247-His27. Multiple sequence alignment with closely related homologues (Figure 1B) revealed the catalytic nucleophile Ser located in the typical family V conserved motif G-X-S-X-G-G; and the PTL motif varied among the enzyme sequences. Regarding the catalytic site of aspartate, the enzymes EST5, Est16 and ORF2 showed the conserved motif DPL.

2.2. Obtaining Protein EST5

The full-length *est5* sequence was amplified and cloned into the expression vector pET-28a(+) with an N-terminal 6× His tag. and expressed in *E. coli* BL21(DE3). The highest amount of protein was achieved at 30 °C for 4 h. Following induction, the encoded EST5 was expressed in active form in the soluble fraction of the host cells. Molecular mass of the recombinant enzyme was about 37 kDa, as analyzed by polyacrylamide gel electrophoresis (SDS-PAGE) (Figure 2A), in accordance with information predicted from amino acid sequence. Figure 2B shows the purification of the enzyme, in which two steps were combined: Ni-NTA affinity chromatography, where the protein was eluted mostly at 50 mM imidazole, and size exclusion chromatography, to proceed with the circular dichroism (CD) and fluorescence analysis.

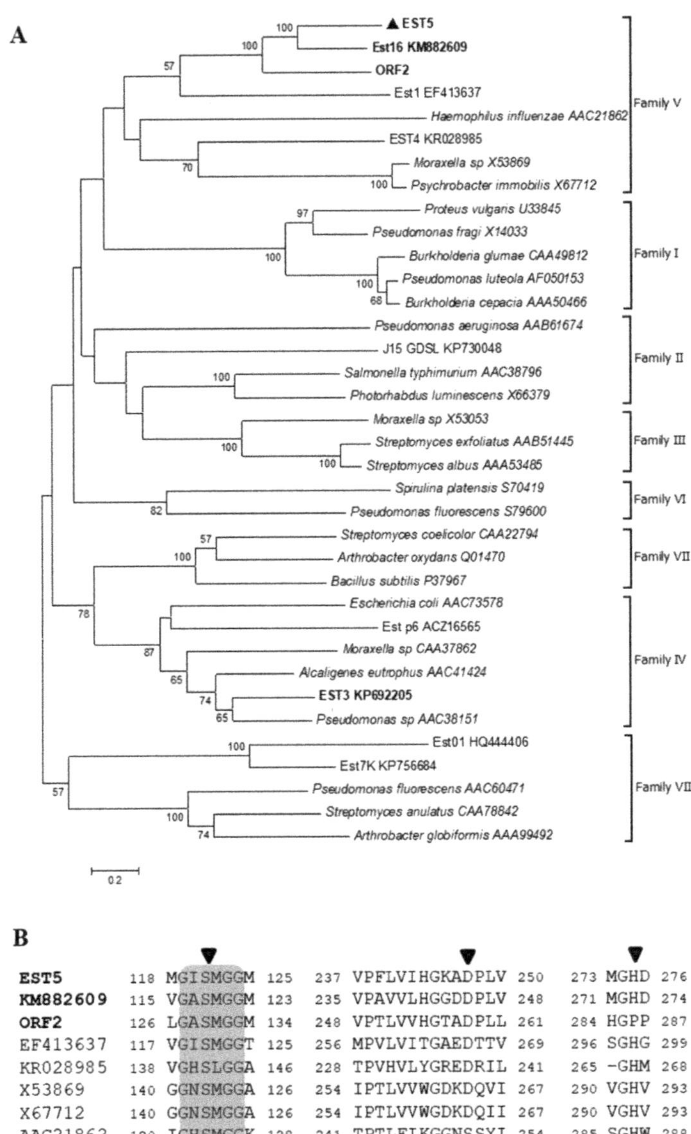

Figure 1. Sequence analysis of EST5. (**A**) Phylogenetic relationship of EST5 (filled triangle) and other lipolytic enzymes. Lipolytic enzymes from the same metagenomic source as EST5 are in black. Only bootstrap values higher than 50% are shown. The scale bar represents 0.2 changes per amino acid. (**B**) Multiple sequence alignment of conserved regions of lipolytic enzymes belonging to family V. Conserved motif is indicated with a gray box. The catalytic triad is indicated with black triangles. Shaded areas (gray) represent the conserved pentapeptide of the lipolytic enzymes and the residues of the catalytic triad are marked with an inverted triangle (▼). Identical (*) or similar residues (. and :) are symbolized below the alignment. The numbers next to the sequence indicate the positions of the amino acids.

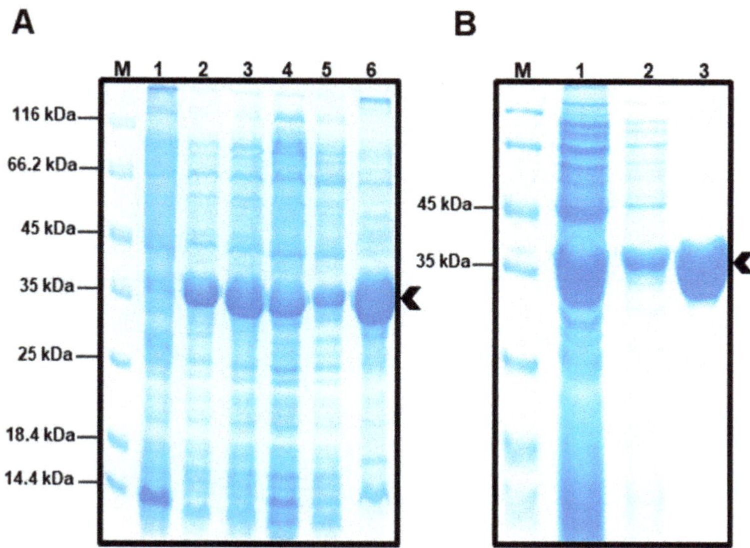

Figure 2. SDS-PAGE analysis of EST5 purification. (**A**) Ni-NTA affinity chromatography purification fractions of EST5 from *E. coli* BL21(DE3) cells carrying the pET28a-*est5* vector. M: molecular weight standard (Thermo Scientific). Lane 1: crude extract of non-induced cells. Lane 2: crude extract of induced cells. Lane 3: soluble extract. Lane 4: flow-through fraction from affinity chromatography. Lanes 5 and 6: eluted fractions with 20 and 50 mM imidazole, respectively. (**B**) EST5 crude extract (Lane 1), after affinity chromatography (Lane 2) and size exclusion chromatography (Lane 3). M represents the molecular weight standard (Thermo Scientific).

2.3. *Enzyme Activity and Its Affecting Factors*

2.3.1. Effect of pH on Activity and Structure of EST5

The effect of pH on the enzymatic activity of EST5 was investigated using *p*NP-C4 as substrate in a pH range from 3 to 10. The enzyme exhibited activity from pH 6.0 to 9.5, with maximum activity at pH 7.5. EST5 retained more than 50% of its activity from pH 6.5 to 9, and the same activity among both bounds (Figure 3A). To assess if the secondary structure and protein folding was correlated to pH variation, CD spectra were measured at differing pH. The Far-UV CD spectra indicated that EST5 displays stronger negative bands at 208 and 222 nm, mostly at different pH values 7.0 and 8.0. On changing the pH to 5.6 and 9.0, there was a decrease of helical content, as shown by the slightly shallower signal in those wavelengths (Figure 3B) [40,41], in agreement with the pH that the enzyme exhibited lower activity. Still monitoring the structure of EST5 face the variation of pH, intrinsic tryptophan fluorescence was measured at pH 7.0, 8.0 and 9.0. EST5 has one tryptophan (W41); exposure of the protein to more basic pH caused a decrease in fluorescence intensity (Figure 3C), suggesting conformational changes of the protein. These data suggest that, at pH 8.0 and, mostly, at pH 9.0, some changes occur in the protein structure that may affect its activity.

To date, no studies regarding the influence of pH on esterase structure using both circular dichroism and intrinsic tryptophan fluorescence as a tool were reported. In our previous study the esterase Est16 exhibited maximal activity at pH 9.0, although the loss of activity in lower pH was not related to differences in the secondary structure, as evinced by CD analysis performed at pH 7.0, 8.0 and 9.0 [38].

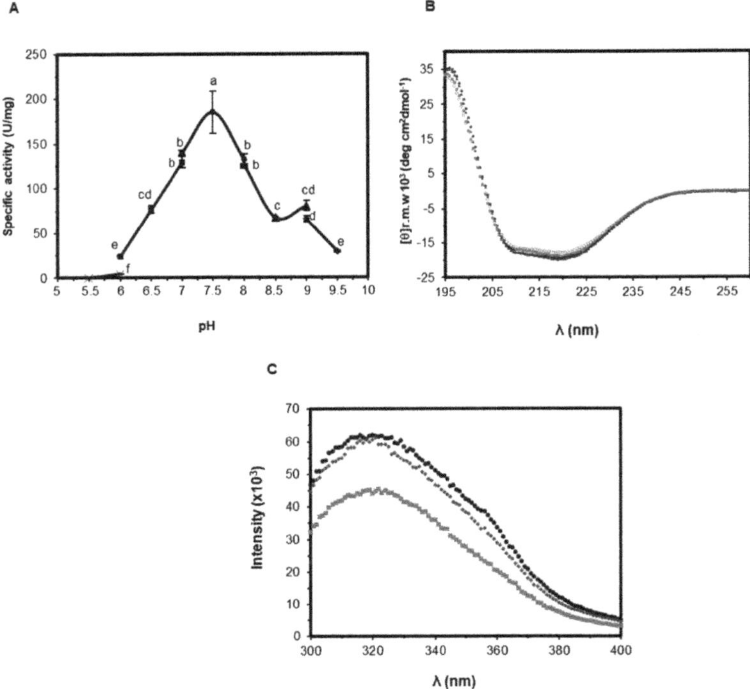

Figure 3. Effect of pH on EST5 activity and structure. (**A**) The effect of pH on enzyme activity. The pH profile was determined in 50 mM Citrate (x), Phosphate (■), HEPES (•), Tris-HCl (▲) and CAPSO (♦) buffers. The amount of released *p*-nitrophenol from *p*NP-butyrate was monitored at 348 nm. Measurements were performed in triplicate assays and error bars represent standard deviation. Small letters on top indicate the significant difference between each condition performed in the experiment, according to ANOVA and Tukey's test at 5% probability. (**B**) Circular dichroism spectra at pH 5.6 (■), 7.0 (•), 8.0 (▲) and 9.0 (◊). The spectrum was obtained in the range from 195 to 260 nm with 2.9 μmol of protein. (**C**) Intrinsic tryptophan fluorescence recorded at 340 nm at pH 7.0 (•), 8.0 (♦) and 9.0 (■).

2.3.2. Effect of Temperature on the Activity and Stability of EST5

The effect of temperature variation on the enzymatic activity of EST5 was investigated using 50 mM sodium phosphate buffer, pH 7.5. EST5 displayed high activity in a temperature range of 30 °C to 45 °C with highest activity at 45 °C (Figure 4A). To test EST5 thermostability, the enzyme was preincubated at 10, 30, 37 and 45 °C and its residual activity was assayed. At 10 °C, 30 °C and 37 °C, the activity was not affected after 15 min of incubation, and was reduced to 74% at 45 °C. After 1 h, EST5 retained 60% of its activity at 45 °C; at 10 °C and 30 °C, the activity was higher than the control (no incubation) and remained higher up to the maximum incubation period (4 h) (Figure 4B).

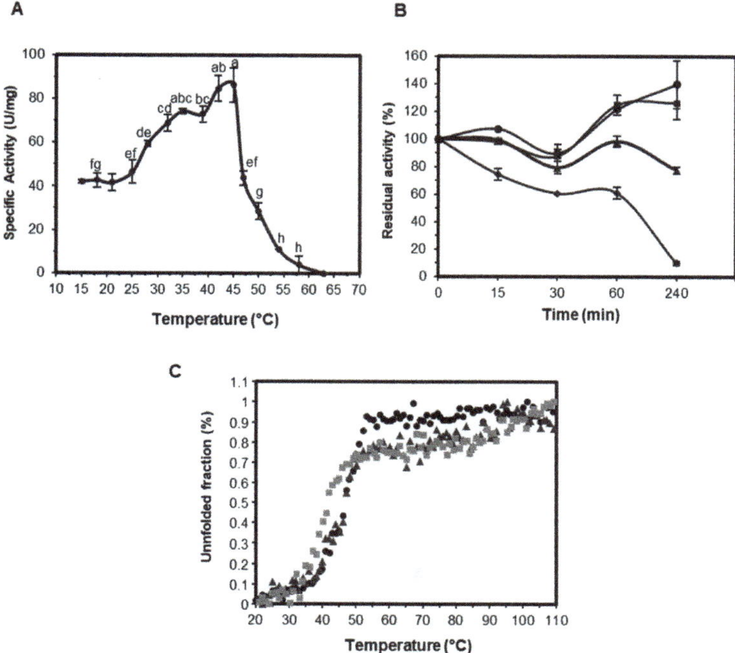

Figure 4. Effect of temperature on EST5 activity and stability. (**A**) Enzyme activity determined from 15 to 65 °C. (**B**) Thermostability. The activity of EST5 was evaluated at different periods of incubation at 10 °C (■), 30 °C (•), 37 °C (▲) and 45 °C (♦). The assays in A and B were performed in 50 mM sodium phosphate pH 7.5 using *p*NP-butyrate as substrate. (**C**) Thermal denaturation profile of EST5 monitored by following the ellipticity at 222 nm from 20 to 110 °C at pH 7,0 (•), 8,0 (▲) and 9,0 (■). Measurements were performed in triplicate assays and error bars represent standard deviation. Small letters on top indicate the significant difference between each condition performed in the experiment, according to ANOVA and Tukey's test at 5% probability.

The effect of temperature on stability of EST5 was also determined by CD spectra. The T*m* value of the enzyme was 47 °C when evaluated at pH 7.0 and 8.0, and 42 °C using buffer at pH 9.0 (Figure 4C). These data confirm low activity of the enzyme above 47 °C inferred from enzymatic activity measurements and reinforce the lower activity of the protein in more basic pH, as conformational changes were observed through intrinsic tryptophan fluorescence tests. Some studies on lipolytic enzymes from metagenomic sources involving circular dichroism or fluorescence have already been reported: a lipase from a metagenomic library of a hot spring soil sample was evaluated based on variation of temperature. Circular dichroism assays revealed distortion in secondary structure at temperatures above 35 °C. However, the study of intrinsic fluorescence data revealed that even with the loss of secondary structure, its tertiary structure was retained [42]. The esterase rEst97 from a high Arctic intertidal zone sediment metagenomic library was considered a cold-adapted enzyme, by having optimum activity at mild temperature (35 °C) and loss of native protein structure at 35–40 °C, as indicated by CD and calorimetry analysis [43]. Est16 protein from our previous study showed two steps for complete unfolding when evaluated at pH 9.0 (40 °C and 95 °C) [38].

The T*m* parameter of EST5 also indicated greater stability at the pH 7.0 and 8.0 in comparison to pH 9.0. By contrast, the esterase Est16 exhibited maximal activity at pH 9.0, although the loss of activity in lower pH was not related to differences in the secondary structure, as evinced by CD analysis performed at pH 7.0, 8.0 and 9.0.

Characterization of EST5 indicated highest hydrolytic activity at 45 °C. However, it is interesting that the enzyme remained at least 50% active at low temperatures. These results show that even though EST5 is a mesophilic enzyme, it has high activity at lower temperatures. Recently, many cold-active or cold-resistant esterases have been isolated, mostly from marine environments or psychrophilic organisms [44–48]. To our knowledge, EST5 is the first cold-active esterase that did not come from low-temperature environment. Moreover, thermal stability analysis showed that incubation of the enzyme up to 30 °C was responsible to increase its activity to 240%.

2.3.3. Substrate Specificity and Kinetic Parameters of EST5

The substrate preference of EST5 was determined under standard assay conditions at pH 7.5 and 45 °C, using *p*NP-esters with different chain lengths (C2–C14) as substrates. The enzyme could hydrolyze all evaluated substrates, as shown in Figure 5. EST5 exhibited higher activity toward *p*NP-C5 and 78% of the maximum activity against *p*NP-C8. The activities toward *p*NP-C2 and *p*NP-C10 were the same, 30% of the relative activity, whereas activities for longer-chain *p*NP-esters (C12 and C14) declined considerably.

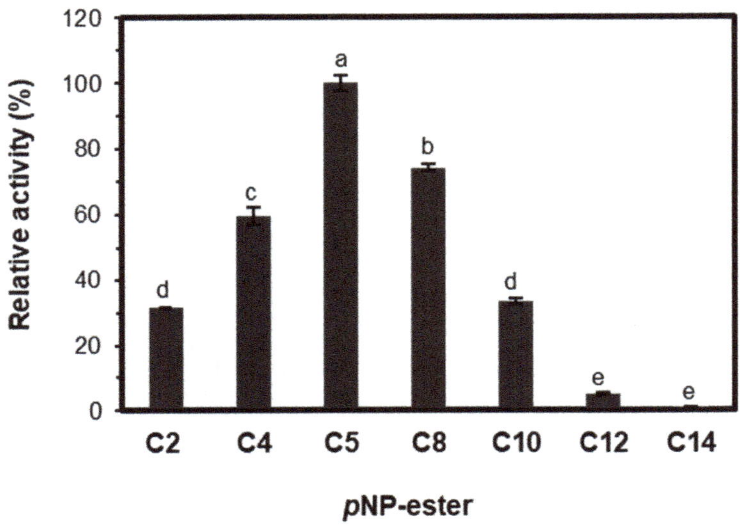

Figure 5. Substrate specificities EST5 on *p*NP-esters. *p*NP-esters of different chain lengths (C2-C14) were assayed at 35 °C in 50 mM Tris-HCl buffer, pH 8.0. Measurements were performed in triplicate assays and error bars represent standard deviation. Small letters on top indicate the significant difference between each condition performed in the experiment, according to ANOVA and Tukey's test at 5% probability.

Kinetic parameters of EST5 are summarized in Table 1. The enzyme showed higher ratio of K_{cat}/K_m against *p*NP-C5 (1533.27 $s^{-1}{\cdot}mM^{-1}$), suggesting that this was the most favored substrate among the *p*NP-esters tested. The V_{max} and K_{cat}/K_m values are, respectively, 1.8- and 2.3-fold higher for C8 in comparison to C4. The substrates C2 and C10 had similar values of catalytic efficiency (18.1 and 17.5 s^{-1} mM^{-1}, respectively).

Table 1. Kinetic parameters of esterase EST5 on *p*NP-esters.

Substrate	Vmax (μM·s^{-1})	Km (mM)	Kcat (s^{-1})	K_{cat}/K_m (s^{-1} mM^{-1})
*p*NP-acetate (C2)	0.5 ± 0.03	1.1 ± 0.13	18.97 ± 1.18	18.1
*p*NP-butyrate (C4)	0.78 ± 0.04	0.4 ± 0.05	29.0 ± 1.46	67.7
*p*NP-valerate (C5)	2.4 ± 0.04	0.1 ± 0.01	88.9 ± 1.02	1533.27
*p*NP-octanoate (C8)	1.4 ± 0.03	0.3 ± 0.04	52.2 ± 1.28	158.24
*p*NP-decanoate (C10)	0.3 ± 0.02	0.73 ± 0.1	12.79 ± 0.74	17.5

Despite EST5 ability to hydrolyze substrates with different acyl chains, the enzyme displayed relative activity markedly lower when the acyl chain length exceeded C10. In view of these findings, the lipolytic enzyme EST5 can be classified as esterase rather than a lipase, since esterases show preference for shorter triacylglycerols, while lipases act more efficiently on esters with longer carbon chain lengths [49].

To date, only a few esterases from family V have been biochemically characterized. Among them, esterase Est1 from metagenomic library derived from sediments of hot spring, which was capable of hydrolyzing *p*NP-esters from C2 to C16 [50]. As EST5, this enzyme showed higher activity toward C5, however K_{cat}/K_m value about 1.6-fold lower. The esterase LC-Est1 from a leaf-branch compost metagenome library exhibited higher catalytic efficiency for C4, with K_{cat}/K_m value of 19.33 s^{-1}·mM^{-1} [51]. Mostly, esterases from metagenomic libraries from different sources exhibited higher activity on *p*NP-C4 [32,52,53]. Still comparing the catalytic efficiency of EST5 with esterases from family V, Est16 [38] exhibited higher kinetic parameters values for almost all evaluated substrates, although EST5 showed 2.3-fold higher catalytic efficiency for C5.

2.3.4. Effect of Additives on EST5 Activity

Many biotechnological processes occur in the presence of some ions in the reaction media that can modify enzyme activity [54]. Thus, the activity of EST5 was evaluated in the presence of ions and salts in the concentrations of 0.5 and 1 mM (Figure 6A). The activity increased in the presence of almost all ions and salts tested at 0.5 mM, mostly Mn^{2+}, K^+ and Ca^{2+}, which were responsible for an increase of around 65%; Fe^{+2} and Na^+ did not affect the enzyme activity, and neither Mn^{2+}, Ni^{2+}, Ca^{2+} and Al^{+3} at 1 mM. Mg^{+2} increased EST5 activity in both tested concentrations, with an increase of about 30%. By increasing the concentration of Na^+ to 1 mM, the relative activity increased to 141%. Some metal ions can increase the solubility of substrates, increasing their availability for the enzyme [55]. In contrast, the enzyme was moderately inhibited by K^+, Li^+, Fe^{+2} and Co^{2+} (Figure 6A). Cu^{+2} inhibited the activity of EST5 at 0.5 mM to 15%, and the enzyme showed no activity in the presence of this ion at 1 mM. EST5 showed similar behavior with other esterases reported in the literature, responding differently to the presence of certain metal ions [56–58]. At 0.5 mM, EDTA increased the enzyme activity but showed slight influence when evaluated at 1 mM. The activity of the lipolytic enzyme Lp_3562 from *Lactobacillus plantarum* was enhanced in the presence of EDTA, showing relative activity of 121%, and also in the presence of Mg^{+2}, Ni^{+2} and Mn^{+2} [59] similarly to the esterase EST5.

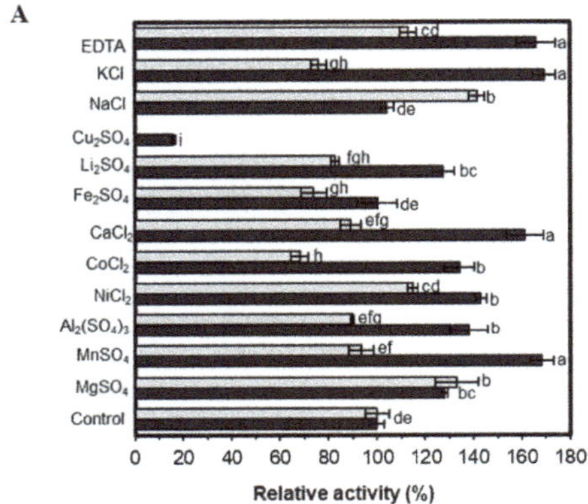

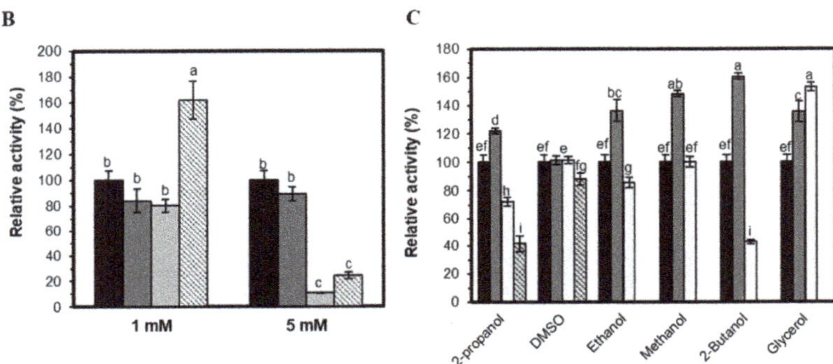

Figure 6. Effect of additives on EST5 activity. (**A**) Effects of metal ions and salts. The enzyme activity was determined in the presence of 0.5 mM (light gray) and 1 mM (dark gray) metal ions, salts and EDTA. (**B**) Effect of detergents at 1 mM and 5 mM: Control (black), Tween 20 (dark gray), Triton X-100 (light gray) and CTAB (striped). (**C**) Effect of organic solvents and glycerol at 1% (dark gray), 10% (light gray), and 20% (striped). Controls are in black bars. Reactions were performed at 35 °C using *p*NP-valerate in 50 mM sodium phosphate buffer, pH 7.5. All measurements were performed in triplicate assays and error bars represent standard deviation. Small letters on top of the bars indicate the significant difference between each condition performed in the experiment, according to ANOVA and Tukey's test at 5% probability.

Detergents are typically added to the reaction media of lipolytic enzymes to improve the quality of the emulsion; however, high concentrations can denature the enzyme [60]. EST5 activity remained stable in the presence of Triton X-100 (1 mM) and Tween 20 (1 mM and 5 mM). 5 mM Triton X-100 inhibited the activity of the enzyme, remaining only 10% of relative activity and Tween 80 caused complete inhibition. Regarding ionic detergents, SDS completely inhibited enzyme activity; 1 mM CTAB increased by almost 70% the activity of EST5 and, at 5 mM, the remaining activity was 24% (Figure 6B).

The effect of organic solvents and glycerol at 1, 10 and 20% on the activity of EST5 is exposed in Figure 6C. The enzyme was almost unaffected in the presence of dimethyl sulfoxide (DMSO), in the three concentrations evaluated, or in the presence of 10% methanol, since no significant difference was detected in comparison to the control. Methanol, ethanol, 2-butanol, and 2-propanol increased enzyme activity at 1% concentration, however, except for methanol, 10% of these substances inhibit the activity of EST5, resulting in 70, 83 and 41% of relative activity, respectively. At 20% concentration of organic solvents, the activity was only detected in DMSO (86%) and 2-propanol (40%). In the presence of 1% and 10% glycerol, the activity of EST5 increased to 132% and 147%, respectively, being inhibited in the presence of 20% of this substance. It is known that, despite the cryoprotectant effect, high concentrations of glycerol have negative effects on the enzymatic activity due to reduced diffusion of the hydrophobic substrate in the active site of the enzyme [61]. The same pattern was observed for esterase EstOF4, which remained stable in 10% DMSO, was inhibited in the presence of 2-propanol and 2-butanol and was slightly stimulated by 10% glycerol [62].

2.4. EST5 Showed Esterification Activity

The esterification activity of EST5 was evaluated using butyric acid and methanol as substrates. The enzyme exhibited esterification activity of 127.04 U·mg^{-1}. The esterase in the present study exhibited the same or even higher ability to be used on esterification reactions as the lipases described in Table 2. The commercial lipases Novozyme 435 (lipase B from *Candida antarctica*) and Lipozyme TLIM (lipase from *Thermomyces lanuginosus*) were analyzed using lauric acid and dodecanol, exhibiting activities of 9.89 and 3.54 U mg^{-1}, respectively. The other commercial lipase CL from *Candida rugosa* showed an activity of 15.83 U·mg^{-1} when assayed with butyric acid and propanol. The esterification activity of EST5 was 1.47-fold higher than LIP4 from *Candida rugosa* and 14.32-fold higher than immobilized *Pseudomonas cepacia* lipase (PS). Higher activities values were achieved for LIP2 (from *Candida rugosa*), which shows preference for butyric acid as substrate, and for the lipase from yeast *Sporidiobolus pararoseus*. The free *Candida rugosa* lipase Lipomod 34MDP showed higher conversions in the esterification reactions of free fatty acids and polyols than two of the most popular commercially available immobilized lipases [63], showing that there is always a demand for new enzymes.

Table 2. Specific esterification activity of lipolytic enzymes.

Enzyme	Substrate	Esterification (U·mg^{-1})	Reference
EST5	Butyric acid/Methanol	127.04	This study
LIP4	Butyric acid/Propanol	86.5	[64,65]
CL	Butyric acid/Propanol	15.83	[65]
LIP2	Butyric acid/Propanol	166	[65]
Immobilized Lipase PS	Lauric acid/Dodecanol	8.87	[66]
Novozyme 435	Lauric acid/Dodecanol	9.89	[66]
Lipozyme TLIM	Lauric acid/Dodecanol	3.54	[66]
YLL	Lauric acid/Propanol	0.35	[67]
Sporidiobolus pararoseus lipase	Oleic Acid/Ethanol	489.65	[68]

The fungal esterase RmEstA in free form produced butyl butyrate with an esterification efficiency of 56%, 35% less than the immobilized form of the enzyme [69].

Most currently commercially available esters are obtained by chemical synthesis using strong acids or bases as catalysts. Even though a considerable amount of current studies is focused on the production of aromas and flavors, only a few are obtained biotechnologically in industrial scale [70].

The ability to produce methyl butyrate makes EST5 attractive for possible application in the synthesis of flavor esters, especially fruity flavors used in food and pharmaceutical products [71,72]. Although complementary studies would be necessary, these preliminary studies of EST5 application shows promising results. Probably, after optimization of reaction parameters, higher conversion rates can be achieved.

Since EST5 showed preference for the *p*NP-esters C4, C5 and C8, valeric and caprylic acids could be possible substrates for its esterification reactions. Short- and medium-chain esters have known application as flavoring agents. Recently, environmental issues have driven research on sustainable sources for fuels. The valeric acid is a lignocellulosic material, considered a cheap and renewable feedstock. The esterification product of valeric acid and ethanol—ethyl valerate—was considered a possible second generation biofuel [73,74].

3. Materials and Methods

From the fosmid metagenomic library isolated from a microbial consortium specialized in diesel oil degradation, 30 positive clones were selected based on the formation of clear halos around individual colonies, which indicated hydrolysis of tributyrin in Luria–Bertani (LB) agar plates [38]. Recently, we sequenced the clone PL14.H10, which assembled four Open Reading Frames (ORFs) identified as putative genes encoding lipolytic enzymes [39]. One of the four potential lipolytic genes was selected to be characterized in this work.

3.1. Sequence and Phylogenetic Analysis

Homology searches were performed using BLAST analysis at the National Center for Biotechnology Information (NCBI, Bethesda, MA, USA). Signal peptides were predicted using SignalP 4.0 (http://www.cbs.dtu.dk/services/SignalP/) [75]. The molecular mass and pI of the encoded protein were analyzed via ProtParam tool in the ExPASy website (http://www.expasy.ch/tools/protparam.html).

Lipases/esterases from the eight families proposed by Arpigny and Jaeger (1999) as well as those novel enzymes recently described were used to construct a phylogenetic tree: EST3 [39], Est16 [38] and ORF2 [37] from the same metagenomic library as EST5 (this study); Est_p6 [76]; J15 GDSL [77]; Est7k [78]; Est1 [50] and EST4 [32]. The retrieved sequences were aligned in Clustal W [79] in the BioEdit Sequence Alignment Program (version 7.0.5.3, North Carolina State University, NC, USA, 1998). Phylogenetic tree was constructed by the neighbor-joining algorithm, as implemented in MEGA 6 [80] program package, with bootstrapping [81] based on 1,000 replicates.

3.2. Expression and Purification of the Recombinant Enzyme

For enzyme overexpression, the full length *est5* gene was amplified using oligonucleotides forward 5′-GATGAATTCCCCATGACCGTCAACAT-3′ and reverse 5′-AGGTCTCGAGCCTTATTCCGCGG-3′ with the respective EcoRI and XhoI restriction sites (underlined). PCR was performed using Pfu DNA polymerase (Thermo Scientific, Waltham, MA, USA) according to the manufacturer recommendations. The purified PCR product was digested and cloned into pET-28a(+) vector to obtain the recombinant plasmid pET28a-*est5*. DNA sequencing confirmed the constructed plasmid integrity. Then, the recombinant plasmid was transformed into *E. coli* BL21(DE3) competent cells for heterologous expression of the protein. Recombinant *E. coli* cells were cultivated in LB liquid medium supplemented with 50 μg·mL^{-1} kanamycin and agitated at 200 rpm, 37 °C. For optimization tests, when the optical density OD_{600} of the culture reached an absorbance of 0.6, 0.1 mM IPTG (isopropyl-b-d-thiogalactopyranoside) was added to induce protein expression at different temperatures (22–37 °C) and time (2–8 h) in rotary shaking at 200 rpm. After induction, the culture was centrifuged at 11,953× *g* for 10 min at 4 °C and the pellet was then homogenized in Buffer A (50 mM Tris-HCl, pH 8.0; 100 mM NaCl and 10% glycerol). Lysozyme was added at a final concentration of 4 μg·mL^{-1} and the material was incubated at an ice bath for 1 h. Cells were disrupted by sonication in ice using the Branson sonifier equipment (Branson, CT, USA) with six cycles of 10 pulses at 30% amplitude with 20 s intervals and then centrifuged at 38,724× *g* for 20 min at 4 °C to pellet the cellular debris and collect the supernatant containing the targeted protein.

The enzyme was purified by immobilized metal ion chromatography and size-exclusion chromatography for characterization purposes. The recovered supernatant from the centrifugation step was mixed with Ni-NTA agarose resin (Qiagen, Venlo, The Netherlands) previously equilibrated with Buffer A containing 20 mM imidazole and loaded to a gravity flow column (Qiagen). The target protein was eluted and then concentrated with Vivaspin 20 30,000 molecular weight cut-off (MWCO) (Sartorius Stedim, Aubagne, France) under centrifugation at 200× *g*. Size-exclusion chromatography was used as an additional purification step in a HiLoad 16/60 Superdex 200 column (GE Healthcare, Little Chalfont, UK), at 4 °C using the elution buffer (20 mM Tris-HCl pH 8.0, 50 mM NaCl and

5% glycerol). Purity and concentration was analyzed by SDS-PAGE [82] and Nanodrop ND-1000 spectrophotometer (Thermo Scientific), respectively.

3.3. Biophysical Analysis

The effect of temperature and pH on enzyme conformation was monitored by the circular dichroism (CD) and fluorescence spectroscopic techniques. Circular dichroism measurements were performed with JASCO J-810 spectropolarimeter fitted with a Jasco Peltier-type temperature controller (PFD425S). Far-UV CD spectra were recorded at 10 °C in the wavelength region of 195–260 nm at 0.5 nm intervals, path length of 1 mm. 2.9 µmol of protein was used in buffers: 5 mM sodium citrate pH 5.6; 5 mM sodium phosphate, pH 7.0 and 8.0; and 5 mM N-Cyclohexyl-2-aminoethanesulfonic (CHES), pH 9.0. The thermal denaturation curve of the protein was obtained by monitoring the change in CD values at 222 nm as the temperature increased from 20 °C to 110 °C with spectra collected at 1 °C intervals. The midpoint temperature (Tm) of the unfolding transition was analyzed.

Intrinsic tryptophan fluorescence measurements were carried out on K2 multifrequency spectrofluorometer (ISS Inc., Champaign, IL, USA). Fluorescence spectra were measured at a protein concentration of 1 µM in 5 mM Tris-HCl at pH 7.0, 8.0 and 9.0. The excitation wavelength was set at 295 nm and the emission spectra were recorded between 300 to 400 nm.

3.4. Enzyme Characterization

Enzyme activity was measured at 405 nm with a SpectraMax M2e spectrophotometer (Molecular Devices, Sunnyvale, CA, USA) equipped with a temperature controller. Standard assays, unless otherwise indicated, were performed at 35 °C for 5 min in 100 µL mixture containing: 27.01 nM of the purified enzyme, 50 mM Tris-HCl buffer, pH 8.0, with 0.3% (*v*/*v*) Triton X-100 and 1 mM *p*-nitrophenyl butyrate dissolved in isopropanol/acetonitrile (4:1 *v*/*v*). The amount of *p*-nitrophenol produced by the reaction was determined from the absorbance with a molar extinction coefficient value of 17,000 $M^{-1}\cdot cm^{-1}$. One enzyme unit is the amount of enzyme that releases 1 µmol of product from substrate per minute under standard assay conditions.

All measurements were carried out in triplicate and a blank reaction without enzyme was included for each experiment. Data was analyzed using the R software (R Foundation for Statistical Computing, Vienna, Austria, 2016). ANOVA and Tukey's test at 5% probability were used to make comparisons among the different conditions evaluated.

3.4.1. Substrate Selectivity

Substrate specificity of the enzyme was carried out using *p*NP-esters: *p*NP-acetate (C2), *p*NP-butyrate (C4), *p*NP-valerate (C5), *p*NP-caprylate (C8), *p*NP-caprate (C10), *p*NP-laurate (C12), *p*NP-myristate (C14) and *p*NP-palmitate (C16) (Sigma, St. Louis, MO, USA). Assays were performed according to standard enzyme assay method. Initial reaction velocities were measured with *p*NP-esters at a concentration range from 0.04 mM to 2.5 mM. Kinetic parameters were obtained by nonlinear regression of the data on Michaelis–Menten equation using GraphPad Prism software version 5.0 (GraphPad Software, Inc, San Diego, CA, USA).

3.4.2. Effect of pH and Temperature on Enzyme Activity

To investigate the effect of pH on enzymatic activity, enzymatic assays were performed under standard assay conditions and the following buffers were used at 50 mM: citrate (pH 3.0 to 6.0), sodium phosphate (pH 6.0 to 8.0), 4-(2-Hydroxyethyl)piperazine-1-ethanesulfonic acid (HEPES) (6.5 to 8.0), Tris-HCl (pH 7.5 to 9.0), and 3-(Cyclohexylamino)-2-hydroxy-1-propanesulfonic acid (CAPSO) (pH 9.0 to 10). The temperature effects on enzyme activity were determined by assessing the enzyme activity from 15 to 65 °C, under standard conditions, with 50 mM HEPES buffer, pH 7.5. To study thermostability, the enzyme was incubated from 10 to 60 °C for up to 4 h. After the enzyme solution was heated to appointed temperature, the residual activity was measured under standard assay conditions.

3.4.3. Effect of Additives on Enzyme Activity

To evaluate the effect of detergents, metal ions, organic solvents and chelating agents, the enzyme was incubated in their presence for 5 min at 4 °C. The influence of metal ions (Na^+, K^+, Mg^{2+}, Ca^{2+}, Mn^{2+}, Co^{2+}, Cu^{+2}, Fe^{+2}, Ni^{+2}, Al^{+3} or Li^+) and the chelating agent EDTA (ethylenediaminetetraacetic acid) was investigated at final concentrations of 0.5 and 1 mM (*w*/*v*). To measure the effect of detergents on the enzyme activity Tween 20, Tween 80, Triton X-100, sodium dodecyl sulfate (SDS) or hexadecyltrimethylammonium bromide (CTAB) was added in the final concentrations of 1 and 5 mM (*w*/*v*). Stability of the enzyme in the presence of organic solvents was tested with 1%, 10%, and 20% (*v*/*v*) of methanol, ethanol, 2-propanol, 1-butanol and dimethyl sulfoxide (DMSO). The presence of glycerol was also investigated in the same concentrations. Enzyme activity without additives was defined as 100%.

3.5. Esterification Assays

Esterification activity was assayed according to method described previously [68]. A volume of 0.0395 μmol of protein in solution (10 mM Tris-HCl, pH 8.0, and 50 mM NaCl) was added to the reaction mixture containing 3.66 g of butyric acid and 0.7 g of methanol. The mixtures were incubated for 40 min at 40 °C and shaken at 150 rpm. After incubation time, reactions were stopped by adding 20 mL of an acetone/ethanol solution (1:1, *v*/*v*). Unreacted fatty acids in the mixture were titrated with 0.035 M NaOH up to pH 11.0. One unit of activity was defined as the amount of enzyme able to consume one μmol of fatty acid per minute under assay conditions.

4. Conclusions

Characterization of the esterase in this study provided useful information about its function and structure. EST5 showed activity toward a wide range of substrates and temperatures, was stable in the presence of some detergents and organic solvents, and incubation at 10 °C and 30 °C for up to 4 h increased its activity. In addition, EST5 had higher esterification activity than some previously reported lipases, including commercial enzymes. EST5-catalyzed esterification reactions can be conducted at lower temperatures in the presence of organic solvents and in solvents-free media. Thus, this enzyme displays a high potential for biotechnological applications.

Author Contributions: Conceptualization, T.C.M. and E.G.M.L.; Formal analysis, T.C.M., M.R.P., A.M.G.M. and J.P.B.; Funding acquisition, T.C.M. and E.G.M.L.; Investigation, T.C.M., M.R.P., A.M.G.M. and J.P.B.; Methodology, T.C.M. and E.G.M.L.; Writing—original draft, T.C.M.; Writing—review & editing, M.R.P., A.M.G.M., J.P.B., P.A.M.P. and E.G.M.L. All authors have read and agreed to the published version of the manuscript.

Funding: This research was funded by the São Paulo Research Foundation—FAPESP, grant numbers 2011/09064-6 and 2013/03568-8.

Acknowledgments: We are grateful for the Spectroscopy and Calorimetry Facility (Brazilian Biosciences National Laboratory-LNBio, Campinas, Brazil) for enabling the CD and fluorescence measurements. We thank the infrastructure offered by the São Paulo State University, Jaboticabal Campus, Brazil and the University of São Paulo, São Paulo, Brazil. We also thank Professor Eleni Gomes and the Institute of Biosciences, Languages and Exact Sciences, Department of Biology, which helped with the esther synthesis analysis.

Conflicts of Interest: The authors declare no conflict of interest.

References

1. Sheldon, R.A.; van Pelt, S. Enzyme immobilisation in biocatalysis: Why, what and how. *Chem. Soc. Rev.* **2013**, *42*, 6223–6235. [CrossRef] [PubMed]
2. Bezborodov, A.M.; Zagustina, N.A. Lipases in catalytic reactions of organic chemistry. *Appl. Biochem. Microbiol.* **2014**, *50*, 313–337. [CrossRef]
3. De Miranda, A.S.; Miranda, L.S.M.; de Souza, R.O.M.A. Lipases: Valuable catalysts for dynamic kinetic resolutions. *Biotechnol. Adv.* **2015**, *33*, 372–393. [CrossRef] [PubMed]

4. Chapman, J.; Ismail, A.E.; Dinu, C.Z. Industrial applications of enzymes: Recent advances, techniques, and outlooks. *Catalysts* **2018**, *8*, 238. [CrossRef]
5. Littlechild, J.A. Improving the 'tool box' for robust industrial enzymes. *J. Ind. Microbiol. Biotechnol.* **2017**, *44*, 711–720. [CrossRef] [PubMed]
6. Bornscheuer, U. Microbial carboxyl esterases: Classification, properties and application in biocatalysis. *FEMS Microbiol. Rev.* **2002**, *26*, 73–81. [CrossRef]
7. Ferrer, M.; Bargiela, R.; Martínez-Martínez, M.; Mir, J.; Koch, R.; Golyshina, O.V.; Golyshin, P.N. Biodiversity for biocatalysis: A review of the α/β-hydrolase fold superfamily of esterases-lipases discovered in metagenomes. *Biocatal. Biotransform.* **2016**, *2422*, 1–15. [CrossRef]
8. Hasan, F.; Shah, A.A.; Hameed, A. Industrial applications of microbial lipases. *Enzyme Microb. Technol.* **2006**, *39*, 235–251. [CrossRef]
9. Kanmani, P.; Aravind, J.; Kumaresan, K. An insight into microbial lipases and their environmental facet. *Int. J. Environ. Sci. Technol.* **2015**, *12*, 1147–1162. [CrossRef]
10. Carvalho, A.; Fonseca, T.; Mattos, M.; Oliveira, M.; Lemos, T.; Molinari, F.; Romano, D.; Serra, I. Recent Advances in Lipase-Mediated Preparation of Pharmaceuticals and Their Intermediates. *Int. J. Mol. Sci.* **2015**, *16*, 29682–29716. [CrossRef]
11. Trono, D. Recombinant Enzymes in the Food and Pharmaceutical Industries. In *Advances in Enzyme Technology*; Elsevier: Amsterdam, The Netherlands, 2019; pp. 349–387.
12. Samoylova, Y.V.; Sorokina, K.N.; Piligaev, A.V.; Parmon, V.N. Application of Bacterial Thermostable Lipolytic Enzymes in the Modern Biotechnological Processes: A Review. *Catal. Ind.* **2019**, *11*, 168–178. [CrossRef]
13. Khambhaty, Y. Applications of enzymes in leather processing. *Environ. Chem. Lett.* **2020**, *18*, 747–769. [CrossRef]
14. Zhang, Z.; Lan, D.; Zhou, P.; Li, J.; Yang, B.; Wang, Y. Control of sticky deposits in wastepaper recycling with thermophilic esterase. *Cellulose* **2017**, *24*, 311–321. [CrossRef]
15. Sahay, S.; Chouhan, D. Study on the potential of cold-active lipases from psychrotrophic fungi for detergent formulation. *J. Genet. Eng. Biotechnol.* **2018**, *16*, 319–325. [CrossRef]
16. Kumar, A.; Gudiukaite, R.; Gricajeva, A.; Sadauskas, M.; Malunavicius, V.; Kamyab, H.; Sharma, S.; Sharma, T.; Pant, D. Microbial lipolytic enzymes—Promising energy-efficient biocatalysts in bioremediation. *Energy* **2020**, *192*, 116674. [CrossRef]
17. López-López, O.; Cerdán, M.E.; Gonzalez-Siso, M.I. New Extremophilic Lipases and Esterases from Metagenomics. *Curr. Protein Pept. Sci.* **2014**, *15*, 445–455. [CrossRef]
18. Markets and Markets. *Lipase Market by Source (Microbial Lipases, Animal Lipases), Application (Animal Feed, Dairy, Bakery, Confectionery, Others), & by Geography (North America, Europe, Asia-Pacific, Latin America, RoW)—Global Forecast to 2020*; Markets and Markets: Pune, India, 2015.
19. Fojan, P.; Jonson, P.H.; Petersen, M.T.N.; Petersen, S.B. What distinguishes an esterase from a lipase: A novel structural approach. *Biochimie* **2000**, *82*, 1033–1041. [CrossRef]
20. Arpigny, J.L.; Jaeger, K.-E. Bacterial lipolytic enzymes: Classification and properties. *Biochem. J.* **1999**, *343*, 177–183. [CrossRef]
21. Salgado, V.; Fonseca, C.; Lopes da Silva, T.; Roseiro, J.C.; Eusébio, A. Isolation and Identification of Magnusiomyces capitatus as a Lipase-Producing Yeast from Olive Mill Wastewater. *Waste Biomass Valor.* **2020**, *11*, 3207–3221. [CrossRef]
22. Chen, C.C.; Chi, Z.; Liu, G.L.; Jiang, H.; Hu, Z.; Chi, Z.M. Production, purification, characterization and gene cloning of an esterase produced by Aureobasidium melanogenum HN6.2. *Process Biochem.* **2017**, *53*, 69–79. [CrossRef]
23. Rade, L.L.; da Silva, M.N.P.; Vieira, P.S.; Milan, N.; de Souza, C.M.; de Melo, R.R.; Klein, B.C.; Bonomi, A.; de Castro, H.F.; Murakami, M.T.; et al. A Novel Fungal Lipase With Methanol Tolerance and Preference for Macaw Palm Oil. *Front. Bioeng. Biotechnol.* **2020**. [CrossRef] [PubMed]
24. Turati, D.F.M.; Almeida, A.F.; Terrone, C.C.; Nascimento, J.M.F.; Terrasan, C.R.F.; Fernandez-Lorente, G.; Pessela, B.C.; Guisan, J.M.; Carmona, E.C. Thermotolerant lipase from Penicillium sp. section Gracilenta CBMAI 1583: Effect of carbon sources on enzyme production, biochemical properties of crude and purified enzyme and substrate specificity. *Biocatal. Agric. Biotechnol.* **2019**, *17*, 15–24. [CrossRef]
25. Lin, X.-S.; Zhao, K.-H.; Zhou, Q.-L.; Xie, K.-Q.; Halling, P.J.; Yang, Z. Aspergillus oryzae lipase-catalyzed synthesis of glucose laurate with excellent productivity. *Bioresour. Bioprocess.* **2016**, *3*, 2. [CrossRef]

26. Bhardwaj, K.K.; Dogra, A.; Kapoor, S.; Mehta, A.; Gupta, R. Purification and Properties of an Esterase from Bacillus licheniformis and it's Application in Synthesis of Octyl Acetate. *Open Microbiol. J.* **2020**, *14*, 113–121. [CrossRef]
27. Noor, H.; Satti, S.M.; Din, S.U.; Farman, M.; Hasan, F.; Khan, S.; Badshah, M.; Shah, A.A. Insight on esterase from Pseudomonas aeruginosa strain S3 that depolymerize poly(lactic acid) (PLA) at ambient temperature. *Polym. Degrad. Stab.* **2020**, *174*, 109096. [CrossRef]
28. Bharathi, D.; Rajalakshmi, G.; Komathi, S. Optimization and production of lipase enzyme from bacterial strains isolated from petrol spilled soil. *J. King Saud Univ. Sci.* **2019**, *31*, 898–901. [CrossRef]
29. Boran, R.; Ugur, A.; Sarac, N.; Ceylan, O. Characterisation of Streptomyces violascens OC125-8 lipase for oily wastewater treatment. *3 Biotech* **2019**, *9*, 5. [CrossRef]
30. Kumar, A.; Dhar, K.; Kanwar, S.S.; Arora, P.K. Lipase catalysis in organic solvents: Advantages and applications. *Biol. Proced. Online* **2016**, *18*, 2. [CrossRef]
31. Lai, O.-M.; Lee, Y.-Y.; Phuah, E.-T.; Akoh, C.C. Lipase/Esterase: Properties and Industrial Applications. In *Encyclopedia of Food Chemistry*; Elsevier: Amsterdam, The Netherlands, 2019; pp. 158–167.
32. Gao, W.; Wu, K.; Chen, L.; Fan, H.; Zhao, Z.; Gao, B.; Wang, H.; Wei, D. A novel esterase from a marine mud metagenomic library for biocatalytic synthesis of short-chain flavor esters. *Microb. Cell Fact.* **2016**, *15*, 41. [CrossRef]
33. Brault, G.; Shareck, F.; Hurtubise, Y.; Lépine, F.; Doucet, N. Short-chain flavor ester synthesis in organic media by an E. coli whole-cell biocatalyst expressing a newly characterized heterologous lipase. *PLoS ONE* **2014**, *9*, e91872. [CrossRef]
34. SÁ, A.G.A.; Meneses, A.C.D.; Araújo, P.H.H.D.; Oliveira, D.D. A review on enzymatic synthesis of aromatic esters used as flavor ingredients for food, cosmetics and pharmaceuticals industries. *Trends Food Sci. Technol.* **2017**, *69*, 95–105. [CrossRef]
35. Zhong, X.L.; Tian, Y.Z.; Jia, M.L.; Liu, Y.D.; Cheng, D.; Li, G. Characterization and purification via nucleic acid aptamers of a novel esterase from the metagenome of paper mill wastewater sediments. *Int. J. Biol. Macromol.* **2020**, *153*, 441–450. [CrossRef] [PubMed]
36. Almeida, J.M.; Martini, V.P.; Iulek, J.; Alnoch, R.C.; Moure, V.R.; Müller-Santos, M.; Souza, E.M.; Mitchell, D.A.; Krieger, N. Biochemical characterization and application of a new lipase and its cognate foldase obtained from a metagenomic library derived from fat-contaminated soil. *Int. J. Biol. Macromol.* **2019**, *137*, 442–454. [CrossRef] [PubMed]
37. Garcia, R.A.M.; Pereira, M.R.; Maester, T.C.; de Macedo Lemos, E.G. Investigation, Expression, and Molecular Modeling of ORF2, a Metagenomic Lipolytic Enzyme. *Appl. Biochem. Biotechnol.* **2015**, *175*, 3875–3887. [CrossRef]
38. Pereira, M.R.; Mercaldi, G.F.; Maester, T.C.; Balan, A.; Lemos, E.G.M. Est16, a new esterase isolated from a metagenomic library of a microbial consortium specializing in diesel oil degradation. *PLoS ONE* **2015**, *10*, e0133723. [CrossRef]
39. Maester, T.C.; Pereira, M.R.; Machado Sierra, E.G.; Balan, A.; de Macedo Lemos, E.G. Characterization of EST3: A metagenome-derived esterase with suitable properties for biotechnological applications. *Appl. Microbiol. Biotechnol.* **2016**, *100*, 5815–5827. [CrossRef]
40. Greenfield, N.J. Using circular dichroism collected as a funcion of temperature to determine the thermodynamics of protein unfolding and binding interactions. *Nat. Protoc.* **2009**, *1*, 2527–2535. [CrossRef]
41. Greenfield, N.J. Biomacromolecular Applications of Circular Dichroism and ORD BT—Encyclopedia of Spectroscopy and Spectrometry. In *Encyclopedia of Spectroscopy and Spectrometry*, 2nd ed.; Lindon, J.C., Tranter, G.E., Holmes, J.L., Eds.; Academic Press: San Diego, CA, USA, 1999; Volume 1, pp. 153–165.
42. Sharma, P.K.; Singh, K.; Singh, R.; Capalash, N.; Ali, A.; Mohammad, O.; Kaur, J. Characterization of a thermostable lipase showing loss of secondary structure at ambient temperature. *Mol. Biol. Rep.* **2012**, *39*, 2795–2804. [CrossRef]
43. Fu, J.; Leiros, H.K.S.; De Pascale, D.; Johnson, K.A.; Blencke, H.M.; Landfald, B. Functional and structural studies of a novel cold-adapted esterase from an Arctic intertidal metagenomic library. *Appl. Microbiol. Biotechnol.* **2013**, *97*, 3965–3978. [CrossRef]
44. Dong, J.; Gasmalla, M.A.A.; Zhao, W.; Sun, J.; Liu, W.; Wang, M.; Han, L.; Yang, R. Characterisation of a cold adapted esterase and mutants from a psychotolerant Pseudomonas sp. strain. *Biotechnol. Appl. Biochem.* **2016**, *64*, 686–699. [CrossRef]

45. Tchigvintsev, A.; Tran, H.; Popovic, A.; Kovacic, F.; Brown, G.; Flick, R.; Hajighasemi, M.; Egorova, O.; Somody, J.C.; Tchigvintsev, D.; et al. The environment shapes microbial enzymes: Five cold-active and salt-resistant carboxylesterases from marine metagenomes. *Appl. Microbiol. Biotechnol.* **2015**, *99*, 2165–2178. [CrossRef] [PubMed]
46. Jiang, H.; Zhang, S.; Gao, H.; Hu, N. Characterization of a cold-active esterase from Serratia sp. and improvement of thermostability by directed evolution. *BMC Biotechnol.* **2016**, *16*, 7. [CrossRef] [PubMed]
47. Borchert, E.; Selvin, J.; Kiran, S.G.; Jackson, S.A.; O'Gara, F.; Dobson, A.D.W. A novel cold active esterase from a deep sea sponge stelletta normani metagenomic library. *Front. Mar. Sci.* **2017**, *4*, 287. [CrossRef]
48. De Santi, C.; Altermark, B.; Pierechod, M.M.; Ambrosino, L.; De Pascale, D.; Willassen, N.P. Characterization of a cold-active and salt tolerant esterase identified by functional screening of Arctic metagenomic libraries. *BMC Biochem.* **2016**, *17*, 1. [CrossRef] [PubMed]
49. Jaeger, K.-E.; Dijkstra, B.W.; Reetz, M.T. Bacterial Biocatalysts: Molecular Biology, Three-Dimensional Structures, and Biotechnological Applications of Lipases. *Annu. Rev. Microbiol.* **1999**, *53*, 315–351. [CrossRef]
50. Tirawongsaroj, P.; Sriprang, R.; Harnpicharnchai, P.; Thongaram, T.; Champreda, V.; Tanapongpipat, S.; Pootanakit, K.; Eurwilaichitr, L. Novel thermophilic and thermostable lipolytic enzymes from a Thailand hot spring metagenomic library. *J. Biotechnol.* **2008**, *133*, 42–49. [CrossRef]
51. Okano, H.; Hong, X.; Kanaya, E.; Angkawidjaja, C.; Kanaya, S. Structural and biochemical characterization of a metagenome-derived esterase with a long N-terminal extension. *Protein Sci.* **2015**, *24*, 93–104. [CrossRef]
52. Zarafeta, D.; Szabo, Z.; Moschidi, D.; Phan, H.; Chrysina, E.D.; Peng, X.; Ingham, C.J.; Kolisis, F.N.; Skretas, G. EstDZ3: A new esterolytic enzyme exhibiting remarkable thermostability. *Front. Microbiol.* **2016**, *7*, 1779. [CrossRef]
53. Sukul, P.; Lupilov, N.; Leichert, L.I. Characterization of ML-005, a novel metaproteomics-derived esterase. *Front. Microbiol.* **2018**, *9*, 2716. [CrossRef]
54. Bofill, C.; Prim, N.; Mormeneo, M.; Manresa, A.; Javier Pastor, F.I.; Diaz, P. Differential behaviour of Pseudomonas sp. 42A2 LipC, a lipase showing greater versatility than its counterpart LipA. *Biochimie* **2010**, *92*, 307–316. [CrossRef]
55. Kumar, S.; Mathur, A.; Singh, V.; Nandy, S.; Khare, S.K.; Negi, S. Bioremediation of waste cooking oil using a novel lipase produced by Penicillium chrysogenum SNP5 grown in solid medium containing waste grease. *Bioresour. Technol.* **2012**, *120*, 300–304. [CrossRef] [PubMed]
56. Maqbool, Q.U.A.; Johri, S.; Rasool, S.; Riyaz-ul-Hassan, S.; Verma, V.; Nargotra, A.; Koul, S.; Qazi, G.N. Molecular cloning of carboxylesterase gene and biochemical characterization of encoded protein from Bacillus subtilis (RRL BB1). *J. Biotechnol.* **2006**, *125*, 1–10. [CrossRef] [PubMed]
57. Wu, C.; Sun, B. Identification of novel esterase from metagenomic library of Yangtze River. *J. Microbiol. Biotechnol.* **2009**, *19*, 187–193. [PubMed]
58. Jeon, J.H.; Kim, J.T.; Lee, H.S.; Kim, S.J.; Kang, S.G.; Choi, S.H.; Lee, J.H. Novel lipolytic enzymes identified from metagenomic library of deep-sea sediment. *Evid. Based Complement. Altern. Med.* **2011**, *2011*, 271419. [CrossRef]
59. Esteban-Torres, M.; Mancheño, J.M.; de las Rivas, B.; Muñoz, R. Characterization of a halotolerant lipase from the lactic acid bacteria Lactobacillus plantarum useful in food fermentations. *LWT Food Sci. Technol.* **2015**, *60*, 246–252. [CrossRef]
60. Glogauer, A.; Martini, V.P.; Faoro, H.; Couto, G.H.; Müller-Santos, M.; Monteiro, R.A.; Mitchell, D.A.; de Souza, E.M.; Pedrosa, F.O.; Krieger, N. Identification and characterization of a new true lipase isolated through metagenomic approach. *Microb. Cell Fact.* **2011**, *10*, 1–15. [CrossRef]
61. Xu, Y.; Nordblad, M.; Nielsen, P.M.; Brask, J.; Woodley, J.M. In situ visualization and effect of glycerol in lipase-catalyzed ethanolysis of rapeseed oil. *J. Mol. Catal. B Enzym.* **2011**, *72*, 213–219. [CrossRef]
62. Rao, L.; Xue, Y.; Zheng, Y.; Lu, J.R.; Ma, Y. A Novel Alkaliphilic Bacillus Esterase Belongs to the 13th Bacterial Lipolytic Enzyme Family. *PLoS ONE* **2013**, *8*, e60645. [CrossRef]
63. Cavalcanti, E.D.C.; Aguieiras, É.C.G.; da Silva, P.R.; Duarte, J.G.; Cipolatti, E.P.; Fernandez-Lafuente, R.; da Silva, J.A.C.; Freire, D.M.G. Improved production of biolubricants from soybean oil and different polyols via esterification reaction catalyzed by immobilized lipase from Candida rugosa. *Fuel* **2018**, *215*, 705–713. [CrossRef]
64. Tang, S.J.; Sun, K.H.; Sun, G.H.; Chang, T.Y.; Lee, G.C. Recombinant expression of the Candida rugosa lip4 lipase in Escherichia coli. *Protein Expr. Purif.* **2000**, *20*, 308–313. [CrossRef]

65. Lee, G.-C.; Lee, L.-C.; Sava, V.; Shaw, J.-F. Multiple mutagenesis of non-universal serine codons of the Candida rugosa LIP2 gene and biochemical characterization of purified recombinant LIP2 lipase overexpressed in Pichia pastoris. *Biochem. J.* **2002**, *366*, 603–611. [CrossRef]
66. Liu, T.; Vora, H.; Khosla, C. Quantitative analysis and engineering of fatty acid biosynthesis in *E. coli*. *Metab. Eng.* **2010**, *12*, 378–386. [CrossRef]
67. De Oliveira, D.; Feihrmann, A.C.; Dariva, C.; Cunha, A.G.; Bevilaqua, J.V.; Destain, J.; Oliveira, J.V.; Freire, D.M.G. Influence of compressed fluids treatment on the activity of Yarrowia lipolytica lipase. *J. Mol. Catal. B Enzym.* **2006**, *39*, 117–123. [CrossRef]
68. Smaniotto, A.; Skovronski, A.; Rigo, E.; Tsai, S.M.; Durrer, A.; Foltran, L.L.; Paroul, N.; Di Luccio, M.; Vladimir Oliveira, J.; de Oliveira, D.; et al. Concentration, characterization and application of lipases from Sporidiobolus pararoseus strain. *Braz. J. Microbiol.* **2014**, *45*, 294–301. [CrossRef] [PubMed]
69. Liu, Y.; Xu, H.; Yan, Q.; Yang, S.; Duan, X.; Jiang, Z. Biochemical characterization of a first fungal esterase from Rhizomucor miehei showing high efficiency of ester synthesis. *PLoS ONE* **2013**, *8*, e77856. [CrossRef] [PubMed]
70. Radzi, S.M.; Mustafa, W.A.F.; Othman, S.S.; Noor, H.M. Green Synthesis of Butyl Acetate, A Pineapple Flavour via Lipase-Catalyzed Reaction. *World Acad. Sci. Eng. Technol.* **2011**, *132*, 7038–7042.
71. Pires-Cabral, P.; da Fonseca, M.M.R.; Ferreira-Dias, S. Esterification activity and operational stability of Candida rugosa lipase immobilized in polyurethane foams in the production of ethyl butyrate. *Biochem. Eng. J.* **2010**, *48*, 246–252. [CrossRef]
72. Kaur, M.; Mehta, A.; Gupta, R. Synthesis of methyl butyrate catalyzed by lipase from Aspergillus fumigatus. *J. Oleo Sci.* **2019**, *68*, 989–993. [CrossRef]
73. Dong, L.-L.; He, L.; Tao, G.-H.; Hu, C. High yield of ethyl valerate from the esterification of renewable valeric acid catalyzed by amino acid ionic liquids. *RSC Adv.* **2013**, 3, 4806. [CrossRef]
74. Paes, F.C.; Kanda, L.R.S.; Voll, F.A.P.; Corazza, M.L. Liquid-liquid equilibrium of ternary systems comprising ethyl valerate(1), water(2), ethanol(3) and valeric acid(4). *J. Chem. Thermodyn.* **2017**, *111*, 185–190. [CrossRef]
75. Petersen, T.N.; Brunak, S.; von Heijne, G.; Nielsen, H. SignalP 4.0: Discriminating signal peptides from transmembrane regions. *Nat. Methods* **2011**, *8*, 785–786. [CrossRef] [PubMed]
76. Peng, Q.; Wang, X.; Shang, M.; Huang, J.; Guan, G.; Li, Y.; Shi, B. Isolation of a novel alkaline-stable lipase from a metagenomic library and its specific application for milkfat flavor production. *Microb. Cell Fact.* **2014**, *13*, 1–9. [CrossRef] [PubMed]
77. Shakiba, M.H.; Ali, M.S.M.; Rahman, R.N.Z.R.A.; Salleh, A.B.; Leow, T.C. Cloning, expression and characterization of a novel cold-adapted GDSL family esterase from Photobacterium sp. strain J15. *Extremophiles* **2016**, *20*, 45–55. [CrossRef] [PubMed]
78. Lee, H.W.; Jung, W.K.; Kim, Y.H.; Ryu, B.H.; Doohun Kim, T.; Kim, J.; Kim, H. Characterization of a novel alkaline family viii esterase with S-enantiomer preference from a compost metagenomic library. *J. Microbiol. Biotechnol.* **2016**, *26*, 315–325. [CrossRef] [PubMed]
79. Thompson, J.D.; Higgins, D.G.; Gibson, T.J. CLUSTAL W: Improving the sensitivity of progressive multiple sequence alignment through sequence weighting, position-specific gap penalties and weight matrix choice. *Nucleic Acids Res.* **1994**, *22*, 4673–4680. [CrossRef] [PubMed]
80. Tamura, K.; Stecher, G.; Peterson, D.; Filipski, A.; Kumar, S. MEGA6: Molecular evolutionary genetics analysis version 6.0. *Mol. Biol. Evol.* **2013**, *30*, 2725–2729. [CrossRef] [PubMed]
81. Felsenstein, J. Confidence limits on phylogenies: An approach using the bootstrap. *Evolution* **1985**, *39*, 783–791. [CrossRef]
82. Laemmli, U.K. Cleavage of structural proteins during the assembly of the head of bacteriophage T4. *Nature* **1970**, *227*, 680–685. [CrossRef]

Article

Counterbalance of Stability and Activity Observed for Thermostable Transaminase from *Thermobaculum terrenum* in the Presence of Organic Solvents

Ekaterina Yu. Bezsudnova [1,*], Alena Yu. Nikolaeva [2], Sergey Y. Kleymenov [1,3], Tatiana E. Petrova [4], Sofia A. Zavialova [1], Kristina V. Tugaeva [1], Nikolai N. Sluchanko [1] and Vladimir O. Popov [1,2]

1 Bach Institute of Biochemistry, Federal Research Center of Biotechnology of the Russian Academy of Sciences, Leninsky Ave. 33, bld. 2, 119071 Moscow, Russia; kleymenov@gmail.com (S.Y.K.); s.zavyalova@fbras.ru (S.A.Z.); kri94_08@mail.ru (K.V.T.); nikolai.sluchanko@mail.ru (N.N.S.); vpopov@inbi.ras.ru (V.O.P.)

2 Kurchatov Complex of NBICS-technologies, National Research Centre "Kurchatov Institute", Akad. Kurchatova sqr 1, 123182 Moscow, Russia; aishome@mail.ru

3 Koltzov Institute of Developmental Biology of Russian Academy of Sciences, 26 Vavilov Street, 119334 Moscow, Russia

4 Institute of Mathematical Problems of Biology, RAS, Branch of Keldysh Institute of Applied Mathematics of the Russian Academy of Sciences, 1, Professor Vitkevich St., 142290 Pushchino, Russia; tania.petrova.ru@gmail.com

* Correspondence: eubez@inbi.ras.ru

Received: 11 August 2020; Accepted: 4 September 2020; Published: 6 September 2020

Abstract: Pyridoxal-5′-phosphate-dependent transaminases catalyze stereoselective amination of organic compounds and are highly important for industrial applications. Catalysis by transaminases often requires organic solvents to increase the solubility of reactants. However, natural transaminases are prone to inactivation in the presence of water-miscible organic solvents. Here, we present the solvent tolerant thermostable transaminase from *Thermobaculum terrenum* (*TaTT*) that catalyzes transamination between L-leucine and alpha-ketoglutarate with an optimum at 75 °C and increases the activity ~1.8-fold upon addition of 15% dimethyl sulfoxide or 15% methanol at high but suboptimal temperature, 50 °C. The enhancement of the activity correlates with a decrease in the thermal denaturation midpoint temperature. The blue-shift of tryptophan fluorescence suggested that solvent molecules penetrate the hydration shell of the enzyme. Analysis of hydrogen bonds in the *TaTT* dimer revealed a high number of salt bridges and surface hydrogen bonds formed by backbone atoms. The latter are sensitive to the presence of organic solvents; they rearrange, conferring the relaxation of some constraints inherent to a thermostable enzyme at low temperatures. Our data support the idea that the counterbalance of stability and activity is crucial for the catalysis under given conditions; the obtained results may be useful for fine-tuning biocatalyst efficiency.

Keywords: biocatalysis; transaminases; enzyme stability; organic solvent; enzyme activation

1. Introduction

Pyridoxal-5′-phosphate (PLP)-dependent transaminases (TAs) catalyze stereoselective amination of keto acids, ketones and aldehydes and are of particular interest as industrially relevant enzymes. Catalysis by TAs requires the application of organic solvents to increase the solubility of reactants and drive the reaction equilibrium towards the amination of ketones [1]. The industrial applications of TAs require an improvement of their thermostability and solvent tolerance as well as a search for new robust enzymes in nature [1]. In the optimized sitagliptin manufacturing process, (*R*)-amination of the pro-sitagliptin ketone is performed by the engineered ATA-117-Rd11 transaminase that tolerates

200 g/L ketone in 50% dimethyl sulfoxide (DMSO) at 40 °C [1,2]. For industrial applications, a search for thermostable TAs is advantageous because of the proven feature of naturally thermostable enzymes to manifest stability under various harsh conditions [3–5]. The enzymes engineered for thermostability showed stability in the presence of organic solvents [6]. The stability of enzyme molecules is largely determined by the non-covalent interactions of different natures and, in particular, by the properties of the surface based on the amino acid composition [3,7]. Electrostatic interactions of various intensity, namely, salt bridges, hydrogen bonds, long-range ion pairs, dipole-dipole interactions, etc., are considered to be the main structural factors responsible for the thermostability of a protein globule [8,9]. A large number of salt bridges on the surface of thermostable enzymes maintain the structural integrity under "hot" water attacks due to a low temperature dependence of electrostatic interactions as well as a lack of geometrical restrictions because of the central symmetry of a charge [10,11]. A high number of charged residues promote the extensive hydrogen-bonding network and tighten both the hydration shell and the interface of the enzyme molecules, thus defining the balance between the structural integrity and the conformational flexibility at any temperatures [4,9,12]. Recent advances in the improvement of enzyme thermostability are based in particular on the insertion of salt bridges into the protein globule [13,14].

The solvent tolerance as well as the enzymatic activity changes in water-organic solvent media are still poorly understood. Somewhat counterintuitively, crystallographic studies of solvent-resistant enzymes did not reveal significant changes in structures obtained from crystals grown (or soaked) in organic solvents and showed almost indistinguishable protein conformation to that obtained in buffers [15–19]. However, several solvent interaction sites were identified as well as changes in the side-chain conformations [19,20]. In particular, it was observed that solvent molecules replaced water molecules on the surface and even in the active site of enzymes [17–21]. It is considered that water-miscible organic solvent–water mixtures destabilize the protein globule: solvent molecules penetrate the hydration shell of the enzyme molecules and strip water from the surface, thus solvating hydrophobic patches and disrupting the surface hydrogen bonds due to the interaction with backbone atoms [7,22–26]. A water-miscible organic solvent can interfere with the activity of an enzyme by substituting catalytic water molecules and breaking conserved hydrogen bonds in the active site [27,28]. Thus, the ability of a protein molecule to maintain the integrity of its hydration shell contributes to organic solvent tolerance [3,16,23]. The effective accumulation of water molecules by the charged residues and the strengthening of Coulomb interactions in water-miscible organic solvent–water media make an excess of charged surface residues a powerful instrument counteracting solvent penetration [3,29–31]. Some enzymes can increase activity in the presence of water-miscible organic solvents. The enhancement of specific activity was observed for proteases, laccases, lipases, etc. [15,22,32–34]. The changes in non-covalent interactions and conformational flexibility as well as the influence of a solvent on particular steps of the catalyzed reactions seem to facilitate the catalysis and underlie the enzyme activation in water–solvent media [15,22,32,35]. Recently, the increase of transaminase activity in cell-free lysates in the presence of organic solvents was demonstrated [36].

The effects of organic cosolvents depend on their physico-chemical characteristics. Non-polar water-immiscible organic solvents do not penetrate the hydration shell, and enzymes preserve structural integrity and the hydration shell in the presence of non-polar cosolvents [3,5]. The penetration potential and water-stripping capacity of water-miscible organic solvents depend on such physical parameters as hydrophobicity, polarity index, dipole moment, and the hydrogen bond donating/accepting abilities [22,37,38]. The enzymes' responses to the addition of organic solvents are complex and depend on the concentration and properties of the organic solvents as well as the enzyme's characteristics.

Here, we focus on the enhancement of the activity of recombinant thermostable transaminase from *Thermobaculum terrenum* (*TaTT*) in water–methanol and water–DMSO media. Both DMSO and methanol are water-miscible organic solvents and are employed in biocatalytic applications. These organic cosolvents were selected for their different physico-chemical characteristics: they differed by hydrophobicity (logP for DMSO is −1.35 and for methanol is −0.74), by dipole moment (μ for DMSO is 3.96 D and for methanol is 1.70 D) and by hydrogen bond formation ability. Methanol is a hydrogen bond donating cosolvent, whereas DMSO is a hydrogen-bond-accepting cosolvent. We found that *TaTT* significantly increased its activity in the presence of methanol and DMSO at the expense of the thermal stability, with the impact of methanol being more significant. The hydrogen bonding of *TaTT* in terms of its structural integrity and solvent tolerance is discussed.

2. Results and Discussion

TaTT is a highly thermostable enzyme [39]. The optimal temperature (*Toptm*) for the catalyzed transamination reaction between L-Leu and α-ketoglutarate is 75 °C [39] (Scheme 1).

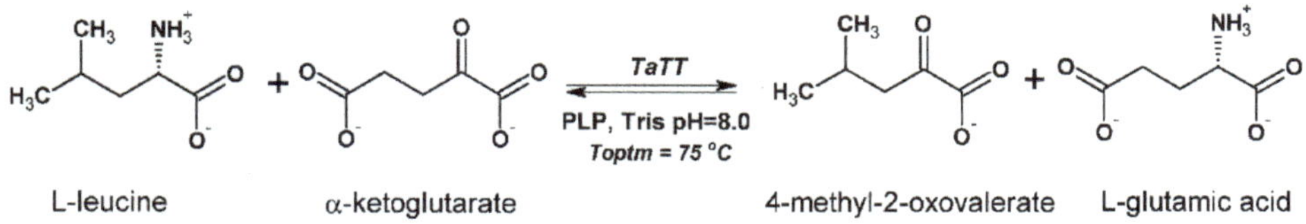

Scheme 1. Transamination reaction between L-Leu and α-ketoglutarate catalyzed by *Thermobaculum terrenum* (*TaTT*).

The differential scanning calorimetry (DSC) profile of *TaTT* is represented by two calorimetric domains with melting points (T_m) of 79.4 and 84.3 °C (Figure 1a). DSC rescanning indicated the irreversibility of the thermal denaturation of *TaTT*. Kinetic stability assay revealed the 50% reduction of *TaTT* only after 40 h incubation at 70 °C and 150 h incubation at 50 °C. *TaTT* retained 100% of the initial activity after 24 h incubation in 50% DMSO and 70% of the activity after 24 h incubation in 50% methanol at 50 °C (Figure 1b,c). The addition of 50% DMSO or 50% methanol in the standard assay decreased the specific activity of *TaTT* by half. The addition of 30% methanol slightly increased the specific activity of *TaTT*; 30% DMSO in the standard assay reduced the specific activity of *TaTT* by 10–15%. By contrast, the addition of 15% methanol and 15% DMSO led to an increase in specific activity by 1.5-1.6 times; at the same time, after 24 h incubation in the presence of 15% methanol and 15% DMSO at 50 °C, *TaTT* retained the initial activity. The latter indicated the absence of *TaTT* denaturation and the reversibility of changes in the *TaTT* molecules at low concentrations of DMSO and methanol.

We examined the effects of 15% methanol and 15% DMSO on *TaTT* activity in the transamination reaction between L-Leu and α-ketoglutarate at different temperatures. The temperature dependences (Figure 2) showed the shift of *Toptm* to lower temperatures and, as a result, an enhancement of the *TaTT* activity at 50 °C. The observed decrease in thermophilicity of *TaTT* in the presence of the cosolvents pointed out small changes in protein molecule as well as possible effects of the co-solvent on the enzymatic reaction at 50 °C. To clarify this issue, we determined the parameters of the transamination reaction in the presence of 15% DMSO (DMSO was a better solvent for the indirect GDH assay, see Material and Method section). The addition of 15% DMSO induced a 1.5-fold increase in maximum velocity and *Km* value at 50 °C, with catalytic efficiency toward L-Leu remaining constant (Table 1). Minor changes of kinetic parameters indicated the similarity of substrate binding and catalytic transformations in both reaction media.

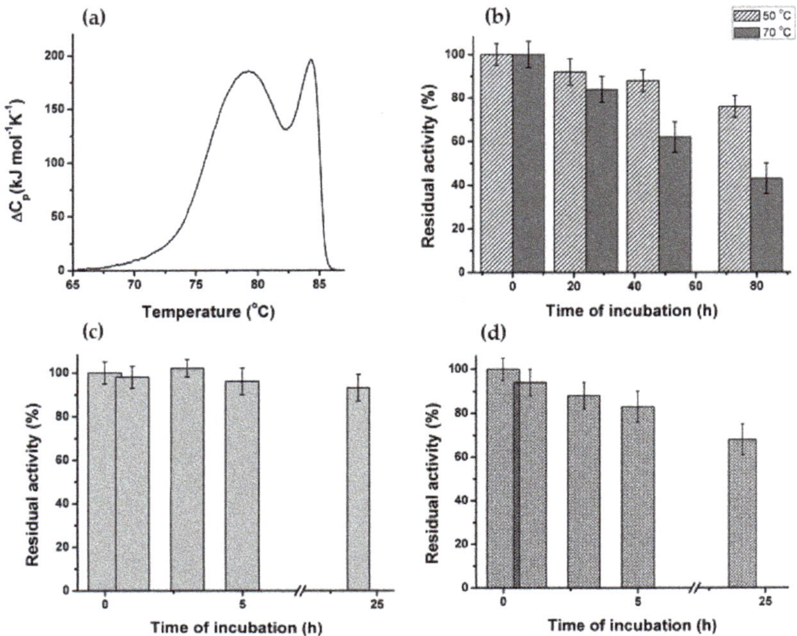

Figure 1. Thermal stability of *TaTT*. (**a**) Differential scanning calorimetry (DSC) profile of 1.0 mg/mL *TaTT*. (**b**)–(**d**) Residual activity of *TaTT* after incubation (**b**) in buffer S (50 mM Tris-HCl, pH 8.0, containing 100 mM NaCl, 60 µM PLP) at 50 °C and 70 °C; (**c**) in buffer S with 50% DMSO (*v/v*) at 50 °C; (**d**) in buffer S with 50% (*v/v*) methanol at 50 °C. 100% corresponds to 40 ± 4 U/mg in the standard assay. Error bars represent standard deviation.

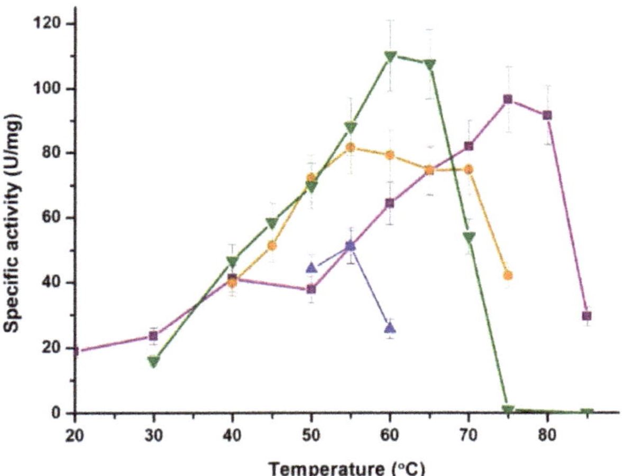

Figure 2. Temperature dependences of the specific activity of *TaTT* in the transamination reaction between L-Leu and α-ketoglutarate in 50 mM Na-phosphate buffer, pH 8.0, containing 50 mM NaCl (purple), and after the addition of 15% DMSO (orange), 15% methanol (olive) or 30% methanol (blue). All measurements were performed at least in triplicates. Error bars represent standard deviation.

Table 1. Steady-state kinetic parameters of the transamination reaction catalyzed by *TaTT* at 50 °C.

Solvent	V_{max}, U/mg	K_m, mM	k_{cat}/K_m, s^{-1} M^{-1}
Buffer	178 ± 23	7.8 ± 2.3	13,700 ± 4400
15% DMSO	280 ± 40	12 ± 4	14,000 ± 5000

To address the organic solvent-induced structural changes, we studied the thermal unfolding of *TaTT* in the presence of 15% cosolvents using intrinsic fluorescence (Figure 3). $T_{0.5}$ decreased from 74.9 °C in the buffer to 68.3 °C and 64.5 °C in the presence of 15% DMSO and 15% methanol, respectively. Notably, the cosolvents expanded the temperature range of the thermal unfolding of *TaTT* from ~10 °C in the buffer to 20.5 and 22.5 °C in the presence of DMSO and methanol, respectively, thereby significantly lowering the cooperativity of thermal transition. This is in line with the idea that water-miscible cosolvents increase structural dynamics of the enzyme, promoting its structural rearrangements in a wide range of temperatures and removing some structural constraints typical of thermostable enzymes at low temperatures.

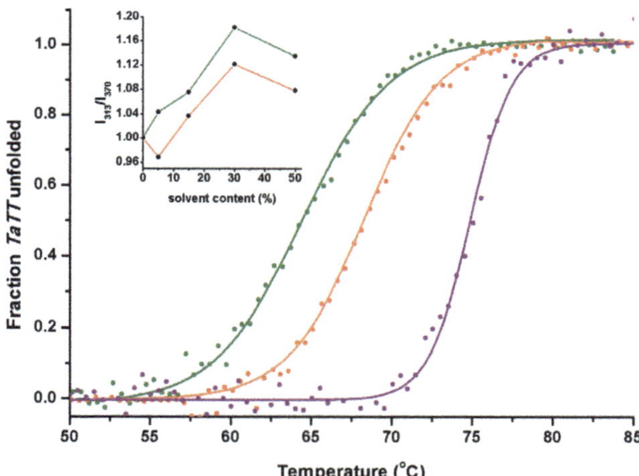

Figure 3. Thermally-induced unfolding curves of *TaTT* in Na-phosphate buffer (purple) and in the presence of 15% DMSO (orange) or 15% methanol (olive). **Inset:** The dependence of I_{313}/I_{370} for 0.1 mg/mL *TaTT* on percentage (*v*/*v*) of methanol (olive) and DMSO (orange) in Na-phosphate buffer at 50 °C. The data are the averages from triplicate experiments.

Additionally, we observed that 0-30% methanol or DMSO caused blue-shifts of the tryptophan fluorescence spectrum of *TaTT* (Inset panel on Figure 3; Figure 4a) at 50 °C, implying an alteration of the environment of tryptophan residues to more hydrophobic [40,41]. The addition of 30–50% of either cosolvent caused a red-shift, which likely reflects the onset of denaturation provoked by an excessive weakening of non-covalent interactions. Among eight tryptophan residues of the *TaTT* subunit, all except one were exposed to the bulk solvent (Figure 4c). The blue-shift is in favor of the notion that solvent molecules penetrate the hydration shell of *TaTT*, causing a decrease in the water density near the hydrophobic indolyl groups and preferential solvation of tryptophan residues (and similarly other exposed hydrophobic groups) by cosolvents [29,30,33]. Sharp growth in light scattering indicated aggregation accompanying the unfolding (Figure 4b) due to the exposure of hydrophobic patches. These changes as well as changes of *TaTT* thermostability and activity were more pronounced in the presence of methanol. This is likely because of a higher hydrophobicity of methanol (logP = −0.74), compared to DMSO (logP = −1.35). Overall, the $T_{0.5}$ decrease coincided with the decrease in *Toptm* of the transamination reaction, suggestive of the common reason underlying the changes. The solvent-induced structural changes can alter the conformational flexibility of the enzyme and adjust the balance between rigidity and flexibility at suboptimal temperatures, i.e., lower than 75 °C.

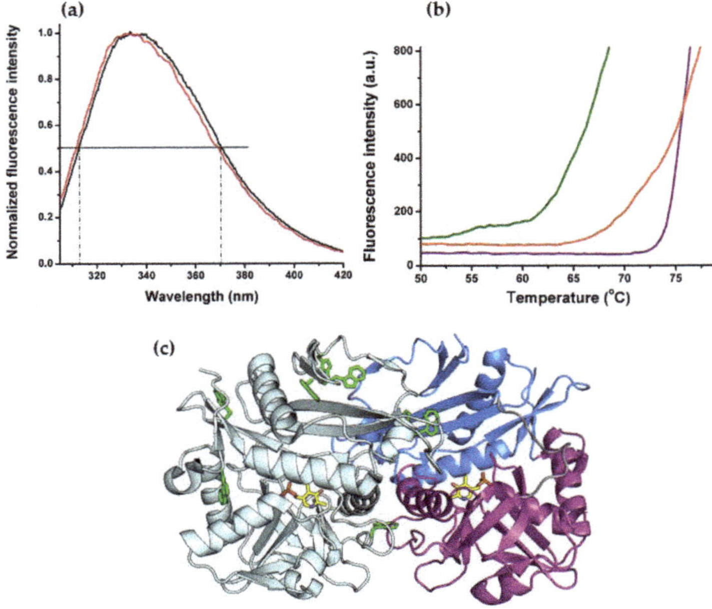

Figure 4. Thermal unfolding of *TaTT* studied by using its intrinsic tryptophan fluorescence. (**a**) Fluorescence spectra of 0.1 mg/mL *TaTT* in the PLP form in 50 mM phosphate buffer, pH 8.0, containing 50 mM NaCl, (black) and in the presence of 30% methanol (pink) at 50 °C. Intensities at 313 nm and 370 nm correspond to the half of the maximum of the *TaTT* fluorescence spectrum. (**b**) Light scattering accompanying thermally-induced unfolding of *TaTT* in Na-phosphate buffer (purple) or the presence of 15% DMSO (orange) or 15% methanol (olive). (**c**) Model of the functional dimer of *TaTT* (PDB ID: 6GKR). Two subunits forming the dimer are shown. In the right subunit, the small domain is colored blue, the large domain is colored purple; in the left subunit, tryptophan residues are shown in green. PLP molecules are colored yellow.

To interpret the effects of the organic solvents on the thermostability and thermophilicity of *TaTT*, we focused on the analysis of hydrogen bonding in *TaTT*, considering hydrogen-bonding network as a crucial structural factor of the stability of enzymes in harsh conditions. According to the recent studies, the penetration of solvent molecules into the hydration shell of an enzyme molecule disturbs the surface hydrogen bonds because of the amphiphilic nature of the cosolvents (dipole strength is 3.96 D for DMSO, 1.70 D for methanol, and 1.85 D for water) [24,26,29]. Both DMSO and methanol can interact with the backbone atoms and outcompete water molecules to form surface hydrogen bonds [24–26,33,39]. Present in the hydration shell, DMSO can interact with the exposed backbone NH-groups via its oxygen atoms [29]. Methanol is instead a hydrogen-bonding donor, which can form H-bonds with the backbone CO-group [25]. At the same time, both DMSO and methanol do not significantly affect side-chain hydrogen bonds and salt bridges [24,25,29]. The analysis of hydrogen bonds in the functional dimers of *TaTT* homologs showed that *TaTT* is distinguished by the excess of surface hydrogen bonds, by hydrogen bonds formed by side chains of charged residues (salt bridges) and by the number of hydrogen bonds per one amino acid residue (Figure 5, Table A1).

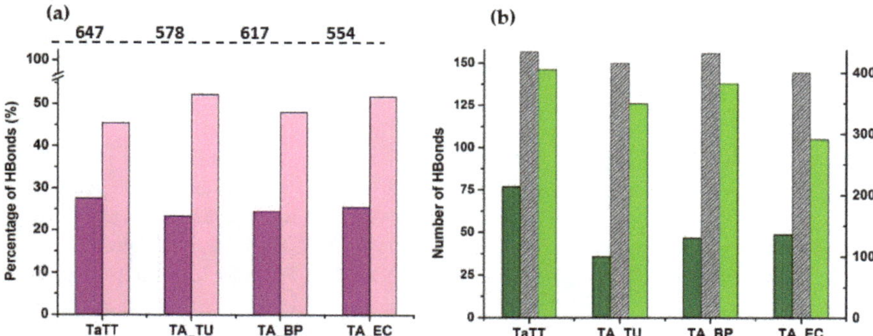

Figure 5. Bar graphs showing the number of hydrogen bonds in functional dimers of *TaTT* (PDB ID: 6GKR) and its homologs: **TA_TU** from *Thermoproteus uzoniensis* (*Toptm* in the standard assay is 95 °C [42], PDB ID: 5CE8,), **TA_BP** from *Burkholderia pseudomallei* (*Toptm* is unknown, the organism grows optimally at 40 °C, PDB ID 3U0G) and **TA_EC** from *Escherichia coli* (*Toptm* in the standard assay is 37 °C [43], PDB ID 1I1K). (**a**) Percentage of Surface–Surface (violet) and Inside–Inside (pink) hydrogen bonds in the total number (100%, underlined by dash line) of hydrogen bonds in the dimers; (**b**) Number of **Charged–Charged** hydrogen bonds (green, left axis), **Charged–Neutral** (light-green, left axis) and **Neutral–Neutral** (grey, right axis) hydrogen bonds in the dimers.

The profound hydrogen-bonding network of the molecule of *TaTT* coincides well with the thermostability and thermophilicity of the enzyme. It is noteworthy that in all TA dimers, the number of **Neutral–Neutral** hydrogen bonds is the highest among all types of hydrogen bonds. While salt bridges tolerate water-organic solvent media and maintain the integrity of both the protein globule and the hydration shell, surface **Neutral–Neutral** and **Neutral–Charged** hydrogen bonds are much more susceptible because of the interactions of organic solvent molecules with the exposed backbone atoms. These solvent-induced disturbances can increase the flexibility of particular surface regions and release extra tension in regions of the enzyme that are important for catalysis. We suggest that the combination of hydrophobic effects and disturbances of hydrogen bonds induced by the addition of organic solvents leads to a weakening of some constraints inherent for *TaTT* as a thermostable enzyme at low temperatures. These changes induce an increase in the conformational flexibility of *TaTT* molecules and enhance the catalysis at lower temperatures, compared with catalysis in buffer systems.

3. Materials and Methods

3.1. Enzyme Production and Activity Assays

Enzyme production was described in [39]. Briefly, the His6TEV-tagged *TaTT* was expressed in *E. coli* BL21(DE3)pLys (Stratagene, California, CA, USA). The recombinant *TaTT* was isolated using subtractive Ni-affinity chromatography and gel filtration. Fractions showing the activity were stored in 50 mM Tris-HCl buffer, pH 8.0, containing 100 mM NaCl, 100 μM PLP and 50% glycerol at −20 °C. The PLP form of *TaTT* was obtained by incubating *TaTT* with the excess of both PLP and α-ketoglutarate overnight, followed by the transfer into the assay buffer using a HiTrap Desalting column (GE Healthcare, Chicago, IL, USA).

In the standard assay, the activity of *TaTT* at 0.5-1.5 μg/mL was measured in the overall transamination reaction with 5 mM L-Leu and 1 mM α-ketoglutarate in 50 mM Tris-HCl buffer, pH 8.0, supplemented with 50 mM NaCl and 60 μM PLP at 50 °C. The amount of L-Glu product was measured by the indirect photometric glutamate dehydrogenase (GluDH (Sigma-Aldrich, St. Louis, MO, USA, cat. N G2626)) assay described in [39]. To evaluate the effects of organic solvents on the overall reaction, the standard assay was carried out in the presence of 15–50% (*v*/*v*) methanol or DMSO.

The effect of solvents on the GluDH reaction was taken into account, and the corrected calibration curves were used accordingly. All measurements were performed at least in duplicate. The data were analyzed using Origin 8.0 software (Origin Lab, Northampton, MA, USA).

3.2. Analysis of *TaTT* Stability

To evaluate the thermal stability, 1.0 mg/mL *TaTT* was incubated at 50 °C or 70 °C in buffer S (50 mM Tris-HCl, pH 8.0, containing 100 mM NaCl, 60 μM PLP). After the thermal treatment, the residual activity was estimated in the standard assay. The solvent stability was estimated by measuring the residual activity after incubation of 1.0 mg/mL *TaTT* in buffer S, containing 15% or 50% (*v*/*v*) organic cosolvents (DMSO, methanol) at 50 °C. After 1, 3, 5 and 24 h, aliquots were taken to determine the residual activity using the standard assay.

Differential scanning calorimetry (DSC) was conducted on a MicroCal VP-Capillary calorimeter (Malvern Instruments, Northampton, MA, USA) with tantalum capillary cells at a heating rate of 1 K/min in 50 mM Na-phosphate buffer, pH 8.0, containing 50 mM NaCl and 1.0 mg/mL *TaTT* in the PLP form.

Steady-state fluorimetry was conducted using a Cary Eclipse spectrofluorimeter (Varian Inc., Victoria, Australia) equipped with a Peltier-controlled cuvette holder and thermosensor; λ_{ex} was 297 nm, λ_{em} was 313 nm and 370 nm (all slits were 5 nm). Intensities at 313 nm and 370 nm correspond to the half of the maximum of the *TaTT* fluorescence spectrum at 50 °C, and were used to obtain thermal unfolding curves upon constant heating of the sample at 1 °C/min [44]. The half-transition temperatures of thermal denaturation ($T_{0.5}$) were calculated using the Boltzmann model. Light scattering at 90 degrees as recorded at λ_{ex} 350 nm and λ_{em} 355 nm (5 nm slits). The buffer was 50 mM Na-phosphate, pH 8.0, containing 50 mM NaCl; the final concentration of *TaTT* in the PLP form was 0.02 mg/mL.

3.3. Analysis of Hydrogen Bonds

Calculations of the number of hydrogen bonds in the dimers of *TaTT* and homologous TAs were performed using the programm HBOND [45] and classification described in Appendix A. For *TaTT* (PDB ID: 6GKR) and its counterparts, PLP and other ligands were removed from the models before the analysis (for details see the Appendix A).

4. Conclusions

In this work, we studied the effects of methanol and DMSO on the activity of thermostable transaminase from *T. terrenum*. Earlier, we characterized this PLP fold type IV transaminase as highly thermostable (*Toptm* = 75 °C) and distinguished by the high rates in the transamination reactions with branched-chain amino acids (maximum velocity 178 and 260 U/mg with L-leucine and L-norvaline at 50 °C, respectively) and aromatic amino acids (maximum velocity 46 and 25 U/mg at 50 °C with L-phenylalanine and L-tryptophan, respectively) and significant activity towards (*R*)-(+)-1-phenylethylamine (0.33 U/mg at 50 °C) [39,46]. Here, we analyzed the solvent tolerance and the activation of *TaTT* in water-miscible organic solvent–water mixtures: *TaTT* increased the activity upon addition of 15% DMSO or 15% methanol 1.5–1.7-fold in the standard assay at high but suboptimal temperature, 50 °C. The enhancement of the activity correlated with a decrease in the thermal denaturation midpoint temperature from 74.9 to 68.3 and 64.5 °C upon the addition of DMSO and methanol, respectively. The blue shift of tryptophan fluorescence suggested the penetration of solvent molecules into the hydration shell of the enzyme. The analysis of the hydrogen bonding of *TaTT* revealed a high number of salt bridges and surface hydrogen bonds formed by the backbone nitrogen and oxygen atoms. We suggested that salt bridges stabilize the protein globule against "hot" water attacks and organic solvent denaturation, but hydrogen bonds formed by the backbone nitrogen and oxygen atoms are susceptible to the presence of solvent molecules and rearrange, underlying a relaxation of some constraints inherent to a thermostable enzyme at low temperatures. According to Botero et al. [47], the gram-positive thermophile *T. terrenum* shows optimal growth at 67 °C, implying that all its metabolic processes are adopted to this temperature. At 50 °C, the balance between integrity and flexibility of the *TaTT* molecule is not optimal for catalysis. The addition of solvents seems to release the extra tension characteristic of thermostable enzymes at suboptimal temperatures and optimize conformational flexibility, thus improving the catalytic activity at a given temperature. At the same time, the inherent stability prevents *TaTT* from denaturation in water-organic solvent mixtures. Despite the challenge of the prediction of co-solvent effects, their great influence on the counterbalance of stability and activity is a useful tool for fine-tuning the efficiency of biocatalytic processes.

Author Contributions: Conceptualization, E.Y.B. and V.O.P.; methodology, N.N.S. and S.Y.K.; software, T.E.P.; validation, K.V.T., A.Y.N., T.E.P. and S.A.Z.; data curation, E.Y.B.; investigation, A.Y.N., T.E.P., S.Y.K., K.V.T. and S.A.Z.; writing—original draft preparation, E.Y.B.; writing—review and editing, E.Y.B and N.N.S.; supervision, V.O.P.; funding acquisition, E.Y.B. All authors have read and agreed to the published version of the manuscript.

Funding: This research was funded by the Russian Science Foundation, grant number 19-14-00164 (in part of fluorescence and structural analysis), and by the Ministry of Science and Higher Education of the Russian Federation.

Conflicts of Interest: The authors declare no conflict of interest.

Appendix A

Procedure for the calculation and analysis of hydrogen bonds:

1. For each analyzed transaminase, a model from the Protein Data Bank (PDB) was taken, and a file (in the pdb format) was prepared, which contained only those monomers that form functional dimer. Water molecules and ligands were removed from the functional dimer model.

2. All potential hydrogen bonds (HBonds) in the model were determined using the program HBOND (http://cib.cf.ocha.ac.jp/bitool/HBOND). The maximal distance between a donor and an acceptor atom was 3.5 Å.

3. The accessible surface area (ASA) for each atom in the model was calculated using the program AREAIMOL of CCP4 package [48]. Atoms with ASA values equal to 0.0 were considered as inner atoms, other atoms were considered as surface atoms.

4. If a protein model contained some residues in multiple conformations, both files (output of HBOND and AREAMOL) were corrected.

5. Both files (output of HBOND and AREAMOL) were merged into one single file. Depending on the ASA value and the side group of the residue, all bonds were categorized into several groups. We determined the number of hydrogen bonds between inner atoms (**Inside–Inside**), between atoms on the surface (**Surface–Surface**), between atoms one of which either belonged to a side group of a neutral residue or was N or O atom of the main chain and the other belonged to a side group of a charged residue (**Charged–Neutral**), between the atoms of the side group of neutral residues and/or N or O atom of the main chain (**Neutral–Neutral**) and between the atoms of the side groups of charged residues (**Charged–Charged**). The difference between the total number of hydrogen bonds and the sum of Inside–Inside and Surface–Surface hydrogen bonds is the number of **Inside–Surface** hydrogen bonds formed between a solvent-accessible atom and an atom buried inside the globule. Note, that **Charged–Charged** hydrogen bonds are nothing but the **salt bridges**.

Table A1. Analysis of hydrogen bonds in functional dimers of *TaTT* and the counterparts: transaminase (TA) from *Thermoproteus uzoniensis* (*Toptm* in the standard assay is 95 °C [42]), TA from *Burkholderia pseudomallei* (*Toptm* is unknown, the organism grows optimally at 40 °C) and TA from *Escherichia coli* (*Toptm* in the standard assay is 37 °C [43]). Two values for the *TaTT* dimer correspond to two dimers composed of different subunits in the 6GKR model.

	Dimer of			
	TaTT	**TA from *T. uzoniensis***	**TA from *B. pseudomalle***	**TA from *E. coli***
Number of residues per subunit	316	295	307	309
Hydrogen bonds				
• Total number in a dimer	647/648	578	617	554
• **Per one a.a. residue in a dimer**	**1.025**	**0.98**	**1.005**	**0.896**
• Number of **Inside–Inside** hydrogen bonds (between atoms inside the globule)	293/296	302	296	287
• Number of **Surface–Surface** hydrogen bonds (between atoms on the globule surface)	177/179	135	151	141
• Number of **Neutral–Neutral** bonds (between atoms of the side group of neutral residues and/or N or O atoms of the main chain)	430/434	416	432	400
• number of **Charged–Charged** bonds (between atoms of the side group of charged residues)	77/68	36	47	49
• Number of **Charged–Neutral** hydrogen bonds	140/146	126	138	105
• Between subunits in the dimer	18/24	14	23	12
PDB ID:	6GKR	5CE8	3U0G	1I1K

References

1. Slabu, I.; Galman, J.L.; Lloyd, R.C.; Turner, N.J. Discovery, engineering, and synthetic application of transaminase biocatalysts. *ACS Catal.* **2017**, *7*, 8263–8284. [CrossRef]
2. Savile, C.K.; Janey, J.M.; Mundorff, E.C.; Moore, J.C.; Tam, S.; Jarvis, W.R.; Colbeck, J.C.; Krebber, A.; Fleitz, F.J.; Brands, J.; et al. Biocatalytic asymmetric synthesis of chiral amines from ketones applied to sitagliptin manufacture. *Science* **2010**, *329*, 305–309. [CrossRef] [PubMed]
3. Doukyu, N.; Ogino, H. Organic solvent-tolerant enzymes. *Biochem. Eng. J.* **2010**, *48*, 270–282. [CrossRef]
4. Bezsudnova, E.Y.; Petrova, T.E.; Popinako, A.V.; Antonov, M.Y.; Stekhanova, T.N.; Popov, V.O. Intramolecular hydrogen bonding in the polyextremophilic short-chain dehydrogenase from the archaeon *Thermococcus sibiricus* and its close structural homologs. *Biochimie* **2015**, *118*, 82–89. [CrossRef]
5. Kumar, A.; Dhar, K.; Kanwar, S.S.; Arora, P.K. Lipase catalysis in organic solvents: Advantages and applications. *Biol. Proced. Online* **2016**, *18*, 2. [CrossRef]
6. Stepankova, V.; Bidmanova, S.; Koudelakova, T.; Prokop, Z.; Chaloupkova, R.; Damborsky, J. Strategies for stabilization of enzymes in organic solvents. *ACS Catal.* **2013**, *3*, 2823–2836. [CrossRef]
7. Feller, G. Protein stability and enzyme activity at extreme biological temperatures. *J. Phys. Condens. Matter* **2010**, *22*, 323101. [CrossRef]
8. Kumar, S.; Tsai, C.-J.; Ma, B.; Nussinov, R. Contribution of salt bridges toward protein thermostability. *J. Biomol. Struct. Dyn.* **2000**, *17*, 79–85. [CrossRef]
9. Berezovsky, I.N.; Shakhnovich, E.I. Physics and evolution of thermophilic adaptation. *Proc. Natl. Acad. Sci. USA* **2005**, *102*, 12742–12747. [CrossRef]
10. Matsui, I.; Harata, K. Implication for buried polar contacts and ion pairs in hyperthermostable enzymes. *FEBS J.* **2007**, *274*, 4012–4022. [CrossRef]
11. Elcock, A.H. The stability of salt bridges at high temperatures: Implications for hyperthermophilic proteins 1 Edited by B. Honig. *J. Mol. Biol.* **1998**, *284*, 489–502. [CrossRef] [PubMed]
12. Dong, Y.; Liao, M.; Meng, X.; Somero, G.N. Structural flexibility and protein adaptation to temperature: Molecular dynamics analysis of malate dehydrogenases of marine molluscs. *Proc. Natl. Acad. Sci. USA* **2018**, *115*, 1274–1279. [CrossRef] [PubMed]

13. Wijma, H.J.; Floor, R.J.; Janssen, D.B. Structure- and sequence-analysis inspired engineering of proteins for enhanced thermostability. *Curr. Opin. Struct. Biol.* **2013**, *23*, 588–594. [CrossRef] [PubMed]
14. Xu, Z.; Cen, Y.K.; Zou, S.P.; Xue, Y.P.; Zheng, Y.G. Recent advances in the improvement of enzyme thermostability by structure modification. *Crit. Rev. Biotechnol.* **2020**, *40*, 83–98. [CrossRef] [PubMed]
15. Wu, M.H.; Lin, M.C.; Lee, C.C.; Yu, S.M.; Wang, A.H.J.; Ho, T.-H.D. Enhancement of laccase activity by pre-incubation with organic solvents. *Sci. Rep.* **2019**, *9*, 9754. [CrossRef] [PubMed]
16. Said, Z.S.A.A.M.; Arifi, F.A.M.; Salleh, A.B.; Rahman, R.N.Z.R.A.; Leow, A.T.C.; Latip, W.; Ali, M.S.M. Unravelling protein -organic solvent interaction of organic solvent tolerant elastase from *Pseudomonas aeruginosa* strain K crystal structure. *Int. J. Biol. Macromol.* **2019**, *127*, 575–584. [CrossRef]
17. Fitzpatrick, P.A.; Steinmetz, A.C.U.; Ringe, D.; Klibanov, A.M. Enzyme crystal structure in a neat organic solvent. *Proc. Natl. Acad. Sci. USA* **1993**, *90*, 8653–8657. [CrossRef]
18. Cianci, M.; Tomaszewski, B.; Helliwell, J.R.; Halling, P.J. Crystallographic analysis of counterion effects on subtilisin enzymatic action in acetonitrile. *J. Am. Chem. Soc.* **2010**, *132*, 2293–2300. [CrossRef]
19. English, A.C.; Groom, C.R.; Hubbard, R.E. Experimental and computational mapping of the binding surface of a crystalline protein. *Protein Eng. Des. Sel.* **2001**, *14*, 47–59. [CrossRef]
20. Gupta, M.N.; Tyagi, R.; Sharma, S.; Karthikeyan, S.; Singh, T.P. Enhancement of catalytic efficiency of enzymes through exposure to anhydrous organic solvent at 70 °C. Three-dimensional structure of a treated serine proteinase at 2.2 Å resolution. *Proteins Struct. Funct. Genet.* **2000**, *39*, 226–234. [CrossRef]
21. Schmitke, J.L.; Stern, L.J.; Klibanov, A.M. The crystal structure of subtilisin Carlsberg in anhydrous dioxane and its comparison with those in water and acetonitrile. *Proc. Natl. Acad. Sci. USA* **1997**, *94*, 4250–4255. [CrossRef] [PubMed]
22. Stepankova, V.; Damborsky, J.; Chaloupkova, R. Organic co-solvents affect activity, stability and enantioselectivity of haloalkane dehalogenases. *Biotechnol. J.* **2013**, *8*, 719–729. [CrossRef] [PubMed]
23. Polizzi, K.M.; Bommarius, A.S.; Broering, J.M.; Chaparro-Riggers, J.F. Stability of biocatalysts. *Curr. Opin. Chem. Biol.* **2007**, *11*, 220–225. [CrossRef] [PubMed]
24. Mohtashami, M.; Fooladi, J.; Haddad-Mashadrizeh, A.; Housaindokht, M.R.; Monhemi, H. Molecular mechanism of enzyme tolerance against organic solvents: Insights from molecular dynamics simulation. *Int. J. Biol. Macromol.* **2019**, *122*, 914–923. [CrossRef]
25. Shao, Q. Methanol concentration dependent protein denaturing ability of guanidinium/methanol mixed solution. *J. Phys. Chem. B* **2014**, *118*, 6175–6185. [CrossRef]
26. Pazhang, M.; Mardi, N.; Mehrnejad, F.; Chaparzadeh, N. The combinatorial effects of osmolytes and alcohols on the stability of pyrazinamidase: Methanol affects the enzyme stability through hydrophobic interactions and hydrogen bonds. *Int. J. Biol. Macromol.* **2018**, *108*, 1339–1347. [CrossRef]
27. Kuper, J.; Wong, T.S.; Roccatano, D.; Wilmanns, M.; Schwaneberg, U. Understanding a mechanism of organic cosolvent inactivation in heme monooxygenase P450 BM-3. *J. Am. Chem. Soc.* **2007**, *129*, 5786–5787. [CrossRef]
28. Roccatano, D. Computer simulations study of biomolecules in non-aqueous or cosolvent/water mixture solutions. *Curr. Protein Pept. Sci.* **2008**, *9*, 407–426. [CrossRef]
29. Srivastava, K.R.; Goyal, B.; Kumar, A.; Durani, S. Scrutiny of electrostatic-driven conformational ordering of polypeptide chains in DMSO: A study with a model oligopeptide. *RSC Adv.* **2017**, *7*, 27981–27991. [CrossRef]
30. Johnson, M.E.; Malardier-Jugroot, C.; Head-Gordon, T. Effects of co-solvents on peptide hydration water structure and dynamics. *Phys. Chem. Chem. Phys.* **2010**, *12*, 393–405. [CrossRef]
31. Markel, U.; Zhu, L.; Frauenkron-Machedjou, V.; Zhao, J.; Bocola, M.; Davari, M.; Jaeger, K.-E.; Schwaneberg, U. Are directed evolution approaches efficient in exploring nature's potential to stabilize a lipase in organic cosolvents? *Catalysts* **2017**, *7*, 142. [CrossRef]
32. Cao, H.; Nie, K.; Xu, H.; Xiong, X.; Krastev, R.; Wang, F.; Tan, T.; Liu, L. Insight into the mechanism behind the activation phenomenon of lipase from *Thermus thermophilus* HB8 in polar organic solvents. *J. Mol. Catal. B Enzym.* **2016**, *133*, S400–S409. [CrossRef]
33. Kamal, M.Z.; Yedavalli, P.; Deshmukh, M.V.; Rao, N.M. Lipase in aqueous-polar organic solvents: Activity, structure, and stability. *Protein Sci.* **2013**, *22*, 904–915. [CrossRef]
34. Batra, R.; Gupta, M.N. Enhancement of enzyme activity in aqueous-organic solvent mixtures. *Biotechnol. Lett.* **1994**, *16*, 1059–1064. [CrossRef]

35. Kudryashova, E.V.; Gladilin, A.K.; Vakurov, A.V.; Heitz, F.; Levashov, A.V.; Mozhaev, V.V. Enzyme-polyelectrolyte complexes in water-ethanol mixtures: Negatively charged groups artificially introduced into α-chymotrypsin provide additional activation and stabilization effects. *Biotechnol. Bioeng.* **1997**, *55*, 267–277. [CrossRef]
36. Gord Noshahri, N.; Fooladi, J.; Syldatk, C.; Engel, U.; Heravi, M.M.; Zare Mehrjerdi, M.; Rudat, J. Screening and comparative characterization of microorganisms from iranian soil samples showing ω-transaminase activity toward a plethora of substrates. *Catalysts* **2019**, *9*, 874. [CrossRef]
37. Sirotkin, V.A.; Zinatullin, A.N.; Solomonov, B.N.; Faizullin, D.A.; Fedotov, V.D. Calorimetric and Fourier transform infrared spectroscopic study of solid proteins immersed in low water organic solvents. *Biochim. Biophys. Acta Protein Struct. Mol. Enzymol.* **2001**, *1547*, 359–369. [CrossRef]
38. Sirotkin, V.A.; Kuchierskaya, A.A. α-chymotrypsin in water-acetone and water-dimethyl sulfoxide mixtures: Effect of preferential solvation and hydration. *Proteins. Struct. Funct. Bioinforma.* **2017**, *85*, 1808–1819. [CrossRef]
39. Bezsudnova, E.Y.; Boyko, K.M.; Nikolaeva, A.Y.; Zeifman, Y.S.; Rakitina, T.V.; Suplatov, D.A.; Popov, V.O. Biochemical and structural insights into PLP fold type IV transaminase from *Thermobaculum terrenum*. *Biochimie* **2019**, *158*, 130–138. [CrossRef]
40. Lakowicz, J.R. *Principles of Fluorescence Spectroscopy*; Lakowicz, J.R., Ed.; Springer US: Boston, MA, USA, 2006; ISBN 978-0-387-31278-1.
41. Bushueva, T.L.; Busel, E.P.; Burstein, E.A. Relationship of thermal quenching of protein fluorescence to intramolecular structural mobility. *Biochim. Biophys. Acta Protein Struct.* **1978**, *534*, 141–152. [CrossRef]
42. Boyko, K.M.; Stekhanova, T.N.; Nikolaeva, A.Y.; Mardanov, A.V.; Rakitin, A.L.; Ravin, N.V.; Bezsudnova, E.Y.; Popov, V.O. First structure of archaeal branched-chain amino acid aminotransferase from Thermoproteus uzoniensis specific for l-amino acids and R-amines. *Extremophiles* **2016**, *20*, 215–225. [CrossRef] [PubMed]
43. Yu, X.; Wang, X.; Engel, P.C. The specificity and kinetic mechanism of branched-chain amino acid aminotransferase from Escherichia coli studied with a new improved coupled assay procedure and the enzyme's potential for biocatalysis. *FEBS J.* **2014**, *281*, 391–400. [CrossRef]
44. Permyakov, E.A.; Burstein, E.A. Some aspects of studies of thermal transitions in proteins by means of their intrinsic fluorescence. *Biophys. Chem.* **1984**, *19*, 265–271. [CrossRef]
45. Available online: http://cib.cf.ocha.ac.jp/bitool/HBOND (accessed on 29 April 2009).
46. Bezsudnova, E.Y.; Popov, V.O.; Boyko, K.M. Structural insight into the substrate specificity of PLP fold type IV transaminases. *Appl. Microbiol. Biotechnol.* **2020**, *104*, 2343–2357. [CrossRef] [PubMed]
47. Botero, L.M.; Brown, K.B.; Brumefield, S.; Burr, M.; Castenholz, R.W.; Young, M.; McDermott, T.R. *Thermobaculum terrenum* gen. nov., sp. nov.: A non-phototrophic gram-positive thermophile representing an environmental clone group related to the Chloroflexi (green non-sulfur bacteria) and *Thermomicrobia*. *Arch. Microbiol.* **2004**, *181*, 269–277. [CrossRef]
48. Collaborative Computational Project, Number 4. The CCP4 suite: Programs for protein crystallography. *Acta Crystallogr. Sect. D Biol. Crystallogr.* **1994**, *50*, 760–763. [CrossRef] [PubMed]

Article

Kinetic Analysis of the Lipase-Catalyzed Hydrolysis of Erythritol and Pentaerythritol Fatty Acid Esters: A Biotechnological Application for Making Low-Calorie Healthy Food Alternatives

Olga A. Gkini [1], Panagiota-Yiolanda Stergiou [1], Athanasios Foukis [1], Panayiotis V. Ioannou [2] and Emmanuel M. Papamichael [1,*]

[1] Enzyme Biotechnology & Genetic Engineering Group, Department of Chemistry, University of Ioannina, 45110 Ioannina, Greece; olgouita@hotmail.com (O.A.G.); stergiou715@gmail.com (P.-Y.S.); afouk717@gmail.com (A.F.)

[2] Laboratory of Inorganic Chemistry, Department of Chemistry, University of Patras, 26500 Patras, Greece; ioannou@chemistry.upatras.gr

* Correspondence: epapamic@uoi.gr; Tel.: +30-26510-08395 or +30-6944-469-829

Received: 3 August 2020; Accepted: 20 August 2020; Published: 23 August 2020

Abstract: Contemporary consumers demand healthier and more nourishing food, and thus, alternative foods that are low-calorie in fats and/or sugars are preferred. These desired properties may be attained by substituting the fatty acid esters of erythritol and pentaerythritol due to their antioxidant action and low toxicity for humans. In this work, the catalyzed hydrolysis of five fatty acid tetraesters of erythritol and/or pentaerythritol by both porcine pancreas type VI-s lipase (PPL) and *Candida antarctica* lipase-B (CALB) were studied kinetically. In all cases, except the hydrolysis of pentaerythritol tetrastearate by CALB, Michaelis–Menten kinetics were observed. In addition, the pK_a values of the fatty acids released due to the catalyzed hydrolysis of the studied tetraesters by CALB were estimated. In the course of the aforementioned procedures, it was found that the CALB-catalyzed hydrolysis was incomplete to various degrees among four of the five studied tetraesters (excluding erythritol tetraoleate), and one or more estimated apparent pK_a values were obtained. These results are novel, and by means of applied methodology, they reveal that erythritol and/or pentaerythritol tetraesters of medium- and long-chain fatty acids are suitable candidates for use as beneficial alternatives to butter and/or sweeteners.

Keywords: erythritol; pentaerythritol; tetraesters; fatty acids; *Candida antarctica* lipase-B; low calorie dietary foods

---

1. Introduction

Erythritol [(2S,3R)-butane-1,2,3,4-tetrol] is a low-calorie compound that occurs in nature and in fermented products, and it can be produced through the bioconversion of starch and/or glucose [1]. It has been reported that erythritol may be helpful in reducing both the caloric value of foods that contain carbohydrates and the side effects of other relative food additives [2]. In addition to erythritol having an almost zero caloric value, it is a sweetener that can improve food taste and texture and the stability of low-calorie foods [1]. These latter properties can meet the demands of contemporary consumers for a healthier lifestyle [3], and erythritol has additional importance due to its confirmed antioxidant action [4].

Tetrol pentaerythritol [2,2-bis(hydroxymethyl)propane-1,3-diol] is well known as an important component in the manufacturing of cosmetics. Pentaerythritol has a low toxicity in the human body, is released in the urine relatively quickly, and does not show mutagenic action against bacteria [5].

Therefore, fatty acid esters of erythritol and pentaerythritol may be healthy, low-calorie alternatives to butter and/or sweeteners; they may not cause caries, too [1].

This work comprises a detailed kinetic study of the lipase-catalyzed hydrolysis of three fatty acid tetraesters of erythritol and two tetraesters of pentaerythritol. These five tetraesters have the potential for being biotechnologically useful in food and in cosmetics applications as compared to the triesters of glycerol [6]. All five studied tetraesters herein were hydrolyzed by two lipases, which contain similar catalytic triads (S^{152}/S^{105}, H^{263}/H^{224}, D^{176}/D^{187}), namely, porcine pancreatic lipase (PPL) and *Candida antarctica* lipase-B (CALB). Conversely, PPL and CALB differ in their catalytic mode. The former requires interfacial activation due to a lid domain, whereas the latter lacks a lid domain and does not require interfacial activation [7,8]. These differences in the catalytic modes of lipases CALB and PPL, along with their very many applications, made these enzymes appropriate for the objectives of this work. Additionally, the various applications of these lipases were based on their catalytic strength in a wide range of experimental conditions (e.g., pH value and temperature). Furthermore, the kinetic mechanisms of action of both CALB and PPL have been investigated extensively [8–13]. Hence, the hydrolytic function of these lipases on the fatty acid tetraester substrates of erythritol and pentaerythritol was considered to be kinetically interesting, and it was essential to investigate their use as low-calorie food substitutes. Subsequently, by means of an automatic titrator system and a relatively new method, the corresponding Michaelis–Menten kinetics and other important parameters were estimated, as well as the pK_a values of the released fatty acids. The calculated pK_a values through the automatic titrator system contributed significantly to the conclusions of this work. In fact, the results of this work converge to suggest that four among the five studied tetraesters may be used as low-calorie dietary food alternatives.

2. Results and Discussion

2.1. Hydrolysis of the Tetraesters of Erythritol and Pentaerythritol by PPL and CALB

The kinetic experimental data were best fitted, in most cases, by the Michaelis–Menten equation (see Supplementary Materials). The hydrolysis of pentaerythritol tetrastearate by CALB seems to follow non-Michaelis–Menten kinetics. From the best fitting of the experimental kinetic data, these results, i.e., the estimated parameter values (k_{cat}, K_m, k_{cat}/K_m, and additional parameters), are summarized in Table 1.

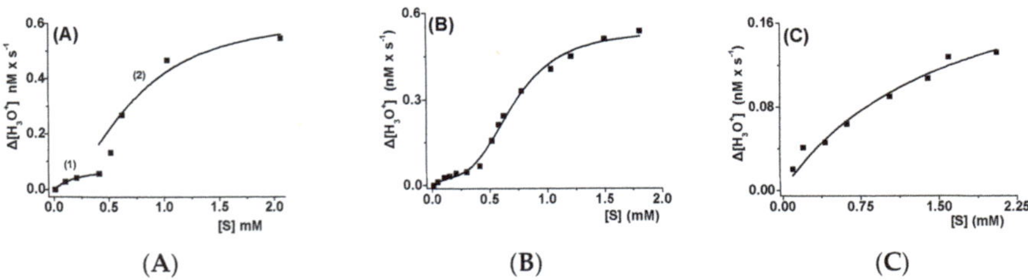

(A) (B) (C)

Figure 1. The kinetic data from the hydrolysis of pentaerythritol tetrastearate by CALB (curves (**A**,**B**)) are discontinuous and may be split into two groups. In curve (**A**) (sub-curves A_1 and A_2), the data were fitted by Equation (2), i.e., are compatible with substrate activation conditions. In curve (**B**), there were more experimental data due to the higher number of parameters, and these fitted with Equation (3) as a whole. In contrast, the experimental data from the hydrolysis of the same substrate by PPL (curve (**C**)) were fitted by the Michaelis–Menten Equation (1); saturation was not achieved due to the limited solubility of tetraester in the corresponding reaction mixture.

Table 1. The estimated parameters from the best fit of experimental data of the hydrolysis of the studied tetraesters by both *Candida antarctica* lipase-B (CALB) and porcine pancreas type VI-s lipase (PPL). The experimental data from the hydrolysis of pentaerythritol tetrastearate by CALB were best fitted by Equations (2) and (3).

SUBSTRATE	k_{cat} (s^{-1})		K_m (mM)		k_{cat}/K_m ($M^{-1} \times s^{-1}$)	
	CALB	PPL	CALB	PPL	CALB	PPL
Erythritol tetraolate	9.9×10^{-4}	1.2	1.7	3.5	0.6	342.9
Erythritol tetrapalmitate	7.0×10^{-4}	1.5	1.8	2.1	0.4	714.3
Erythritol tetralaurate	2.8×10^{-4}	2.9	0.8	5.8	0.4	500.0
Pentaerythritol tetrapalmitate	6.2×10^{-3}	7.3	5.1	4.8	1.2	1520.8
Pentaerythritol tetrastearate	(A_1) $4.7 \times 10^{-5\,a}$ (A_1) $5.5 \times 10^{-4\,b}$ (A_2) $3.7 \times 10^{-4\,a}$ (A_2) $4.3 \times 10^{-4\,b}$ (B) 3.2×10^{-4}	(C) 0.3	(A_1) 0.3 [c] (A_1) 11.2 [d] (A_2) 0.7 [c] (A_2) 0.3 [d] (B) 1.9	(C) 1.6	(A_1) 0.2 [e] (A_1) 0.1 [f] (A_2) 0.5 [e] (A_2) 1.4 [f] (B) 0.2	(C) 187.5
	(B) Virial coefficients $\Rightarrow V_1 = -3.9$; $V_2 = 9.4$					

All results were rounded to one decimal digit. In all cases of fitting, the goodness-of-fit (R^2 value) was estimated to be $\leq 0.998 \pm 0.010$. (A_1), (A_2), (B), and (C) refer to the corresponding curves of Figure 1. Subscripts a, b, c, d, e, and f correspond to k_{cat1}, k_{cat2}, K_m, K_{SS}, k_{cat1}/K_m, and k_{cat2}/K_m, respectively, of Equation (2).

The kinetic analysis of hydrolysis of pentaerythritol tetrastearate by CALB was based on the observation that the experimental points are apparently divided into two groups. The graphs of these results, along with the best fit of the kinetics of the hydrolysis of pentaerythritol tetrastearate by PPL, are depicted in Figure 1.

The kinetically discontinuous behavior of pentaerythritol tetrastearate (Figure 1A) could be explained by both its structure and the relatively high concentration of CALB in the reaction mixture, as compared to those of PPL (1700/0.884 ≈ 1923). Therefore, as the concentration of pentaerythritol tetrastearate increases in the reaction mixture, a discontinuous saturation of the enzyme molecules by substrate occurs, which is not uncommon. The enzyme–substrate complex was dispersed as an emulsion where Traube's Rule is valid [14] and was kinetically expressed as substrate activation. The additional best fit of the experimental data of the hydrolysis of pentaerythritol tetrastearate by CALB was performed by means of Equation (3), which is based on an expansion similar to the virial one concerning real fluids. Therefore, an increase in the powers of [S] ($>K_m$) is included in Equation (3), depending on the mean value of all the possible enzyme and substrate contacts and the conformational changes of the ES complex. Additionally, when the first and second derivatives of the polynomial $[S](1+A[S]+[B][S]^2)$, in the numerator and denominator of Equation (3), take on a zero-value simultaneously, one bending point is observed [15].

2.2. The Estimated pK_a Values of the Released Fatty Acids

In most of the examined cases, more than one pK_a value was calculated during the titration procedures of the fatty acids released through the CALB hydrolysis of the studied tetraesters. However, only one pK_a value (5.12) was estimated (Figure 2A) from the hydrolysis of erythritol tetraoleate by CALB and the release of oleic acid; for oleic acid, a pK_a of 5.02 in aqueous solution has been reported [16]. As opposed to the former result, four pK_a values (i.e., 4.54, 6.38, 6.76, and 7.76) were estimated when using the pentaerythritol tetrastearate (Figure 2B), while for stearic acid, a p*K*a of 4.78 has been reported [17].

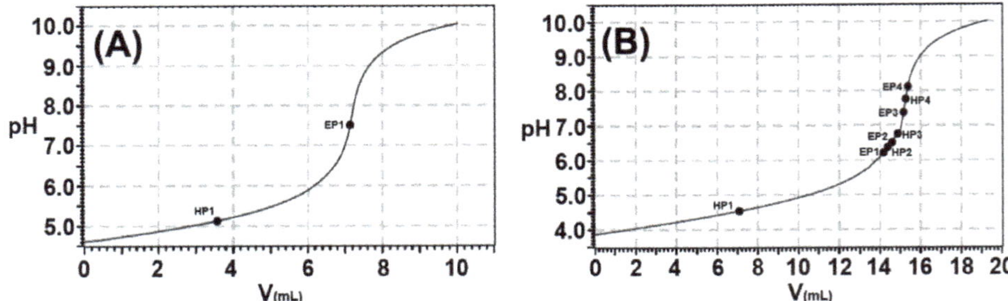

Figure 2. Automatically generated titration curves for the estimation of pKa values (CALB-catalyzed hydrolysis). Curve (**A**) corresponds to the titration of erythritol tetraoleate (one pKa = 5.12), while curve (**B**) corresponds to the titration of pentaerythritol tetrastearate (four pKa = {4.54, 6.38, 6.76, and 7.76}).

Three pK_a values were estimated in the case of erythritol tetrapalmitate, i.e., 4.45, 6.77, and 7.51 (Figure 3A), whereas two pK_a values were estimated in both cases of erythritol tetralaurate (i.e., 5.01 and 7.82) and pentaerythritol tetrapalmitate (i.e., 4.61 and 6.95), as depicted in Figure 3B,C, respectively. Values of $pK_a = 4.75$ and $pK_a = 5.30$ have been reported for palmitic and lauric acid, respectively [18,19]. Figure 2A,B as well as Figure 3A–C were generated automatically by the utilized software. The pK_a values (HP in Figures 2 and 3) have been marked on the titration curves by considering half the volume required to neutralize each of the released acids (EP—equivalence point in Figures 2 and 3). Therefore, as monoprotic acids are titrated, the Henderson–Hasselbalch equation is degenerated to $pH = pK_a$ [20].

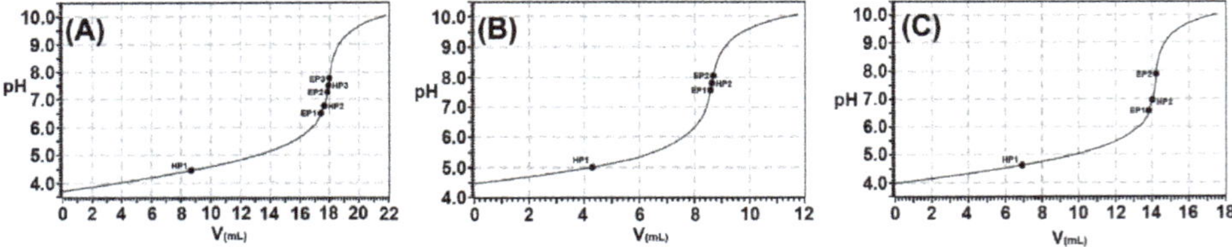

Figure 3. Automatically generated titration curves for the estimation of the pKa values (CALB-catalyzed hydrolysis). Curve (**A**) corresponds to the titration of erythritol tetrapalmitate (three pKa = {4.45, 6.77, and 7.51}), curve (**B**) corresponds to the titration of erythritol tetralaurate (two pKa = {5.01 and 7.82), and curve (**C**) corresponds to the titration of pentaerythritol tetrapalmitate (two pKa = {4.61, 6.95}).

Solvent/solute interactions affect the equilibrium between the solute's ionized/non-ionized forms and influence the estimated pK_a values [21]. In this work, the more likely to be released tri-, bi-, and mono-esters and the corresponding fatty acid molecules may form micellar bilayers according to Traube's Rule, as well as through autocatalysis [14,22], where the solute's ionized/non-ionized forms are present in equal concentrations. Additionally, four out of the five studied substrates were subjected to further hydrolysis during the titration process for the estimation of the pK_a values by varying pH values in a range of about six units. Hence, the more likely to be formed molecular species may yield the observed apparent pK_a values (pK_a^{app}). The aforementioned argument explains why, in this work, involving medium- and long-chain fatty acids, we observed these pK_a^{app} values (Figures 2 and 3).

Consequently, it seems likely that in the CALB-catalyzed hydrolysis of erythritol and pentaerythritol tetraesters, a complete hydrolysis was only achieved in the case of erythritol tetraoleate. In the case of pentaerythritol tetrastearate, a less complete hydrolysis was observed, because three pK_a^{app} values were estimated. Intermediate cases were observed in the hydrolysis of erythritol tetrapalmitate (two pK_a^{app} values) and in the hydrolyses of both erythritol tetralaurate and pentaerythritol tetrapalmitate (one pK_a^{app} value).

3. Materials and Methods

3.1. Materials

All analytical grade chemicals and reagents, including gum Arabic, PPL, as well as CALB (recombinant, expressed in *Aspergillus niger*—62288), were purchased from Sigma-Aldrich (St. Louis, MO, USA). Argon of high purity (≥99.99%) was purchased from Linde Hellas Ltd. (Linde Group). The herein utilized tetraesters, i.e., erythritol tetraoleate, erythritol tetrapalmitate, erythritol tetralaurate, pentaerythritol tetrapalmitate, and pentaerythritol tetrastearate, were synthesized as described previously [6]. Because the yield of pentaerythritol tetrapalmitate, obtained from the acylation of pentaerythritol by palmitoyl chloride in the presence of $BF_3{\cdot}Et_2O$, was only 47% [6], we describe in the Supplementary Materials the RCOCl/pyridine method for acylation, which gave 90% yield.

3.2. Solutions and Devices

Stock solutions of 0.1 M in dimethyl sulfoxide (DMSO) were prepared for all the utilized tetraester substrates. Aqueous stock solution (5% *w/v*) of gum Arabic was prepared by dissolving 5 g in 96 mL of bidistilled water and 4 mL of absolute ethanol to avoid formation of agglomerates. Standard solutions of 0.005 N NaOH, 0.500 N NaOH, and 0.500 N HCl were also prepared. All stock and standard solutions, including the bidistilled water, were degassed and stored under an argon atmosphere in bottles equipped with suitable sealing valves. The stock solution of PPL was prepared by dissolving the enzyme in 5% *w/v* solution of gum Arabic, as described previously [8], whereas the stock solution of CALB was prepared by dissolving the enzyme in bidistilled water [8,23]. The enzymatic stock solutions were degassed and stored under an argon atmosphere. All measurements were performed by means of an automatic titrator system Metrohm 907 Titrando (Metrohm AG, Herisau, Switzerland), which was computer-driven through appropriate software (Tiamo). The titrator system and the peripheral devices are illustrated graphically in Figure 4.

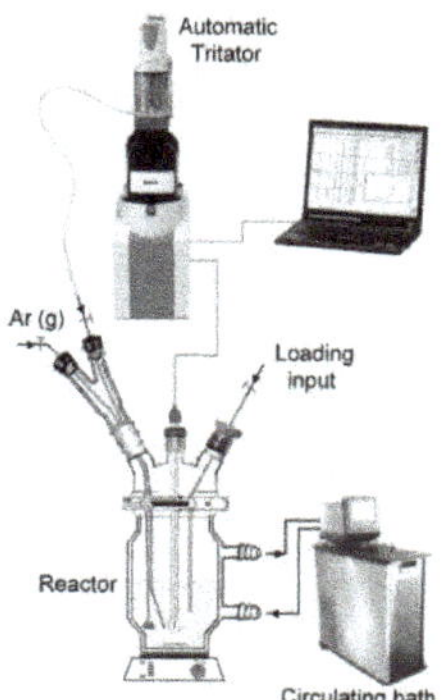

Figure 4. Representation of the devices for the titrimetry of the lipase-catalyzed hydrolysis of the five tetraester substrates employed in this work; the T-shapes indicate sealing valves. All reagents were added to the reactor through the loading input.

3.3. Kinetic Measurements

These measurements were performed by recording the decrease in the pH value of the reaction mixture vs. time due to the increased fatty acid concentration as a result of the lipase-catalyzed hydrolysis of the aforementioned erythritol and pentaerythritol tetraesters. In all measurements, the active concentrations of the employed lipases were kept stable in the reaction mixtures, i.e., 0.884 nM and 1700 nM in the cases of PPL and of CALB, respectively.

In a typical kinetic run, which was similar for both lipases, 27 mL of bidistilled water was transferred into the reactor (Figure 4), followed by the addition of 3 mL of 5% *w/v* gum Arabic stock solution, 20 μL of the appropriate lipase stock solution, and varied volumes of DMSO (<2.98 mL) so that the sum of volumes of DMSO and substrate combined to be 2.98 mL. Then, the reactor was sealed, the thermostat was fixed at 37 °C by means of a circulating bath, and argon gas was bubbled through the reaction mixture under continuous stirring (150 rpm). The driving software (Tiamo) of the automatic titrator was activated in order to set an initial pH value of 8.0 in the reaction medium (by adding 0.005 N NaOH standard solution) before starting the lipase-catalyzed hydrolyses. All of the hydrolytic procedures were initialized by the addition of a suitable volume of the studied tetraester substrate stock solution in the reaction mixture, followed by the end of the argon gas bubbling. The final volume of DMSO was always 6% in a 33 mL total reaction volume.

The initial velocities of the hydrolytic reactions were recorded as $\Delta[H_3O^+]$ nM × s^{-1} during the continuous decrease in the pH value of the reaction mixture. In all cases, the results were adjusted by performing blank measurements at an initial pH value of 7.0 in the reaction medium, where the volume of the studied substrate was replaced by an equal volume of DMSO. Each individual kinetic run was performed three times depending upon the standard deviation, which was set to be less than or equal to 5% of the mean value; otherwise, the three measurements were repeated. The employed concentrations of the tetraester substrates varied mainly from 0.10 mM up to 2.06 mM; however, in several cases, higher concentrations of a substrate were used according to its solubility in the reaction mixture. Active site titrations of the stock solutions with both lipases (PPL and CALB) were carried out under conditions similar to those of the kinetic runs, using the irreversible inhibitor phenylmethylsulfonyl fluoride (PMSF), as previously described [8], and then the active enzyme concentrations $[E]_0$ were estimated.

3.4. *Estimation of the* pK_a *Values*

The pKa values of the released fatty acids were estimated over the course of the CALB-catalyzed hydrolysis of each of the studied tetraester substrates, when a high substrate concentration was used in the corresponding kinetic run. The CALB-catalyzed hydrolysis was chosen due to the fact that the corresponding Michaelis–Menten curves approached saturation much more rapidly than those of the PPL-catalyzed hydrolysis in almost all cases. Therefore, after the end of each kinetic run, a suitable driving software subroutine of the automatic titrator was activated, and the reaction mixture was titrated by a standard solution of 0.500 N HCl up to a pH value of 3.50, along with the bubbling of argon gas and continuous stirring. Subsequently, the bubbling was stopped, and the reaction mixture was titrated by a 0.500 N NaOH standard solution through a suitable software routine of the automatic titrator, up to a pH value of 10.00. Subsequently, the corresponding pK_a values and the equivalent points were recorded automatically.

3.5. *Analysis and Curve Fitting of the Experimental Results*

In most cases, the experimental results derived from the kinetic runs were best fitted by the well-known Michaelis–Menten equation:

$$\mathrm{v} = \frac{k_{cat}[E]_0[S]}{K_m + [S]} \quad (1)$$

However, in the case of the hydrolysis of pentaerythritol tetrastearate by CALB, both Equations (2) and (3) best fitted the experimental data [15,24]:

$$\mathrm{v} = \frac{k_{cat1}[E]_0 + \frac{k_{cat2}[E]_0\,[S]}{K_{SS}}}{1 + \frac{K_m}{[S]} + \frac{[S]}{K_{SS}}} \quad (2)$$

$$v = \frac{k_{cat}[E]_0\,[S](1+V_1[S]+V_2[S]^2)}{K_m+[S](1+V_1[S]+V_2[S]^2)} \tag{3}$$

In Equations (1)–(3), the parameters k_{cat} and K_m have well-known meanings, whereas K_{SS} in Equation (2) is the equilibrium constant of the reaction ES + S ⇔ SES (substrate inhibition), a parameter similar to K_m. In Equation (3), V_1 and V_2 are the virial coefficients [15]. All fitting procedures were achieved by means of the OriginPro 2019 trial version, while different weighted least-squares tests were considered as convergence criteria [13].

4. Conclusions

In this work, we studied the kinetics of the lipase-catalyzed (PPL and CALB; the latter being widely used in biotechnological applications) hydrolysis of three fatty acid tetraesters of erythritol, as well as two fatty acid tetraesters of pentaerythritol. The values of Michaelis–Menten parameters k_{cat}, K_m, and k_{cat}/K_m were estimated during the hydrolysis of four of the aforementioned tetraesters by both lipases. Although the PPL-catalyzed hydrolysis of pentaerythritol tetrastearate followed Michaelis–Menten kinetics, in contrast, the CALB-catalyzed hydrolysis of this tetraester showed apparently discontinuous non-Michaelis–Menten kinetics; subsequently, novel important parameters were estimated. This latter kinetic behavior was explained in terms of substrate activation, as well as through Traube's Rule. Moreover, during the attempt to estimate the pK_a values of the released fatty acids as a result of the hydrolysis of the studied tetraesters by CALB, novel findings appeared. During these procedures, it was found that, excluding erythritol tetraoleate, the other four tetraesters were incompletely hydrolyzed by CALB, as between one and three apparent pK_a values (pK_a^{app}) were estimated (Figures 2 and 3). Likewise, the calculated k_{cat} and k_{cat}/K_m values were relatively low, and discontinuous kinetics of the hydrolysis of the pentaerythritol tetrastearate were detected. These findings indicate that the four tetraesters (erythritol tetrapalmitate, erythritol tetralaurate, pentaerythritol tetrapalmitate, and pentaerythritol tetrastearate), which were hydrolyzed by CALB incompletely, can be considered to be biotechnologically interesting. These compounds may be low-calorie foods and healthy alternatives to butter and/or sweeteners, augmented by the low toxicity of both erythritol and pentaerythritol in humans. Further research is necessary to synthesize substrates with similar structures that show much lower k_{cat} and k_{cat}/K_m values but with much higher K_m ones when catalytically hydrolyzed by PPL.

Supplementary Materials: The following are available online at http://www.mdpi.com/2073-4344/10/9/965/s1, Figure S1: The best fitting (Michaelis-Menten) of the experimental kinetic data concerning the hydrolyses by both lipases, i.e., PPL (curves A, C, E, G) and CALB (curves B, D, F, H), of the tetraesters erythritol tetraoleate (A,B), erythritol tetrapalmitate (C,D), erythritol tetralaurate (E,F), and pentaerythritol tetrapalmitate (G,H), respectively. Due to the limited solubility of several tetraesters, in the reaction mixtures, the corresponding Michaelis-Menten curves did not reach saturation, Figure S2. (A) IR (KBr), and (B) 1H NMR (CDCl3, TMS) spectra of pentaerythritol palmitate.

Author Contributions: O.A.G. performed the experimental work. A.F. supported the suitable programming of the software Tiamo. P.-Y.S. supported the suitable programming of the software Tiamo and revised the manuscript. P.V.I. synthesized the tetraesters of erythritol and pentaerythritol and revised the manuscript. E.M.P. designed the whole work, prepared the manuscript, and edited its final form. All authors have read and agreed to the published version of the manuscript.

Funding: This research received no external funding.

Conflicts of Interest: The authors declare no conflict of interest.

Ethical Approval: This article does not contain any studies with human participants or animals performed by any of the authors.

References

1. Röper, H.; Goossens, J. Erythritol, a New Raw Material for Food and Non-food Applications. *Starch/Stärke* **1993**, *45*, 400–405. [CrossRef]
2. Moon, H.J.; Jeya, M.; Kim, I.W.; Lee, J.K. Biotechnological production of erythritol and its applications. *Appl. Microbiol. Biotechnol.* **2010**, *86*, 1017–1025. [CrossRef] [PubMed]

3. Hartog, G.J.M.D.; Boots, A.W.; Adam-Perrot, A.; Brouns, F.; Verkooijen, I.W.C.M.; Weseler, A.R.; Haenen, G.R.M.M.; Bast, A. Erythritol is a sweet antioxidant. *Nutrition* **2010**, *26*, 449–458. [CrossRef]
4. Regnat, K.; Mach, R.L.; Mach-Aigner, A.R. Erythritol as sweetener-wherefrom and whereto? *Appl. Microbiol. Biotechnol.* **2018**, *102*, 587–595. [CrossRef]
5. Bringmann, G.; Kuhn, R. Results of toxic action of water pollutants on Daphnia magna Straus tested by an improved standardized procedure. In Z. *Wasser Abwasser-Forsch*; Aqua Publishing: Richmond, BC, Canada, 1982; Volume 15, pp. 1–6.
6. Ioannou, P.V.; Lala, M.A.; Tsivgoulis, G.M. Preparation and properties of fully esterified erythritol. *Eur. J. Lipid Sci. Technol.* **2011**, *113*, 1357–1362. [CrossRef]
7. Cygler, M.; Schrag, J.D. Structure as basis for understanding interfacial properties of lipases. *Meth. Enzymol.* **1997**, *284*, 3–27. [CrossRef]
8. Kokkinou, M.; Theodorou, L.G.; Papamichael, E.M. Aspects on the Catalysis of Lipase from Porcine Pancreas (type VI-s) in Aqueous Media: Development of Ion-pairs. *Braz. Arch. Biol. Technol.* **2012**, *55*, 231–236. [CrossRef]
9. Stauch, B.; Fisher, S.J.; Cianci, M. Open and closed states of *Candida antarctica* lipase B: Protonation and the mechanism of interfacial activation. *J. Lipid Res.* **2015**, *56*, 2348–2358. [CrossRef] [PubMed]
10. Abdul Rahman, R.N.Z.R.; Salleh, A.B.; Basri, M. Lipases: Introduction. In *New Lipases and Proteases*; Salleh, A.B., Abdul Rahman, R.N.Z.R., Basri, M., Eds.; Nova Science Publishers Inc.: New York, NY, USA, 2006; pp. 1–22. ISBN 1-60021-068-6.
11. Dulęba, J.; Czirson, K.; Siódmiak, T.; Marszałł, M.P. Lipase B from *Candida antarctica*—The wide applicable biocatalyst in obtaining pharmaceutical compounds. *Med. Res. J.* **2019**, *4*, 174–177. [CrossRef]
12. Stergiou, P.-Y.; Foukis, A.; Filippou, M.; Koukouritaki, M.; Parapouli, M.; Theodorou, L.G.; Hatziloukas, E.; Afendra, A.; Pandey, A.; Papamichael, E.M. Advances in lipase-catalyzed esterification reactions. *Biotechnol. Adv.* **2013**, *31*, 1846–1859. [CrossRef] [PubMed]
13. Foukis, A.; Gkini, O.A.; Stergiou, P.-Y.; Papamichael, E.M. New insights and tools for the elucidation of lipase catalyzed esterification reaction mechanism in n-hexane: The synthesis of ethyl butyrate. *Mol. Catal.* **2018**, *455*, 159–163. [CrossRef]
14. Strohmeyer, G.; Martin, G.A.; Khingsmuller, V. Changes of alpha-ketoglutaric acid, pyruvic acid and diphosphopyridine nucleotide in chronic liver diseases. *Klin. Wochenschr.* **1957**, *35*, 385–390. [CrossRef] [PubMed]
15. Lymperopoulos, K.; Kosmas, M.; Papamichael, E.M. A Formulation of Different Equations Applied in Enzyme Kinetics. *J. Sci. Ind. Res. India* **1998**, *57*, 604–606.
16. Jenke, D. Establishing the pH of Extraction Solvents Used to Simulate Aqueous Parenteral Drug Products during Organic Extractables Studies. *Pharm. Outsourcing* **2014**, *15*. [CrossRef]
17. Chemical Book. Available online: https://www.chemicalbook.com/ProductMSDSDetailCB4853859_EN.htm (accessed on 27 July 2020).
18. Available online: https://pubchem.ncbi.nlm.nih.gov/compound/985#section=Dissociation-Constants&fullscreen=true (accessed on 27 July 2020).
19. Available online: https://pubchem.ncbi.nlm.nih.gov/compound/3893#section=pKa&fullscreen=true (accessed on 27 July 2020).
20. Po, H.N.; Senozan, N.M. The Henderson-Hasselbalch Equation: Its History and Limitations. *J. Chem. Educ.* **2001**, *78*, 1499–1503. [CrossRef]
21. Namazian, M.; Halvani, S. Calculations of pKa values of carboxylic acids in aqueous solution using density functional theory. *J. Chem. Thermodyn.* **2006**, *38*, 1495–1502. [CrossRef]
22. Salentinig, S.; Sagalowicz, L.; Glatter, O. Self-Assembled Structures and pKa Value of Oleic Acid in Systems of Biological Relevance. *Langmuir* **2010**, *26*, 11670–11679. [CrossRef] [PubMed]

23. Foukis, A.; Gkini, O.A.; Stergiou, P.-Y.; Sakkas, V.A.; Dima, A.; Boura, K.; Koutinas, A.A.; Papamichael, E.M. Sustainable production of a new generation biofuel by lipasecatalyzed esterification of fatty acids from liquid industrial waste biomass. *Bioresour. Technol.* **2017**, *238*, 122–128. [CrossRef] [PubMed]
24. Papamichael, E.M.; Lymperopoulos, K. Elastase and Cathepsin-G from Human PMN activated by PAF: Relation between their Kinetic Behavior and Pathophysiological Role. In *Advances in Biotechnology*, 1st ed.; Pandey, A., Ed.; Educational Publishers & Distributers: New Delhi, India, 1998; pp. 221–234. ISBN 81-87198-03-6.

Article

Influence of Carrier Structure and Physicochemical Factors on Immobilisation of Fungal Laccase in Terms of Bisphenol A Removal

Kamila Wlizło [1,2,*], Jolanta Polak [2,*], Justyna Kapral-Piotrowska [3], Marcin Grąz [2], Roman Paduch [4] and Anna Jarosz-Wilkołazka [2]

1 Department of Industrial and Environmental Microbiology, Institute of Biological Sciences, Maria Curie-Skłodowska University, Akademicka 19, 20-033 Lublin, Poland

2 Department of Biochemistry and Biotechnology, Institute of Biological Sciences, Maria Curie-Skłodowska University, Akademicka 19, 20-033 Lublin, Poland; graz@poczta.umcs.lublin.pl (M.G.); anna.wilkolazka@poczta.umcs.lublin.pl (A.J.-W.)

3 Department of Functional Anatomy and Cytobiology, Institute of Biological Sciences, Maria Curie-Skłodowska University, Akademicka 19, 20-033 Lublin, Poland; justyna.kapral-piotrowska@poczta.umcs.lublin.pl

4 Department of Virology and Immunology, Institute of Biological Sciences, Maria Curie-Skłodowska University, Akademicka 19, 20-033 Lublin, Poland; rpaduch@poczta.umcs.lublin.pl

* Correspondence: kamila.wlizlo@poczta.umcs.lublin.pl (K.W.); jpolak@poczta.umcs.lublin.pl (J.P.); Tel.: +48-815-375-958 (K.W.); +48-815-375-051 (J.P.)

Received: 22 July 2020; Accepted: 17 August 2020; Published: 20 August 2020

Abstract: Laccase from *Pleurotus ostreatus* was immobilised on porous Purolite® carriers and amino-functionalised ultrafiltration membranes. The results indicated a correlation between the carrier structure and the activity of laccase immobilised thereon. The highest activity was obtained for carriers characterised by a small particle size and a larger pore diameter (the porous carriers with an additional spacer (C_2 and C_6) and octadecyl methacrylate beads with immobilised laccase activity of 5.34 U/g, 2.12 U/g and 7.43 U/g, respectively. The conditions of immobilisation and storage of immobilised laccase were modified to improve laccase activity in terms of bisphenol A transformation. The highest laccase immobilisation activity was obtained on small bead carriers with a large diameter of pores incubated in 0.1 M phosphate buffer pH 7 and for immobilisation time of 3 h at 22 °C. The immobilised LAC was stable for four weeks maintaining 80–90% of its initial activity in the case of the best C_2, C_6, and C_{18} carriers. The immobilised laccase transformed 10 mg/L of BPA in 45% efficiency and decreased its toxicity 3-fold in the Microtox tests. The effectiveness of BPA transformation, and the legitimacy of conducting this process due to the reduction of the toxicity of the resulting reaction products have been demonstrated. Reusability of immobilised LAC has been proven during BPA removal in 10 subsequent batches.

Keywords: laccase; immobilisation; porous carriers; membranes; bisphenol A; environmental toxicity

---

1. Introduction

Laccases are well-known biocatalysts for industrial purposes. Given their substrate promiscuity and the mild reaction conditions, this group of enzymes is widely studied in terms of green catalysis and especially improvement of their properties, which allows adaptation of laccases to specific biotechnological process conditions. Among modern approaches such as direct evolution and rational design, immobilisation of enzymes contributes to increasing laccase stability in adverse reaction conditions, such as extreme pH and temperature values or the presence of organic solvents [1]. Application of an immobilised enzyme in biocatalysis simplifies its separation from the reaction product and reuse in the continuous system of the biotechnological process. This reduces both the cost of production of the biocatalyst and the process itself; hence, immobilisation is the most promising method for industrial application of laccase and enzymes in general [2–4]. Many techniques and carriers for immobilisation have been developed over the years; however, there is no universal method of immobilisation or an ideal support for all enzymes and applications. Both have advantages and disadvantages, which need to be balanced through the optimisation process. Regarding immobilisation techniques, the choice is between the activity and stability of laccase, whereas the dilemma for the support lies between the limitations of substrate diffusion and the amount of laccase loaded on the support [5]. Therefore, the improvement of the immobilisation method in terms of the conditions of laccase attachment and extension of preparation's shelf life is still desired as well as searching for new, cheap supports. An interesting group of supports among many materials used for immobilisation, such as carbon and silica nanostructures, are acrylic resins due to their broad range of structural diversity and ease of modification [6,7]. Depending on the structure, they can be applied as supports for laccase immobilisation through covalent bonds and adsorption giving an efficient and repeatedly used biocatalyst [8]. An alternative for porous carriers may be ultrafiltration membranes characterised by low porosity, which are used especially for covalent attachment of laccase with simultaneous low substrate adsorption. Promising results were reported for laccase encapsulated in alginate, as high activity of immobilised laccase and simultaneous low substrate adsorption were noted [9,10]. However, there may be possible leaking of physically immobilised enzyme from the carrier and limited mass transfer, which are unfavourable phenomena [5,10].

Several groups of porous acrylic carriers (Purolite®) and amine-activated ultrafiltration membranes were tested as potential supports for immobilisation of fungal laccase from *Pleurotus ostreatus*. The supports differed in terms of the material they were made of, porosity, and immobilisation technique. The goals of this study were to carry out (1) evaluation of the correlation between the structure of the styrene and methacrylate carriers with the porosity ranged from 300 to 1800 Å as well as cellulosic and polyethersulfonic membranes and the activity of immobilised laccase, (2) analysis of the impact of the physicochemical factors of the immobilisation process and storage conditions on the activity of immobilised laccase and long-term stability, (3) transformation of bisphenol A (BPA) as a model endocrine chemical compound and analysis of the process efficiency, and (4) evaluation of the process legitimacy based on the results of BPA transformation and the toxicity of its transformation products.

2. Results and Discussion

2.1. Influence of the Carrier Structure

The carrier structure and immobilisation technique are major factors limiting the process of enzyme bonding to the carriers and, consequently, the applicability of immobilised biocatalysts. The 18 tested supports represented two different groups, i.e., porous carriers and membranes. The former group included ready-to-use (1) supports dedicated for covalent immobilisation and possessing a short (C_2) or long (C_6) spacer arm, (2) supports used for adsorption and possessing octadecyl groups (C_{18}), and (3) supports used for adsorption without any functional groups (S_0) (Figure 1).

The latter group comprised polyethersulfone (PES) and cellulose (CEL) membranes as potential supports for covalent immobilisation, which were additionally functionalised through silanisation. The scheme of LAC immobilisation on various types of carriers through covalent bonds and hydrophobic interactions is summarized in Figure 2.

Figure 1. Pictures of carrier's beads and membranes used in BPA transformation; (**A**) C2C carrier, (**B**) C6C carrier, (**C**) C18C carrier, (**D**) CEL100 membrane.

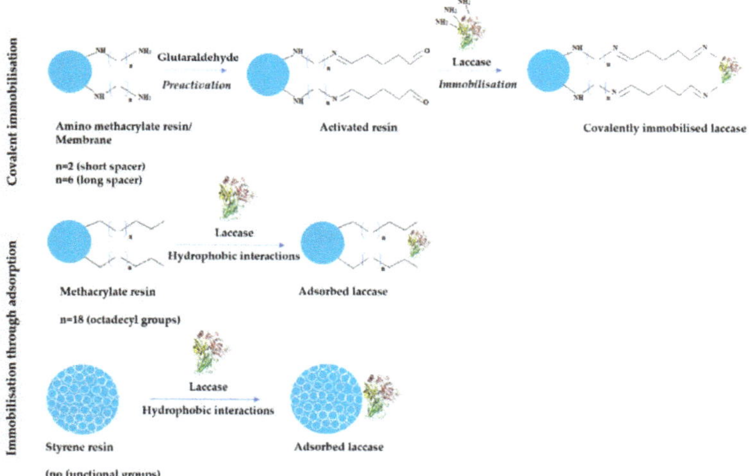

Figure 2. Schematic structure of tested carriers and the mechanism of laccase immobilisation covalently and through adsorption.

The present study showed a high impact of the carrier structure on the activity of immobilised laccase (LAC) from the *Pleurotus ostreatus* strain (Table 1). Both the LAC activity and the yield of protein bonded on porous carriers were significantly higher than in the case of membranes (PES and CEL). This was probably caused by the poor porosity and smaller surface area of the membranes, resulting in low protein loading confirmed by the Bradford method.

As described in several papers, acrylic beads with high porosity provided 100% of protein loading; hence, supports with high porosity are recommended for immobilisation of enzymes [11]. The highest LAC activity in the case of porous carriers with an additional C_2 and C_6 spacer and octadecyl methacrylate beads (C_{18}) was obtained for those characterised by a small particle size and a larger pore diameter (acronyms ending in a letter C), i.e., a structure representing an optimal setup of carrier capacity and flow properties [11]. Interestingly, in the case of the C_2 and C_6 carriers, the decrease in the pore diameter reduced the immobilised LAC activity (C2B, C6B).

This may be the result of laccase immobilisation mainly on the carrier's surface. Therefore, LAC is not protected from external factors, which can be observed in the case of high porosity carriers [11].

Moreover, substrate and product dispersion can be limited due to the smaller pore size; hence, the detected activity of laccase is underestimated.

Table 1. Influence of the type of carrier and immobilisation technique on the activity of immobilised LAC from *Pleurotus ostreatus* [U/g] and yield of protein immobilisation [%]—screening test.

Immobilisation Technique	Carrier	A_{imm} [U/g]	Yield [%]
Covalent	C2A	2.47 ± 0.1	40.5 ± 5.4
	C2B	1.58 ± 0.3	51 ± 1
	C2C	5.34 ± 0.4	80.5 ± 2.5
	C2D	1.86 ± 0.1	55.5 ± 5.5
	C6A	2.75 ± 0.2	76 ± 3.6
	C6B	2.12 ± 0	62 ± 2.1
	C6C	2.12 ± 0.3	68 ± 0.1
	C6D	2.65 ± 0.2	92 ± 0.5
	CEL10	0.30 ± 0	1.23 ± 0.1
	CEL100	0.33 ± 0	1.54 ± 0.1
	PES10	0.55 ± 0	1.0 ± 0.5
	PES100	0.88 ± 0	1.15 ± 0
Adsorption	C18A	1.04 ± 0.1	100 ± 0
	C18C	7.43 ± 0.3	100 ± 0.1
	C18D	2.55 ± 0.2	100 ± 0.1
	S0A	5.80 ± 0.9	96 ± 0.4
	S0B	1.05 ± 0.1	85 ± 3.5
	S0D	3.87 ± 0.3	97 ± 2.1

The length of the spacer arm had an imperceptible impact on the immobilisation process. This conclusion contradicts the general information that an optimal carrier should have a small spacer [5]. Indeed, a short spacer arm promotes multipoint laccase binding, thereby preventing its desorption, but may hinder the mobility of the bound enzyme molecules in contrast to those bound on the longer spacer arm [12,13]. Such bonding and stiffening of the LAC structure may be more desirable in terms of its long-term stability rather than activity.

The immobilisation of LAC was also possible on the tested membranes; however, the activity obtained was much lower than on the porous carriers. Nevertheless, there was a correlation between the pore diameter and the activity. Higher values were obtained after the immobilisation on membranes with 100 kDa porosity, compared to 10 kDa, and the highest LAC activity was observed on the 100 kDa cellulose membranes (CEL100).

Considering the immobilisation technique used, the highest activity of LAC was obtained via adsorption on the C18C support, with a concomitant 100% protein bonding yield, whereas the lowest activity of LAC was detected for the covalent attachment of LAC to the PES membranes. The yields of protein bonding (covalently) to the C_2 and C_6 carriers were in the range from 40% to 92%, whereas a 100% yield was noted in the case of the C_{18} and S_0 supports (adsorption). This phenomenon was associated with the actual activity of immobilised LAC; however, the enzyme-carrier interactions were important. The cause of the lower activity of covalently immobilised LAC in groups C_2, C_6, and membranes was not only the lower amount of bonded protein but also the stabilisation of the LAC quaternary structure by strong covalent bonds and multipoint attachment to the carrier. This contributed to LAC rigidity and, consequently, loss of activity, which is a well-known disadvantage of this technique [14,15]. On the other hand, adsorption of LAC on C_{18} and S_0 carriers is possible through weak hydrogen, hydrophobic, and ionic bonds as well as van der Waals forces, allowing sufficiently strong bonds without any distortion of the structure of the enzyme, which maintains its activity [12].

The contribution of different types of bonds formed during the covalent LAC immobilisation was calculated based on the activity of unbonded LAC in eluates obtained during the carrier washing steps

(Figure 3). Covalent bonds represented the main type of immobilisation on both groups of carriers, whereas unspecific bonds accounted for up to 12%. The percentage of unspecific bonds ranged from 3.8% to 12.1% in the C_2 group and from 1.8% to 4.5% in the C_6 group. The two groups of carriers differed in terms of the proportion of unspecific bonds. There was a clear predominance of adsorption over the other ones in the C_2 group, whereas adsorption in the C_6 group was negligible.

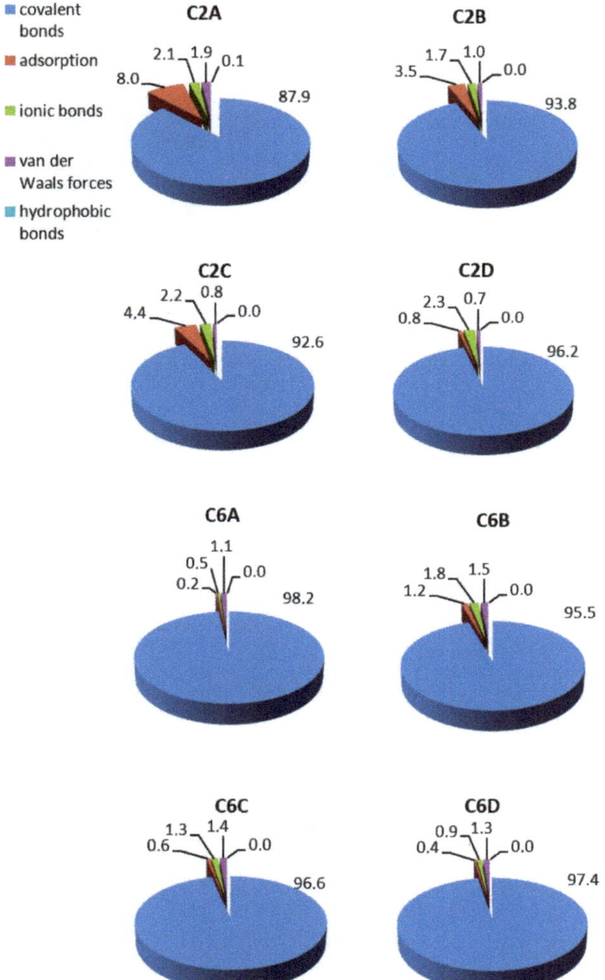

Figure 3. Percentage of bonds formed during immobilisation of LAC on porous carriers with C_2 and C_6 spacer arms.

In the case of the membranes, most LAC was present in the decanted solution remaining after immobilisation, due to the very low bonding on the poorly developed specific surface of the membranes. In the case of the C_{18} and S_0 groups, the adsorption of LAC on the porous carriers was so strong that there was no LAC in the solution decanted over the carriers after the immobilisation. Consequently, both factors contributed to the lack of LAC activity in the eluates during washing of the C_{18} and S_0

carriers and the membranes and thus it was not possible to determine the types of bonds formed during LAC immobilisation thereon.

The differences in the immobilised LAC activity depended on the carrier structure and the immobilisation technique, whereas the stability of LAC during long-term storage was not as obvious. Generally, the activity of laccase immobilised covalently on any of the tested carriers remained above 50% activity after four weeks of storage (Figure 4).

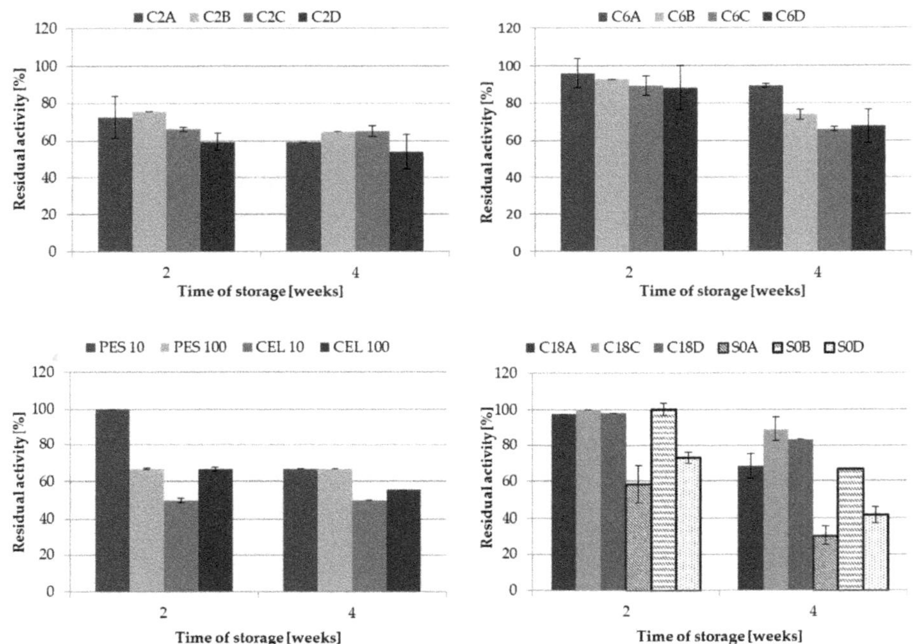

Figure 4. Stability of LAC immobilised on different carriers via covalent bond formation and adsorption during 4 weeks of storage at 4 °C.

LAC immobilised on the C6 supports exhibited the highest stability, as its activity exceed 66% of the initial value. The results obtained on each carrier in its group were comparable, whereas the following differences in the stability were noted between the groups: C_6 group > C_2 group > membranes. High stability of LAC obtained through adsorption was noted in the case all the C18 carriers, with the value in the range from 66% to 89% of the initial LAC activity after four weeks of storage. The largest discrepancies in the results were recorded for the S_0 group. Only for S0B, the carrier retained LAC activity above 60%, which decreased below 50% in the case of the S0A and S0D carriers. Moreover, the dense packing of LAC molecules on the S0A and S0D carriers may have favoured desorption of loosely bound molecules, which could not have been stabilised due to the lack of any functional group on the carriers.

2.2. Immobilisation Conditions

Based on the screening test, six carriers were selected for further studies, i.e., two carriers made from macroporous styrene (S0A and S0D), three made from methacrylate (C2C, C6C, C18C), and one from cellulose membrane (CEL100).

The impact of four physicochemical factors on immobilisation efficiency was evaluated to obtain high activity of immobilised LAC that allows maximising its biocatalytic potential in biotechnological processes. The tested factors included pH values and the concentration of phosphate buffer as an environment of LAC immobilisation, temperature and time of LAC incubation with the support, and the LAC concentration. In this step of the study, the activity of immobilised laccase expressed in U/g was monitored.

2.2.1. Influence of Phosphate Buffer pH Value

Most papers described the use of 0.1 M phosphate buffer pH 7 for protein immobilisation [16,17]. The present results showed a slight effect of the phosphate buffer pH on the activity of immobilised laccase (Table 2).

Table 2. Influence of phosphate buffer pH value on the activity of LAC [U/g] from *P. ostreatus* attached covalently or through adsorption on different porous carriers.

Immobilisation Technique	Carrier	Buffer pH			
		5	6	7	8
Covalent	C2C	5.41 ± 0.2	4.88 ± 0	4.5 ± 0.3	3.13 ± 0
	C6C	5.26 ± 0.2	4.65 ± 0.2	3.99 ± 0	3.76 ± 0
	CEL100	0.45 ± 0	0.45 ± 0	0.35 ± 0	0.23 ± 0
Adsorption	C18C	4.78 ± 0	4.95 ± 0	5.14 ± 0	4.72 ± 0
	S0A	1.89 ± 0	2.06 ± 0	2.09 ± 0	1.5 ± 0
	S0D	1.56 ± 0	1.53 ± 0	1.32 ± 0	1.07 ± 0

In the case of the covalently attached protein, the highest values were noted after using buffer of pH 5 (C2C and C6C carriers), which decreased as the buffer pH increased, whereas the highest activity of laccase adsorbed on the carrier was determined for the C18C carrier when buffer pH 7 was used (Table 2).

The correlation between buffer pH and LAC activity in the case of the covalent immobilisation can be the result of the impact of pH on functional groups, especially on ionisation of hydrogen in primary lysine amino groups on the surface of laccase participating in multipoint attachment, which is more likely to form at higher pH values [18]. In the case of adsorption on C18C, the elevated pH may have negatively influenced the octadecyl groups and thus limited the availability of these groups for interaction with laccase molecules. In contrast, the effect of buffer pH on adsorption on the S_0 supports was marginal. However, samples of immobilised LAC obtained in the pH 5 buffer turned out to be unstable. The process of LAC washing out from the carriers was monitored with the Bradford method. After 7 days of storage, the activity of unbounded LAC present in the buffer solution was higher in the case of laccase immobilised in the pH 5 buffer than in the samples obtained at pH 7 and pH 8. Similar correlations between the buffer pH value and immobilisation efficiency as well as LAC release from the carrier were observed by Pezzella and co-workers, who noted the highest immobilisation yield of laccase after incubation in phosphate buffer pH 7.5 with no simultaneous desorption of covalently bound laccase [19,20]. Considering all these results, pH 7 buffer was selected for further experiments. This is in agreement with literature data reporting that the phosphate pH 7 buffer is an optimal environment for laccase immobilisation, especially when glutaraldehyde is the bi-functional coupling agent [6].

2.2.2. Temperature and Time of Immobilisation

From the wide range of parameters that are important for immobilisation process, the temperature and time of incubation with the enzyme should be optimised, because they facilitate reduction of the immobilisation process cost. Most of the enzyme-engaging processes occur at the temperature of 4 °C to minimise enzyme degradation. Application of 4 °C for immobilisation of LAC can increase its half-life but can also limit its structure flexibility during attachment to the carrier surface or bond formation. The results of the experiment showed the same correlation for all types of carriers used for covalent immobilisation (Table 3).

Table 3. Influence of the temperature on the activity [U/g] of covalently immobilised LAC from *P. ostreatus* during 24-h incubation.

Carrier	Temperature	Time of LAC Incubation [h]			
		1	3	5	24
C2C	4 °C	3.43 ± 0.4	4.73 ± 0.1	4.26 ± 0.3	4.44 ± 0.3
	22 °C	4.41 ± 0.2	5.18 ± 0.1	4.55 ± 0.2	4.30 ± 0.4
C6C	4 °C	3.91 ± 0.2	4.16 ± 0.2	4.37 ± 0.2	4.38 ± 0.7
	22 °C	4.55 ± 0.2	4.61 ± 0.4	3.72 ± 0.2	4.01 ± 0.6
CEL100	4 °C	0.55 ± 0.1	0.63 ± 0.1	0.61 ± 0	0.83 ± 0.1
	22 °C	0.73 ± 0	0.68 ± 0	0.78 ± 0.1	0.83 ± 0.1
C18C	4 °C	4.02 ± 0.2	3.97 ± 0.2	3.63 ± 0.5	4.44 ± 0.3
	22 °C	4.22 ± 0	4.37 ± 0.2	4.01 ± 0.5	4.59 ± 0.1
S0A	4 °C	2.20 ± 0.1	2.15 ± 0.1	2.26 ± 0.1	2.33 ± 0.3
	22 °C	2.06 ± 0	2.63 ± 0.2	2.13 ± 0.1	2.51 ± 0.2
S0D	4 °C	2.40 ± 0.1	2.44 ± 0.1	2.57 ± 0.1	2.40 ± 0.1
	22 °C	2.33 ± 0	2.99 ± 0.2	2.41 ± 0.2	2.83 ± 0.2

The activity of LAC obtained after incubation at 4 °C was lower than in preparations incubated at 22 °C, especially during the first 3 h of incubation; these differences became less significant in the following hours. The optimal time of carrier incubation with the LAC protein varied and depended on the carrier used: it was 3 h for four out of the six carriers. The LAC activities obtained at 22 °C were higher in comparison to their analogue obtained at 4 °C. This may suggest that, after more than 3 h of incubation, the concentration of protein bound to the carrier was too high and restrained dispersion of the substrate and the product [21]. This effect was also observed in the case of the covalent immobilisation of *Pleurotus florida* laccase, where incubation of the enzyme with the porous silica perlite carrier for more than 4 h had no effect on a further increase in the immobilisation yield and the activity of immobilised LAC [19].

2.2.3. Laccase Concentration

The purified laccase from *Pleurotus ostreatus* was immobilised in optimal conditions in a concentration range from 0.5 mg/g to 4 mg/g of the wet mass of the carrier. As presented in Table 4, the concentration of LAC bound to the carrier had a significant impact on enzymatic activity.

Table 4. Effect of the LAC concentration applied for covalent immobilisation (C2C, C6C, CEL100) and via adsorption (C18C, S0A) on the activity of immobilised LAC [U/g]; optimal conditions were: buffer pH 7, buffer concentration—0.025 M for C2C and 0.1 M for C6C, time and temperature of incubation—3 h and 22 °C.

Carrier	LAC Concentration [mg/g]			
	0.5/0.25 *	1/0.5 *	2/0.75 *	4/1 *
C2C	5.38 ± 0.3	12.18 ± 0	8.14 ± 0.3	19.14 ± 0
C6C	3.82 ± 0.5	5.79 ± 0.2	10.73 ± 1.3	14.09 ± 1.4
CEL100	0.74 ± 0	0.77 ± 0.2	0.92 ± 0.1	0.97 ± 0.1
C18C	5.56 ± 0	7.51 ± 0.6	5.31 ± 0.2	5 ± 0.5
S0A	1.51 ± 0.1	3.13 ± 0.1	3.27 ± 0.2	3.56 ± 0.4

* values of LAC concentration only for CEL100.

The application of the higher concentration of LAC in the solution resulted in higher activity of immobilised LAC (Table 4). The protein overcrowding on the carrier surface, affecting the activity of immobilised laccase, is a well-known phenomenon, which limits the full use of carrier capacity. For example, the declared capacity of most of the tested Purolite® carriers is estimated at 50 mg of protein per 1 g of carrier. The present results of immobilised LAC activity clearly indicated that this concentration was far too high for the majority of carriers and did not affect the activity of the obtained laccase proportionally. A drastic decrease in the efficiency after covalent binding of laccase was observed, where the immobilisation yield dropped from 70% to 50% for preparations with 80 IU/g and 300 IU/g loading, respectively [19]. Overloading of the support is a well-known phenomenon, in which a high number of enzyme molecules limit the dispersion of the substrate and reaction product [22].

2.3. Storage Conditions and Thermostability

2.3.1. Composition and pH Value of Buffer

The impact of the buffer composition and its pH value on the activity of both free and immobilised LAC was evaluated during 4-week storage. The results showed that immobilisation of LAC significantly improved its stability (Figure 5).

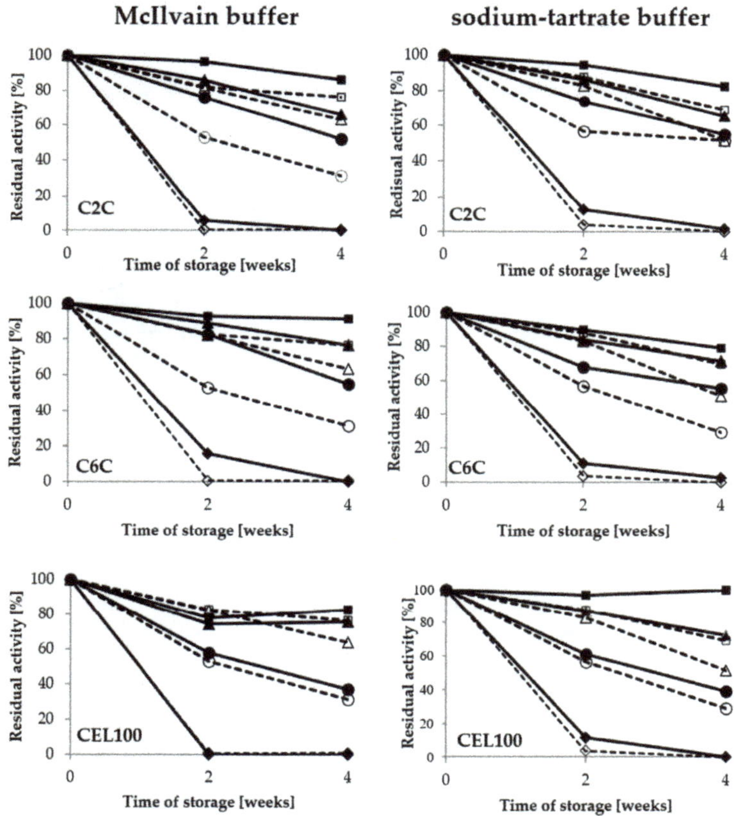

Figure 5. Stability of covalently immobilised laccase activity during 4-week storage in the McIlvain buffer and sodium-tartrate buffer in the range of pH values from 3 to 6; immobilised LAC—filled, free LAC—empty: pH 3—rhombus, pH 4—circle, pH 5—triangle, pH 6—square.

Moreover, it was noted that pH value of buffer had a significant impact on the activity of immobilised LAC during the long-term storage. In the case of the covalent immobilisation, the optimal buffer for storage of both immobilised and free LAC was pH 6, with a minimum influence of the buffer composition. In the case of the porous C2C and C6C carriers, the highest stability of laccase was observed after using the McIlvain buffer, with 87% and 91% of retained activity respectively, after four weeks of storage (Figure 5). Therefore, this buffer was selected for further studies of laccase immobilised on those carriers, while the pH 6 sodium-tartrate buffer was selected for the CEL100 membrane as the best for preservation of the activity of immobilised laccase.

In the case of the adsorbed laccase, the pH value was the most important factor for the enzyme stability as well; however, the composition of the buffer altered the optimal pH value. As shown in Figure 6, the activity of laccase adsorbed on the C18C carrier varied between the different storage buffer types (Figure 6). The highest values for the McIlvain buffer were noted after storage at pH 6, whereas the peak of activity in the sodium-tartrate buffer was observed after storage at pH 5.

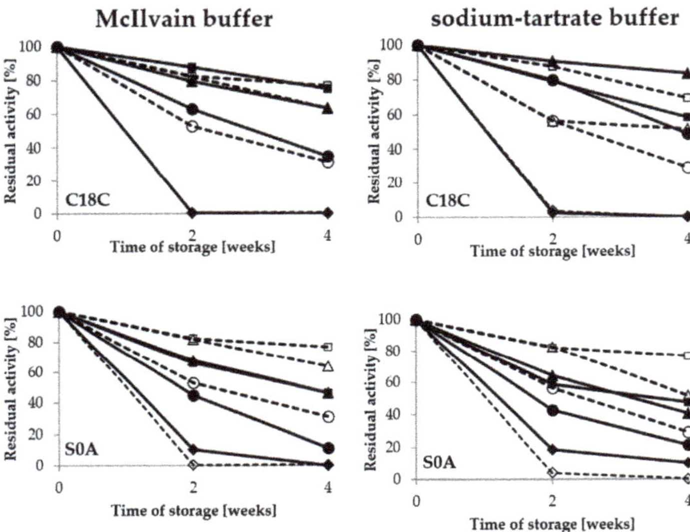

Figure 6. Stability of adsorbed laccase activity during 4-week storage in the McIlvain buffer and the sodium-tartrate in the range of pH from 3 to 6; immobilised LAC—filled, free LAC—empty: pH 3—rhombus, pH 4—circle, pH 5—triangle, pH 6—square.

For laccase adsorbed on the S0A carrier, high activity was also noted in the pH 5 sodium-tartrate buffer, but only after 2 weeks of storage. Immobilisation of laccase preserved its activity in comparison with the free form of the enzyme. The results regarding the beneficial effect of immobilisation on LAC activity preservation in adverse conditions during storage are consistent with those reported for inter alia *Trametes versicolor* laccase immobilised covalently on metal-ion-chelated magnetic microspheres and silica beads [23,24] and laccase adsorbed on vinyl-modified mesoporous fibres or on carbon-based mesoporous magnetic composites [25,26].

2.3.2. Thermostability

One of the main goals of laccase immobilisation is to increase its resistance to adverse environmental conditions such as high temperature. In this experiment, free and immobilised laccase was incubated in the temperature range from 20 °C to 80 °C for 24 h to assess its stability. The results indicated that the immobilisation of laccase improved its stability, compared to the free form (Figure 7). In the case of covalent immobilisation, the highest activity of both immobilised and free laccase was maintained after refrigerated storage and decreased along with the increase in the temperature up to 50 °C, above which protein denaturing conditions were observed. Among the tested covalent carriers, C2C exhibited the highest stability of immobilised laccase in the range of approx. 90% to 50% of the initial activity at the temperature of 20 °C to 40 °C respectively, which constituted a 20% to 11% difference compared to free laccase (Figure 7). The better thermostability results of the C2C samples may have been caused by the formation of multipoint bonds, which are more likely to form when functional groups are mounted on the short spacer and prevent denaturation of the enzyme [18,27]. This feature has been repeatedly described in the literature for different laccases and supports [13,23,27].

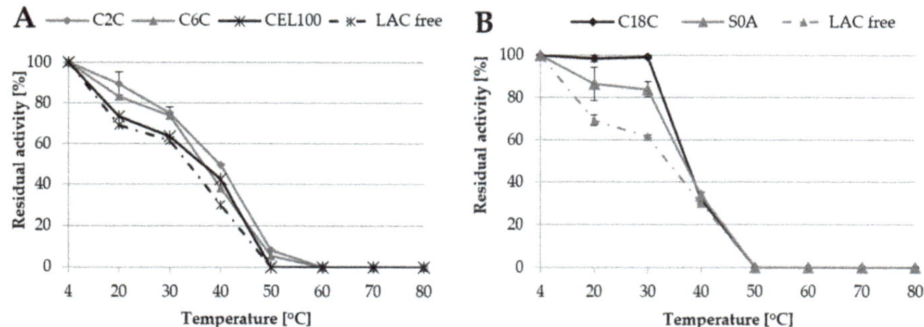

Figure 7. Thermostability of laccase immobilised covalently (**A**) and via adsorption (**B**) in comparison to thermostability of free laccase (LAC free).

The activity of laccase adsorbed on the C18C and S0A carriers was maintained much longer than in the case of the covalent immobilisation. In this case, the results showed approx. 99% and 85% of laccase initial activity at 20 °C and 30 °C, respectively. However, at 40 °C, the stability of the immobilised laccase decreased drastically, even more than that of the free form of the enzyme, which probably resulted from the low stability of the carriers themselves at high temperatures.

2.4. Degradation of Bisphenol A

Bisphenol A (BPA) is a well-known organic pollutant identified mainly in the aquatic environment [28]. Numerous studies have shown that BPA present in drinking water enters the human organism and is contained mainly in human body fluids such as urine, umbilical cord blood, and breast milk [29,30]. Such distribution poses a threat to human health already at the stage of foetal life, since the main harmful effect of BPA is hormonal disorders of the reproductive system [31]. In adults, it leads to infertility in general but also to prostate cancer in men and ovarian cancer in women [32]. Therefore, transformation of BPA is important in terms of limitation of its concentration in wastewater, its distribution in environment, and harmful effects on living organism. Therefore, the potential of immobilised LAC to degrade BPA was assessed.

Samples of immobilised LAC were tested as a potential catalyst for bisphenol A degradation. In the first stage of the experiments, several concentrations of BPA from 0.5 mg/L to 10 mg/L were transformed using immobilised LAC with activity of 5 U/L. After 72 h of transformation, the BPA depletion ranged from 50% to even 100% in the case of the C2C carrier and the C18C carrier, respectively (Figure 8). To distinguish the enzymatic transformation from physical adsorption of BPA, a control experiment regarding BPA adsorption was conducted. The results prompt a conclusion that the depletion of BPA by most of the carriers is mainly due to adsorption of the compound. This was especially visible in the case of the 100% removal by the C18C carrier, probably caused by the presence of octadecyl groups and thus the hydrophobic nature of this carrier, to which BPA as a substance sparingly soluble in water attaches easily. In the case of the C2C and C6C carriers exhibiting the same porosity and particle size, but different functional groups BPA adsorption was no higher than 70%. Interestingly, no BPA adsorption on the CEL100 carrier was observed, and the BPA loss was caused only by the LAC transformation.

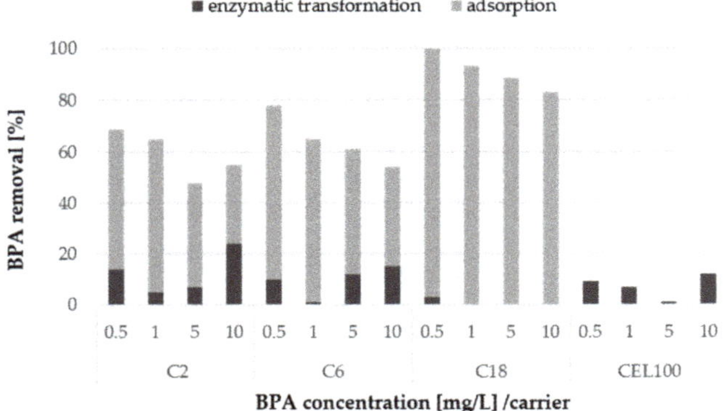

Figure 8. Enzymatic transformation of BPA at a concentration from 0.5 mg/L to 10 mg/L using adsorbed or covalently immobilised laccase and removal of BPA via adsorption on the surface of non-modified selected carriers after 72 h. The standard error of the results did not exceed 5%.

Among the tested BPA concentrations, the highest percent of enzymatic transformation for 10 mg/L BPA was observed in the case of the C2 and C6 carriers with covalently immobilised LAC. It could be expected that the highest percentage of transformation would be observed at a lower concentration of BPA. However, the results suggested that most of the BPA sample in the case of the lower concentration was probably strongly adsorbed on the carrier and thus the transformation was inhibited.

To maximise the BPA transformation using immobilised LAC, the higher activity of 12 U/L was applied. This increased the share of enzymatic transformation over adsorption in the BPA loss up to 39%, 45% and 21% for the C2C carrier, the C6C carrier, and the CEL100 membrane, respectively (Figure 9). The best rate of BPA enzymatic degradation obtained for the C6C carrier was 45%, which corresponds to 0.063 mg/L/h of BPA removal. This result was comparable to results obtained for laccase from *Trametes versicolor* immobilised on *Hippospongia communis*, i.e., 0.08 mg/L/h of BPA removal, and even higher than those reported for *Coriolopsis gallica* laccase immobilised on silica beads (>0.03 mg/L/h of BPA removal) [33,34]. Noteworthy, our results were obtained using nearly 5-times lower LAC activity than the activity of *C. gallica* LAC [33].

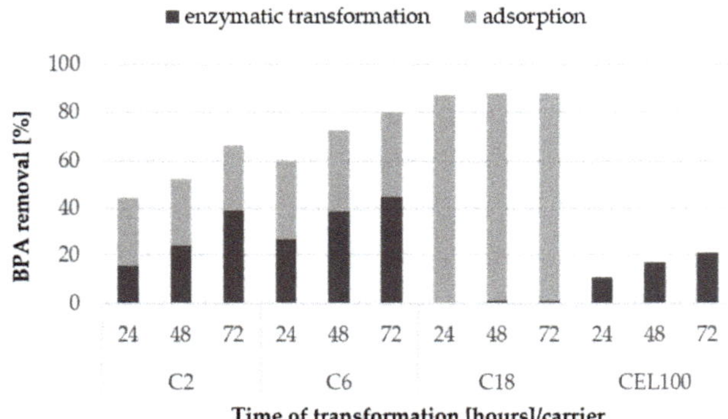

Figure 9. Enzymatic transformation and adsorption of BPA at the concentration of 10 mg/L using laccase immobilised covalently and via adsorption on selected carriers after 24, 48 and 72 h. The standard error of the results did not exceed 5%.

In the case of the C18C carrier, the increase in the LAC activity resulted in lack of transformation, even though the LAC activity on this carrier was the highest and the mass of the carrier used in the transformation was the lowest. Therefore, it may be concluded that carriers with LAC immobilised by adsorption are not suitable in BPA and other partially soluble in water substances transformation, even though they exhibit high activity of LAC attached.

The main advantage of immobilised laccase application is its reusability, what was measured during 10 batches of BPA (final concentration 10 mg/L) transformed by covalently immobilised LAC. From four previously tested carriers, the C18C carrier was excluded due to the high adsorption of the BPA. The highest BPA removal was obtained for C6C carrier in the first five batches comparing to C2C carrier (Figure 10A). In next batches, BPA removal rate was equal for both carriers. The lowest reusability was showed for CEL100 membrane, probably due to low surface area, which is a disadvantage of membranes application [17]. Along with BPA removal activity of immobilised LAC activity also decreased (Figure 10B) and retained about 50% of its initial activity in the fourth batch, which is comparable to activity of laccase immobilised in cross-linked aggregates and metal-ion-chelated magnetic microspheres [24,35].

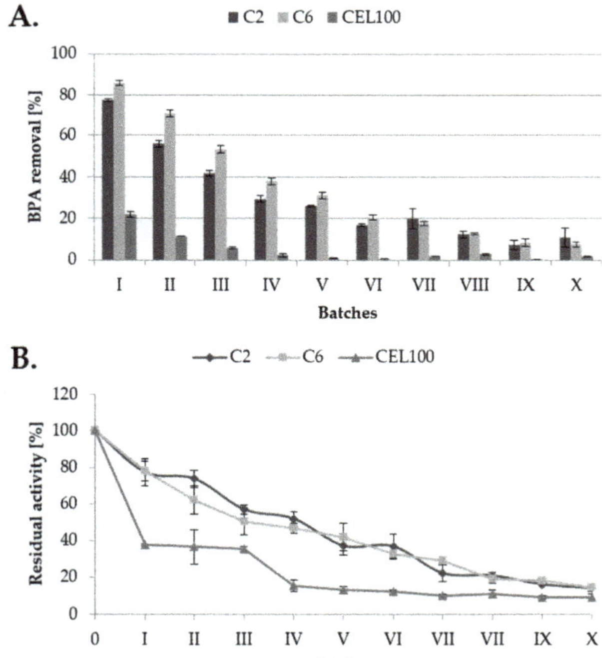

Figure 10. Rate of bisphenol A transformation during ten subsequent batches (**A**) and residual LAC activity (**B**) measured on the tested C2, C6 and CEL100 carriers.

2.5. Toxicity Evaluation

Biological degradation of micropollutants including enzymatic treatment generates reaction intermediates with varied biological activity [36]. Moreover, in some cases, products of reaction may be equally or even more toxic than the treated substrate [37].

Therefore, it is important to ensure that the transformation process is not only effective but also mitigates the hazardous effect of the micropollutant. Since BPA mostly accumulates in the environment, the ecotoxicity of BPA and its transformation products was evaluated and, since it pollutes drinking water and enters the human body through ingestion, its cytotoxicity towards normal colon epithelial cells tests was assessed.

The ecotoxicity was tested using a Microtox® test based on inhibition of bioluminescence of bacterial species *Vibrio fisheri*, which is one of the most common standardised methods used in ecotoxicity assessment [38]. The results of the EC_{50} values indicated that BPA transformation products are ca. 2.5-times less toxic than the BPA substrate (Table 5). These results are consistent with results of BPA degradation with crude laccase from *P. ostreatus*, which reduced BPA toxicity to 4.65% of its initial value [39].

Table 5. Environmental toxicity of BPA and its transformation products obtained using Microtox® test after 5 and 15 min of incubation with *Vibrio fisheri* and expressed as EC_{50}.

Time	EC_{50} [mg/L]	
	BPA	Products
5 min	0.116 ± 0	0.316 ± 0.01
15 min	0.156 ± 0.01	0.375 ± 0.02

The MTT test showed no effect of BPA and its transformation products on mitochondrial activity after 24 h of incubation with colon epithelial cells CCD-841CoTr (Table 6). After 48 h of incubation, a decrease in the toxicity of the product compared to the substrate was noticeable but both values were still low.

Table 6. Cytotoxicity effect of BPA and its transformation products after 24 and 48 h of incubation with human colon cells evaluated using MTT and LDH tests and expressed as % of damaged cells.

Time	MTT		LDH	
	BPA	Products	BPA	Products
24 h	0	0	15 ± 1.82	5 ± 0.23
48 h	10 ± 0.77	7 ± 1.23	17 ± 2.5	3 ± 0.71

Thus, it may be assumed that BPA and its transformation product had a limited influence on cell metabolic activity at the concentration of 4 mg/L; this corresponds to the concentration of 17 mM, which was still over million times higher than that in drinking and surface water [40]. Lan et al. reached similar conclusions after application of the MTT test of BPA in MA-10 cells. They showed that BPA at a concentration lower that 100 μM had no apoptotic effect as well as a long-term use of a low dose BPA below 1 μM [41]. On the contrary, the MTT test showed that a low dose of BPA in the range from 0.1 nM to 10 μM stimulated proliferation of breast cancer MCF-7 and SkBr-3 cells, which is consistent with the suggestion that even such a low dose of BPA exposure may lead to development of breast cancer [42]. On the other hand, the tested concentration was sufficient to induce damage to the plasma membrane demonstrated in the LDH test. After 24 h of incubation, the results for BPA showed a 15% cytotoxic effect, which decreased in the case of the BPA product obtained after the immobilised LAC treatment to 5%. Therefore, it can be argued that cell exposure to BPA causes a certain degree of membrane damage first; however, the metabolic activity of cells is retained for some time.

3. Materials and Methods

3.1. Microorganism and Culture Conditions

The source of extracellular laccase (LAC) was the white rot fungus *Pleurotus ostreatus* (PO13) obtained from the Fungal Collection of the Department of Biochemistry and Biotechnology, Maria Curie-Skłodowska University (Lublin, Poland). The *Pleurotus ostreatus* strain was cultivated on Petri dishes containing maltose agar medium (glucose 20 g/L, maltose extract 20 g/L, peptone 1 g/L, agar 20 g/L), for 10 days in 25 °C and stored in 4 °C afterwards. The four 10 mm plugs of mycelium were transferred to the 250 mL flasks containing 100 mL of PDB medium (glucose 20 g/L, potato extract

4 g/L), and the culture was conducted for 14 days, which was used as *inoculum* for liquid culture of *Pleurotus ostreatus*. The liquid culture was conducted in volume of 7 L, in Lindeberg-Holm medium modified by addition of peptone (0.5 g/L) with idle air supplied (flow 3.5 mL/min). The laccase was obtained after 12 days of cultivation [43]. Culture liquid was concentrated by ultrafiltration using a Millipore polyethersulfone membrane (10 kDa) and purified by ion exchange chromatography (HQ-Sepharose) using 1 M $(NH_4)_2SO_4$ in TRIS-HCl (5 mM) with a linear grade from 0 to 50%. Purified laccase was desalted, concentrated, and stored frozen at −18 °C until use for preparation of working LAC with precise activity.

3.2. Chemicals, Reagents, and Carriers

Glucose, sodium hydroxide, cytric acid, tris(hydroxymethyl)aminomethane (TRIS), and amonium sulphate were purchased from Avantor Performance Materials (Gliwice, Poland). 2,2-Azino-bis (3-ethylbenzthiazoline-6-sulphonic acid) (ABTS), glutaraldehyde, (3-aminopropyl) triethoxysilane bovine serum albumin (BSA), bisphenol A, the MTT dye, the LDH assay kit, and tartaric acid were purchased from Merck-Sigma-Aldrich Company (St. Louis, MO, USA). Porous carriers used for laccase immobilisation were purchased from Purolite® (Gdynia, Poland) whereas polyeterosulphone and cellulose membranes used for LAC immobilisation were purchased from Merck.

3.3. Catalysts

The activity of free or immobilised LAC was determined spectrophotometrically by oxidation of 2.5 mM 2,2′-azino-bis-(3-ethylbenzthiazoline-6-sulphonic acid) (ABTS) in 100 mM Na-tartrate buffer at pH 3. The formation of ABTS+ was monitored spectrophotometrically at 414 nm ($\lambda 414$ = 34,450 M^{-1} cm^{-1}). Laccase activity was expressed in U per litre (U/L). One unit of LAC (U) oxidised 1 µmol of ABTS per 1 min at 25 °C. In the case of immobilised LAC, 5–10 mg samples of the carrier were placed into a stirred vessel containing a three-fold higher volume of buffer and ABTS than the free laccase activity assay; the reaction was monitored using a Cary 50 Bio spectrophotometer (Varian, Pao Alto, CA, USA) equipped with a Peltier stirring module. The activity of immobilised LAC was expressed in Units per gram of the carrier (U/g). The concentration of LAC was evaluated using the Bradford method.

3.4. Immobilisation of LAC

Purified LAC from *Pleurotus ostreatus* was immobilised covalently and through adsorption using the method developed by Bryjak and Rekuć, respectively [44,45]. For covalent immobilisation, acrylic supports with amino groups (C_2 and C_6) as well as silanised polyeterosulphone (PES) and cellulose membranes (CEL) were used. The silanisation procedure involved 30 min shaking of membranes scraps in a 5% acetone silane solution (10 mL per 1 g of membranes scraps) and drying at 45 °Cduring the night. In the case of the adsorption technique, macroporous styrene (S_0) and acrylic beads with octadecyl groups (C_{18}) were used. The characteristics of the carriers and immobilisation techniques are shown in Table 7. The numbers used in the acronyms of the carrier names refer to the number of carbon atoms on the carrier surface. In the case of carriers C_2 and C_6, it is the length of the spacers: short, built of two carbon atoms, and long, built of six carbon atoms, both ending with an amino group. In the case of the C18 carriers, there were 18-carbon chains as octadecyl groups. There were no functional groups on the surface of S_0 carriers. In turn, the numbers used in the acronyms of membranes PES and CEL refer to the pore size: 10 kDa or 100 kDa (Table 7).

Table 7. Characteristics and acronyms of tested carriers.

<table>
<tr><th>Group of Carriers</th><th>Immobilisation Technique</th><th>Type of Carriers</th><th>Functional Groups</th><th>Pore Diameter (Å)</th><th>Particle Size (μm)</th><th>Acronym</th></tr>
<tr><td rowspan="15">Porous</td><td rowspan="8">Covalent (C2 and C6 carriers)</td><td rowspan="4">Amino C2 methacrylate</td><td rowspan="4">NH₂ (short spacer)</td><td rowspan="2">300–600</td><td>150–300</td><td>C2A</td></tr>
<tr><td>300–710</td><td>C2B</td></tr>
<tr><td rowspan="2">1200–1800</td><td>150–300</td><td>C2C</td></tr>
<tr><td>300–710</td><td>C2D</td></tr>
<tr><td rowspan="4">Amino C6 methacrylate</td><td rowspan="4">NH₂ (long spacer)</td><td rowspan="2">300–600</td><td>150–300</td><td>C6A</td></tr>
<tr><td>300–710</td><td>C6B</td></tr>
<tr><td rowspan="2">1200–1800</td><td>150–300</td><td>C6C</td></tr>
<tr><td>300–710</td><td>C6D</td></tr>
<tr><td rowspan="7">Adsorption (C18 and S0 carriers)</td><td rowspan="3">Octadecyl methacrylate</td><td rowspan="3">Octadecyl</td><td>350–450</td><td>150–300</td><td>C18A</td></tr>
<tr><td rowspan="2">500–700</td><td>150–300</td><td>C18C</td></tr>
<tr><td>300–710</td><td>C18D</td></tr>
<tr><td rowspan="3">Macroporous styrene</td><td rowspan="3">None</td><td rowspan="2">900–110</td><td>150–300</td><td>S0A</td></tr>
<tr><td>300–710</td><td>S0B</td></tr>
<tr><td>950–1200</td><td>300–710</td><td>S0D</td></tr>
<tr><td>-</td><td>-</td><td>-</td><td>Pore Diameter (kDa)</td><td>Membrane Diameter (mm)</td><td>Acronym</td></tr>
<tr><td rowspan="4">Membrane</td><td rowspan="4">Covalent (CEL and PES carriers)</td><td rowspan="2">Cellulose (disc)</td><td rowspan="2">NH₂</td><td>10</td><td>47</td><td>CEL10</td></tr>
<tr><td>100</td><td>47</td><td>CEL100</td></tr>
<tr><td rowspan="2">Polyethersulfone (disc)</td><td rowspan="2">NH₂</td><td>10</td><td>47</td><td>PES10</td></tr>
<tr><td>100</td><td>47</td><td>PES100</td></tr>
</table>

3.4.1. Yield of LAC Immobilisation

The amount of LAC immobilised (LAC_{imm}) was calculated based on the difference between the loaded concentration of LAC protein on the support (LAC_{load}) and the concentration of protein present in eluates obtained after washing out of unbound laccase from the support ($LAC_{unbound}$) and expressed in mg of protein per g of wet mass of the support according to Equation (1). The yield of protein immobilisation was expressed in percent (%), based on Equation (2):

$$LAC_{imm}\left(\frac{\text{mg}}{\text{g}}\right) = LAC_{load} - LAC_{unbound} \tag{1}$$

$$Yield\ (\%) = \frac{LAC_{imm}}{LAC_{load}} \times 100\% \tag{2}$$

3.4.2. Activity of Immobilised LAC

The activity of immobilised LAC was measured in sodium-tartrate buffer 0.1 M pH 3, at 25 °C, using ABTS as a substrate. The activity (A_{imm}) was calculated based on formula 3 and expressed in U per gram of wet mass of the support (U/g), which is the amount of enzyme able to oxidise 1 μmol of ABTS, where ΔA_{min} is the absorbance per minute, Vt is the total volume of the sample (2.8 mL), ε_{414} is the ABTS molar absorption coefficient (34.450), t is the time of reaction (1 min), and m is the mass of the support used in the reaction (g):

$$A_{imm} = \frac{A_{min} * Vt}{\varepsilon_{414} * t * m} \tag{3}$$

The optimisation of the LAC immobilisation conditions was started from the screening of the type of support and the immobilisation technique (Table 7). The next step was the optimisation of pH (5–8) of phosphate buffer used for immobilisation. Those steps of the experiments were performed using 0.5 g of wet mass of each carrier on which 1 mg of laccase per 1 g of carrier was loaded. In the last step, a concentration-dependent LAC activity test was performed in the range from 0.5 mg to 4 mg of protein per 1 g of wet mass of the support. The selection was based on the highest measured activity (U/g) of LAC attached to the carrier.

3.4.3. Determination of the Bond Type

Unbound laccase after covalent immobilisation was washed off from the carriers using different buffers to break the unspecific bonds formed between the protein molecule and the carrier. Eluted LAC allowed determination of the type of unspecific bonds, such as adsorption (in phosphate eluates), ionic bonds (in NaCl eluates), van der Waals forces (in phosphate-buffer eluates), and hydrophobic bonds (in water eluates). TRIS-HCl buffer was used to block of unreacted active groups. The wash-off procedure was as follows (10 mL of buffer per g of carrier): 0.1 M phosphate buffer pH 7 (washed four times), 0.5 M NaCl (washed twice for 15 min), phosphate-citrate buffer pH 5 (washed three times for 30 min), 0.5 M TRIS-HCl pH 7 (washed three times), distilled water (washed six times for 10 min), and again 0.1 M phosphate buffer twice. The percent of unspecific bonds allowed calculating the percent of covalent bonds [46].

3.5. Stability of LAC

The selected conditions for long-term storage of immobilised LAC were optimised, i.e., the type (Na-tartaric, McIlvain) and pH values (3.0–6.0) of the buffer and the concentration of Na-tartaric buffer pH 5 and 6 (1–100 mM). Triple preparations of immobilised LAC weighing 0.5 g each were stored at 4 °C and the activity of both immobilised and free LAC was measured every second week. The evaluation of thermostability was performed using immobilised LAC samples (0.1 g) and corresponding activity of free LAC suspended in a buffer that was optimal for the tested carrier. Samples of immobilised and free LAC were incubated at the temperature from 4 °C to 80 °C for 24 h. Changes in LAC activity were expressed as a percentage of initial LAC activity (time 0).

3.6. Transformation of Endocrine Disrupting Chemical

Bisphenol A (BPA) was first transformed in the concentration range from 0.5 mg/L to 10 mg/L using LAC immobilised on four different carriers (C2C, C6C, CEL100, and C18C) with final activity of 5 U/L. The amount of bonded protein differed in the case of each carrier and was 1 mg per 1 g of carriers C2 and C6, 0.25 mg per 1 g of carrier CEL100, and 0.5 mg per 1 g of carrier C18. The reaction was performed in 6-well plates containing a mixture of 0.1 M sodium-tartrate buffer pH 5.5 and the substrate. The reaction mixture was incubated on a rotary shaker (115 rpm) at room temperature. The reaction was conducted in a volume of 10 mL in 6-well plates for 72 h, and samples were taken for analysis every 24 h. Afterwards, transformation of BPA at the concentration of 10 mg/L was performed using preparations of laccase with final activity of 12U/L in analogical conditions. The amount of immobilised protein was increased to 4 mg/g of carriers C2 and C6 or 1 mg/g of carriers CEL100 and C18. In every step, control experiments of substrate absorption were performed using immobilised bovine serum albumin (BSA) at a concentration adequate to the concentration of immobilised laccase of 1 mg/g or 4 mg/g incubated with a mixture of the sodium-tartrate buffer and the substrate.

The reusability of immobilised LAC was performed for carriers C2C, C6C and CEL100 in analogical conditions using LAC final activity of 12 U/L. Between each batch the carriers were washed 5-times using 5 mL of 0.1 M sodium-tartrate buffer pH 5.

3.7. High Pressure Liquid Chromatogrpahy (HPLC)

Transformation mixtures containing samples of BPA were analysed using a Zorbax Eclipse-C18 reverse-phase HPLC column (Agilent®, Santa Clara, CA, USA) and eluent consisting of 70% methanol. The sample volume was 5 μL and the analysis lasted 3 min. A signal was recorded at 275 nm and 225 nm.

3.8. Ecotoxicity Microtox® Test

The environmental toxicity test of BPA and the product of its transformation by immobilised laccase was carried out according to the Microtox® biotest methodology, which consists in analysing the decrease in the bioluminescence of *Vibrio fisheri* bacteria under the influence of the tested compound. The results demonstrated a 50% decrease in bioluminescence intensity designated EC_{50}, i.e., the concentration of tested substances producing a 50% effect. The initial concentration of BPA in this test was 10 mg/L.

3.9. MTT and LDH Cytotoxicity Tests

The cytotoxicity tests examined the effect of BPA and the product of its transformation on the metabolic activity and integrity of the cell membrane in human colon epithelial cells of the CCD-841CoTr line. The solution of BPA and the mixture after its transformation used for the tests were diluted 2.5-times in a 2% solution of foetal bovine serum and a 4 mg/L final concentration solution was obtained, which corresponds to a concentration of 17 mM. Cell viability tests using tetrazolium salt (MTT) and lactate dehydrogenase activity (LDH) methods were performed according to the methodology described in the literature [47].

4. Conclusions

The aim of present work was to describe novel porous Purolite® carriers and ultrafiltration membranes, which had never been tested before as supports for laccase immobilisation, and their utility in BPA degradation through biocatalysis. It is important, that commercially available carriers were tested for this process. The main impact on the activity of immobilised LAC was exerted by the structure of the carrier, which should have a well-developed porosity structure with simultaneously small particle size. The immobilised laccase preparations were stable for one month with certain diversity depending on the immobilisation technique. The selected immobilised LAC preparations proved useful for BPA elimination by both enzymatic conversion and by adsorption of the substrate on the carrier surface. It is reasonable to use enzymatic degradation of BPA for reduction of its ecotoxicity and cytotoxicity, mainly damage to the cell membrane. However, more studies are necessary to determine kinetic parameters of immobilised laccase as well as the operational stability of the most efficient carrier.

Author Contributions: Conceptualization, K.W., J.P., A.J.-W.; Methodology, all; Validation, K.W., J.P.; Investigation, K.W., J.K.-P., M.G., R.P.; Resources, all; Data curation, K.W.; Writing—original draft preparation, K.W.; Writing—review and editing, K.W., J.P., A.J.-W.; Visualization, K.W., J.P.; Supervision, K.W., J.P., A.J.-W.; Project administration, K.W., J.P., A.J.-W.; Funding acquisition, K.W. All authors have read and agreed to the published version of the manuscript.

Funding: This work was supported by National Science Centre (Poland), grant number 2015/17/N/NZ9/03647.

Conflicts of Interest: The authors declare no conflict of interest. The funders had no role in the design of the study; in the collection, analyses, or interpretation of data; in the writing of the manuscript, or in the decision to publish the results.

Abbreviations

LAC	laccase
BPA	bisphenol A

References

1. Mateo, C.; Palomo, J.M.; Fernandez-Lorente, G.; Guisan, J.M.; Fernandez-Lafuente, R. Improvement of enzyme activity, stability and selectivity via immobilization techniques. *Enzym. Microb. Technol.* **2007**, *40*, 1451–1463. [CrossRef]
2. Peralta-Zamora, P.; Pereira, C.M.; Tiburtius, E.R.; Moraes, S.G.; Rosa, M.A.; Minussi, R.C.; Durán, N. Decolorization of reactive dyes by immobilized laccase. *Appl. Catal. B Environ.* **2003**, *42*, 131–144. [CrossRef]

3. Spinelli, D.; Fatarella, E.; Di Michele, A.; Pogni, R. Immobilization of fungal (*Trametes versicolor*) laccase onto Amberlite IR-120 H beads: Optimization and characterization. *Process Biochem.* **2013**, *48*, 218–223. [CrossRef]
4. Gonçalves, M.C.P.; Kieckbusch, T.G.; Perna, R.F.; Fujimoto, J.T.; Morales, S.A.V.; Romanelli, J.P. Trends on enzyme immobilization researches based on bibliometric analysis. *Process Biochem.* **2019**, *76*, 95–110. [CrossRef]
5. Fernández-Fernández, M.; Sanromán, M.Á.; Moldes, D. Recent developments and applications of immobilized laccase. *Biotechnol. Adv.* **2013**, *31*, 1808–1825. [CrossRef]
6. Galliker, P.; Hommes, G.; Schlosser, D.; Corvini, P.F.X.; Shahgaldian, P. Laccase-modified silica nanoparticles efficiently catalyze the transformation of phenolic compounds. *J. Colloid Interface Sci.* **2010**, *349*, 98–105. [CrossRef]
7. Costa, J.B.; Lima, M.J.; Sampaio, M.J.; Neves, M.C.; Faria, J.L.; Morales-Torres, S.; Tavarez, A.P.M.; Silva, C.G. Enhanced biocatalytic sustainability of laccase by immobilization on functionalized carbon nanotubes/polysulfone membranes. *Chem. Eng. J.* **2019**, *355*, 974–985. [CrossRef]
8. Wlizło, K.; Polak, J.; Jarosz-Wilkołazka, A.; Pogni, R.; Petricci, E. Novel textile dye obtained through transformation of 2-amino-3-methoxybenzoic acid by free and immobilised laccase from a *Pleurotus ostreatus* strain. *Enzym. Microb. Technol.* **2020**, *132*, 109398. [CrossRef]
9. Mogharabi, M.; Nassiri-Koopaei, N.; Bozorgi-Koushalshahi, M.; Nafissi-Varcheh, N.; Bagherzadeh, G.; Faramarzi, M.A. Immobilization of laccase in alginate-gelatin mixed gel and decolorization of synthetic dyes. *Bioinorg. Chem. Appl.* **2012**, *2012*, 1–6. [CrossRef]
10. Lassouane, F.; Aït-Amar, H.; Amrani, S.; Rodriguez-Couto, S. A promising laccase immobilization approach for Bisphenol A removal from aqueous solutions. *Bioresour. Technol.* **2019**, *271*, 360–367. [CrossRef]
11. Mohamad, N.R.; Marzuki, N.H.C.; Buang, N.A.; Huyop, F.; Wahab, R.A. An overview of technologies for immobilization of enzymes and surface analysis techniques for immobilized enzymes. *Biotechnol. Biotechnol. Equip.* **2015**, *29*, 205–220. [CrossRef] [PubMed]
12. Nisha, S.; Karthick, S.A.; Gobi, N. A review on methods, application and properties of immobilized enzyme. *Chem. Sci. Rev. Lett.* **2012**, *1*, 148–155.
13. de Bezerra, T.M.S.; Bassan, J.C.; de Santos, V.T.O.; Ferraz, A.; Monti, R. Covalent immobilization of laccase in green coconut fiber and use in clarification of apple juice. *Process Biochem.* **2015**, *50*, 417–423. [CrossRef]
14. Durán, N.; Rosa, M.A.; D'Annibale, A.; Gianfreda, L. Applications of laccases and tyrosinases (phenoloxidases) immobilized on different supports: A review. *Enzym. Microb. Technol.* **2002**, *31*, 907–931. [CrossRef]
15. Asgher, M.; Shahid, M.; Kamal, S.; Iqbal, H.M.N. Recent trends and valorization of immobilization strategies and ligninolytic enzymes by industrial biotechnology. *J. Mol. Catal. B Enzym.* **2014**, *101*, 56–66. [CrossRef]
16. Bautista, L.F.; Morales, G.; Sanz, R. Immobilization strategies for laccase from *Trametes versicolor* on mesostructured silica materials and the application to the degradation of naphthalene. *Bioresour. Technol.* **2010**, *101*, 8541–8548. [CrossRef]
17. Hou, J.; Dong, G.; Ye, Y.; Chen, V. Enzymatic degradation of bisphenol-A with immobilized laccase on TiO_2 sol–gel coated PVDF membrane. *J. Membr. Sci.* **2014**, *469*, 19–30. [CrossRef]
18. Fernández-Lorente, G.; Lopez-Gallego, F.; Bolivar, J.M.; Rocha-Martin, J.; Moreno-Perez, S.; Guisán, J.M. Immobilization of proteins on glyoxyl activated supports: Dramatic stabilization of enzymes by multipoint covalent attachment on pre-existing supports. *Curr. Org. Chem.* **2015**, *19*, 13. [CrossRef]
19. Pezzella, C.; Russo, M.E.; Marzocchella, A.; Salatino, P.; Sannia, G. Immobilization of a *Pleurotus ostreatus* laccase mixture on perlite and its application to dye decolourisation. *BioMed. Res. Int.* **2014**, *2014*, 308613. [CrossRef]
20. Tentori, F.; Bavaro, T.; Brenna, E.; Colombo, D.; Monti, D.; Semproli, R.; Ubiali, D. Immobilization of Old Yellow Enzymes via Covalent or Coordination Bonds. *Catalysts* **2020**, *10*, 260. [CrossRef]
21. Wang, F.; Guo, C.; Yang, L.R.; Liu, C.Z. Magnetic mesoporous silica nanoparticles: Fabrication and their laccase immobilization performance. *Bioresour. Technol.* **2010**, *101*, 8931–8935. [CrossRef] [PubMed]
22. Wang, F.; Guo, C.; Liu, H.Z.; Liu, C.Z. Immobilization of *Pycnoporus sanguineus* laccase by metal affinity adsorption on magnetic chelator particles. *J. Chem. Technol. Biotechnol.* **2008**, *83*, 97–104. [CrossRef]
23. Rahmani, K.; Faramarzi, M.A.; Mahvi, A.H.; Gholami, M.; Esrafili, A.; Forootanfar, H.; Farzadkia, M. Elimination and detoxification of sulfathiazole and sulfamethoxazole assisted by laccase immobilized on porous silica beads. *Int. Biodeterior. Biodegrad.* **2015**, *97*, 107–114. [CrossRef]

24. Lin, J.; Liu, Y.; Chen, S.; Le, X.; Zhou, X.; Zhao, Z.; Ou, Y.; Yang, J. Reversible immobilization of laccase onto metal-ion-chelated magnetic microspheres for bisphenol A removal. *Int. J. Biol. Macromol.* **2016**, *84*, 189–199. [CrossRef]
25. Liu, Y.; Zeng, Z.; Zeng, G.; Tang, L.; Pang, Y.; Li, Z.; Liu, C.; Lei, X.; Wu, M.; Ren, P.; et al. Immobilization of laccase on magnetic bimodal mesoporous carbon and the application in the removal of phenolic compounds. *Bioresour. Technol.* **2012**, *115*, 21–26. [CrossRef]
26. Xu, R.; Si, Y.; Wu, X.; Li, F.; Zhang, B. Triclosan removal by laccase immobilized on mesoporous nanofibers: Strong adsorption and efficient degradation. *Chem. Eng. J.* **2014**, *255*, 63–70. [CrossRef]
27. Sathishkumar, P.; Kamala-Kannan, S.; Cho, M.; Kim, J.S.; Hadibarata, T.; Salim, M.R.; Oh, B.T. Laccase immobilization on cellulose nanofiber: The catalytic efficiency and recyclic application for simulated dye effluent treatment. *J. Mol. Catal. B Enzym.* **2014**, *100*, 111–120. [CrossRef]
28. Michałowicz, J. Bisphenol A—Sources, toxicity and biotransformation. *Environ. Toxicol. Pharmacol.* **2014**, *37*, 738–758. [CrossRef]
29. Rogala, D.; Kulik-Kupka, K.; Śnieżek, E.; Janicka, A.; Moskalenko, O. Bisfenol A—Niebezpieczny związek ukryty w tworzywach sztucznych. *Probl. Hig. Epidemiol.* **2016**, *97*, 213–219.
30. Sifakis, S.; Androutsopoulos, V.P.; Tsatsakis, A.M.; Spandidos, D.A. Human exposure to endocrine disrupting chemicals: Effects on the male and female reproductive systems. *Environ. Toxicol. Pharmacol.* **2017**, *51*, 56–70. [CrossRef]
31. Castro, B.; Sanchez, P.; Torres, J.M.; Preda, O.; del Moral, R.G.; Ortega, E. Bisphenol A exposure during adulthood alters expression of aromatase and 5α-reductase isozymes in rat prostate. *PLoS ONE* **2013**, *8*, e55905. [CrossRef] [PubMed]
32. Konieczna, A.; Rutkowska, A.; Rachoń, D. Health risk of exposure to bisfenol A (BPA). *Rocz. Państ. Zakł. Hig.* **2015**, *66*, 5–11.
33. Nair, R.R.; Demarche, P.; Agathos, S.N. Formulation and characterization of an immobilized laccase biocatalyst and its application to eliminate organic micropollutants in wastewater. *New Biotechnol.* **2013**, *30*, 814–823. [CrossRef] [PubMed]
34. Zdarta, J.; Antecka, K.; Frankowski, R.; Zgoła-Grześkowiak, A.; Ehrlich, H.; Jesionowski, T. The effect of operational parameters on the biodegradation of bisphenols by *Trametes versicolor* laccase immobilized on *Hippospongia communis* spongin scaffolds. *Sci. Total Environ.* **2018**, *615*, 784–795. [CrossRef] [PubMed]
35. Sadeghzadeh, S.; Nejad, Z.G.; Ghasemi, S.; Khafaji, M.; Borghei, S.M. Removal of bisphenol A in aqueous solution using magnetic cross-linked laccase aggregates from *Trametes hirsuta*. *Bioresour. Technol.* **2020**, *306*, 123169. [CrossRef]
36. Dudziak, M. Wpływ warunków środowiska wodnego na rozkład bisfenolu A. *Proc. ECOpolen* **2017**, *11*, 131–139.
37. Varga, B.; Somogyi, V.; Meiczinger, M.; Kováts, N.; Domokos, E. Enzymatic treatment and subsequent toxicity of organic micropollutants using oxidoreductases—A review. *J. Clean. Prod.* **2019**, *221*, 306–322. [CrossRef]
38. Spina, F.; Cordero, C.; Schilirò, T.; Sgorbini, B.; Pignata, C.; Gilli, G.; Bichi, C.; Varese, G.C. Removal of micropollutants by fungal laccases in model solution and municipal wastewater: Evaluation of estrogenic activity and ecotoxicity. *J. Clean. Prod.* **2015**, *100*, 185–194. [CrossRef]
39. de Freitas, E.N.; Bubna, G.A.; Brugnari, T.; Kato, C.G.; Nolli, M.; Rauen, T.G.; Moreira, R.F.P.M.; Peralta, R.A.; Bracht, A.; de Souza, C.G.M.; et al. Removal of bisphenol A by laccases from *Pleurotus ostreatus* and *Pleurotus pulmonarius* and evaluation of ecotoxicity of degradation products. *Chem. Eng. J.* **2017**, *330*, 1361–1369. [CrossRef]
40. Wlizło, K.; Polak, J.; Jarosz-Wilkołazka, A. Związki biologicznie aktywne i metody ich usuwania na drodze biokatalizy. *Postęp. Biochem.* **2017**, *63*, 304–314.
41. Lan, H.C.; Wu, K.Y.; Lin, I.W.; Yang, Z.J.; Chang, A.A.; Hu, M.C. Bisphenol A disrupts steroidogenesis and induces a sex hormone imbalance through c-Jun phosphorylation in Leydig cells. *Chemosphere* **2017**, *185*, 237–246. [CrossRef] [PubMed]
42. Song, H.; Zhang, T.; Yang, P.; Li, M.; Yang, Y.; Wang, Y.; Du, J.; Pan, K.; Zhang, K. Low doses of bisphenol A stimulate the proliferation of breast cancer cells via ERK1/2/ERRγ signals. *Toxicol. In Vitro* **2015**, *302*, 521–528. [CrossRef] [PubMed]
43. Lindeberg, G.; Holm, G. Occurrence of tyrosinase and laccase in fruit bodies and mycelia of some Hymenomycetes. *Physiol. Plant.* **1952**, *5*, 100–114. [CrossRef]

44. Bryjak, J.; Kruczkiewicz, P.; Rekuć, A.; Pęczyńska-Czoch, W. Laccase immobilization on copolymer of butyl acrylate and ethylene glycol dimethacrylate. *Biochem. Eng. J.* **2007**, *35*, 325–332. [CrossRef]
45. Rekuć, A.; Kruczkiewicz, P.; Jastrzembska, B.; Liesiene, J.; Pęczyńska-Czoch, W.; Bryjak, J. Laccase immobilization on the tailored cellulose-based Granocel carriers. *Int. J. Biol. Macromol.* **2008**, *42*, 208–215. [CrossRef]
46. Ginalska, G.; Kowalczuk, D.; Osińska, M. A chemical method of gentamicin bonding to gelatin-sealed prosthetic vascular grafts. *Int. J. Pharm.* **2005**, *288*, 131–140. [CrossRef]
47. Polak, J.; Jarosz-Wilkolazka, A.; Szuster-Ciesielska, A.; Wlizlo, K.; Kopycinska, M.; Sojka-Ledakowicz, J.; Lichawska-Olczyk, J. Toxicity and dyeing properties of dyes obtained through laccase-mediated synthesis. *J. Clean. Prod.* **2016**, *112*, 4265–4272. [CrossRef]

Article

Estimating the Product Inhibition Constant from Enzyme Kinetic Equations Using the Direct Linear Plot Method in One-Stage Treatment

Pedro L. Valencia [1,*], Bastián Sepúlveda [2], Diego Gajardo [2] and Carolina Astudillo-Castro [3]

[1] Department of Chemical and Environmental Engineering, Universidad Técnica Federico Santa María, P.O. Box 110-V, Valparaíso 2390123, Chile

[2] Department of Mathematics, Universidad Técnica Federico Santa María, Valparaíso 2390123, Chile; bastian.sepulveda@sansano.usm.cl (B.S.); diego.gajardom@alumnos.usm.cl (D.G.)

[3] School of Food Engineering, Pontificia Universidad Católica de Valparaíso, Valparaíso 2360100, Chile; carolina.astudillo@pucv.cl

* Correspondence: pedro.valencia@usm.cl

Received: 15 June 2020; Accepted: 8 July 2020; Published: 1 August 2020

Abstract: A direct linear plot was applied to estimate kinetic constants using the product's competitive inhibition equation. The challenge consisted of estimating three kinetic constants, V_{max}, K_m, and K_p, using two independent variables, substrates, and product concentrations, in just one stage of mathematical treatment. The method consisted of combining three initial reaction rate data and avoiding the use of the same three product concentrations (otherwise, this would result in a mathematical indetermination). The direct linear plot method was highly superior to the least-squares method in terms of accuracy and robustness, even under the addition of error. The direct linear plot method is a reliable and robust method that can be applied to estimate K_p in inhibition studies in pharmaceutical and biotechnological areas.

Keywords: direct linear plot; median method; product inhibition; kinetic constants; non-parametric; distribution-free method

1. Introduction

The direct linear plot (DLP) is a graphic method to estimate kinetic constants from enzymatic reactions based on the median as a statistic. The method was developed by Eisenthal and Cornish-Bowden in 1974 [1]. The accuracy and robustness of this method were tested during the estimation of V_{max} and K_{m} from the Michaelis–Menten equation [1,2]. As in this article, all of the recent studies regarding the application of DLP are based on the estimation of these two parameters from the Michaelis–Menten equation. The application of DLP to equations with more than two parameters has not been studied until now [3]. This median-based method was applied herein to estimate three kinetic constants from the substrate-uncompetitive inhibition equation. The results indicated that DLP was not only applicable to equations with more than two parameters, but it was reliable and robust when compared to the least-squares (LS) method. In the present article, we explored the application of DLP to the product-competitive inhibition equation, which is a different type of problem, as will be exposed. Some attempts to estimate inhibition constants from enzyme kinetics using DLP have involved estimating apparent kinetic constants and using secondary plots to estimate the values of these inhibition constants. The application of DLP to estimate inhibition constants was first proposed by Eisenthal and Cornish-Bowden [1]. The literature contains outdated examples of this application because the use of DLP to estimate inhibition constants in one-stage has not been assessed [4–8]. In the literature, DLP has been used to estimate the apparent kinetic constants V_{max} and K_m; however, studies have used secondary plots to estimate the inhibition constant

K_p. The present article studied the application of DLP to estimate the product inhibition constant in one-stage treatment, avoiding the estimation of apparent kinetic constants and secondary plots. As the proper characterization of inhibition in pharmaceutical and biotechnological applications is a major concern, the importance of this proposal is that it opens up the possibility of using a reliable and robust method to estimate product inhibition constants. The purpose of this study is to estimate V_{max}, K_m, and K_p from the competitive inhibition equation in just one stage of calculations, avoiding the apparent constants and secondary plots. This means the values of the product inhibition constants must be obtained from the DLP method, which has never been tried before. The problem is outlined as follows.

The equation for competitive product inhibition involves three parameters—V_{max}, K_m, and K_p—and three variables—substrate concentration, product concentration, and initial reaction rate (Equation (1)).

$$v = \frac{V_{max}S}{K_m + S + \frac{K_m}{K_p}P} \tag{1}$$

This is a different type of problem compared to previous articles where the substrate-uncompetitive inhibition equation involved three parameters—V_{max}, K_m, and K_S—but only two variables—substrate concentration and initial reaction rate (Equation (2)) [3].

$$v = \frac{V_{max}S}{K_m + S + \frac{S^2}{K_s}} \tag{2}$$

The experimental dataset required is traditionally used in the initial reaction rate method, which denotes a series of initial rates vs. substrate concentrations at constant product concentrations, as indicated in the scheme of Figure 1.

$P = P_1$		$P = P_2$		$P = P_m$	
S	v	S	v	S	v
S_1	v_{11}	S_1	v_{21}	S_1	v_{m1}
$\vdots$	$\vdots$	$\vdots$	$\vdots$	$\vdots$	$\vdots$
S_n	v_{1n}	S_n	v_{2n}	S_n	v_{mn}

Figure 1. Scheme of the traditional experimental design based on the initial reaction rate method to estimate inhibition constants in enzyme kinetics.

Equation (1) can be rearranged to Equation (3) in the following form:

$$v_{ij}K_m - S_jV_{max} + \frac{v_{ij}P_iK_m}{K_p} = -v_{ij}S_j \tag{3}$$

The definition of v_{ij} is the rate of product concentration P_i and substrate concentration S_j. As the estimation of three parameters is required, three data pairs and three equations are needed to solve the matrix system in Equation (4).

$$\begin{pmatrix} v_{ii'} & -S_{i'} & v_{ii'}P_i \\ v_{jj'} & -S_{j'} & v_{jj'}P_j \\ v_{kk'} & -S_{k'} & v_{kk'}P_k \end{pmatrix} \begin{pmatrix} K_m \\ V_{max} \\ \frac{K_m}{K_p} \end{pmatrix} = \begin{pmatrix} -v_{ii'}S_{i'} \\ -v_{jj'}S_{j'} \\ -v_{kk'}S_{k'} \end{pmatrix} \tag{4}$$

Likewise, the definition of $v_{ii'} = v(P_i, S_{i'})$, i.e., i values are between 1 and m (the number of product concentrations), and i' values are between 1 and n (the number of substrate concentrations). Alternatively, the solution for Equation (4) is presented in Equation (5) in algebraic form.

$$\begin{aligned} V_{max} &= \frac{v_{ii'}v_{jj'}v_{kk'}\left(P_k\left(S_{i'}-S_{j'}\right)+P_i\left(S_{j'}-S_{k'}\right)+P_j(-S_{i'}+S_{k'})\right)}{\left(P_j-P_i\right)S_{k'}v_{ii'}v_{jj'}+(P_i-P_k)S_{j'}v_{ii'}v_{kk'}+\left(P_j-P_i\right)S_{i'}v_{jj'}v_{kk'}} \\ K_m &= \frac{P_iS_{j'}S_{k'}v_{ii'}\left(v_{kk'}-v_{jj'}\right)+P_kS_{i'}S_{j'}v_{kk'}\left(v_{jj'}-v_{ii'}\right)+P_jS_{i'}S_{k'}v_{jj'}\left(v_{ii'}-v_{kk'}\right)}{\left(P_i-P_j\right)S_{k'}v_{ii'}v_{jj'}+v_{kk'}\left(-P_iS_{j'}v_{ii'}+P_kS_{j'}v_{ii'}+P_jS_{i'}v_{jj'}-P_kS_{i'}v_{jj'}\right)} \\ \frac{K_m}{K_p} &= \frac{S_{j'}S_{k'}v_{ii'}\left(v_{jj'}-v_{kk'}\right)+S_{i'}S_{j'}v_{kk'}\left(v_{ii'}-v_{jj'}\right)+S_{i'}S_{k'}v_{jj'}\left(v_{kk'}-v_{ii'}\right)}{\left(P_i-P_j\right)S_{k'}v_{ii'}v_{jj'}+v_{kk'}\left(-P_iS_{j'}v_{ii'}+P_kS_{j'}v_{ii'}+P_jS_{i'}v_{jj'}-P_kS_{i'}v_{jj'}\right)} \end{aligned} \quad (5)$$

The rank of the matrix in Equation (4) must be 3 for the system to have a solution. For the case when three product concentrations are taken from the same set, $P_i = P_j = P_k$, the system has no solution because the rank of the resulting matrix is 2. This problem can be easily checked by observing the denominators of Equation (5) when replacing P_i, P_j, and P_k with the same value of product concentration. The strategy proposed to solve this system is to take at least two different values of product concentration. In this way, the total number of feasible data combinations will be calculated using Equation (6).

In consequence, calculations of the kinetic constants V_{max}, K_m, and K_p will be obtained from the combination of two P values from the same dataset, one P from a different one, and the combination of three different P values from three datasets. A list of V_{max}, K_m, and K_p values from the total combinations indicated in Equation (6) will be obtained. Finally, the estimators of V_{max}, K_m, and K_p correspond to the median values from the list.

$$\binom{m \cdot n}{3} - \binom{n}{3} \cdot m \quad (6)$$

This article aims to compare the quality of the kinetic constant estimations from the product inhibition equation between the DLP and the LS methods.

2. Results and Discussion

As in our previous article [3], this problem consists of estimating three kinetic constants, but in this new case, three experimental variables will be used. Experimental data of initial rate perturbed with a random error of variance σ^2 were simulated in a dataset of 1000 runs. The dataset, according to Table 1, considers the values of the kinetic constants 1, 1 and 10 for V_{max}, K_m, and K_p, respectively, and the initial rates vs. the substrate concentrations at different product concentrations were plotted in Figure 2.

Table 1. Experimental design of substrate and product concentrations to calculate initial reaction rates.

$P_1 = 0$		$P_2 = 0.5K_p$		$P_3 = K_p$		$P_4 = 2K_p$		$P_5 = 4K_p$	
S	v	S	v	S	v	S	v	S	v
0.1	v_{11}	0.1	v_{21}	0.1	v_{31}	0.1	v_{41}	0.1	v_{51}
0.2	v_{12}	0.2	v_{22}	0.2	v_{32}	0.2	v_{42}	0.2	v_{52}
0.5	v_{13}	0.5	v_{23}	0.5	v_{33}	0.5	v_{43}	0.5	v_{53}
1.0	v_{14}	1.0	v_{24}	1.0	v_{34}	1.0	v_{44}	1.0	v_{54}
2.0	v_{15}	2.0	v_{25}	2.0	v_{35}	2.0	v_{45}	2.0	v_{55}
5.0	v_{16}	5.0	v_{26}	5.0	v_{36}	5.0	v_{46}	5.0	v_{56}
10.0	v_{17}	10.0	v_{27}	10.0	v_{37}	10.0	v_{47}	10.0	v_{57}

These data were used to estimate the kinetic constants using the DLP and LS methods. An example of the dataset obtained for the values of the kinetic constants V_{max}, K_m, and K_p with a normal distribution of error and variance 0.01 is plotted in Figure 3.

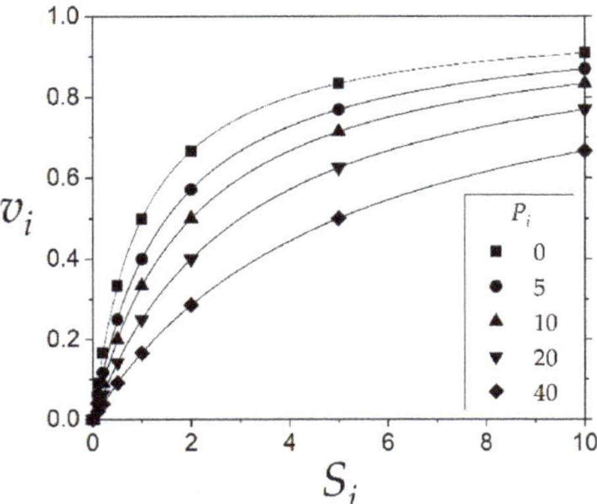

Figure 2. Initial reaction rates vs. substrate concentration for the different product concentrations according to the experimental design presented in Table 1. Data plotted without error for illustration purposes. The kinetic constant values for V_{max}, K_{m}, and K_{p} are 1, 1, and 10, respectively.

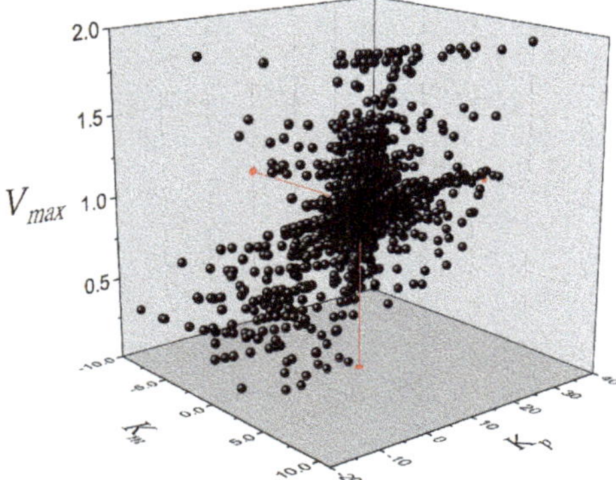

Figure 3. Three-dimensional projection of the direct linear plot (DLP) values obtained according to the experimental design presented in Table 1. The kinetic constant values for V_{max}, K_m, and K_p are 1, 1, and 10, respectively. The median of each kinetic constant (red dots) is projected on the planes. Some data are outside of the layers for illustration purposes only.

The number of kinetic constant values plotted in Figures 3 and 4 was 6370. This amount of data was obtained avoiding the use of the same three product concentrations during the calculations with Equation (4). The dispersion of the data was enormous compared to the value of the kinetic constant, as can be observed in Figure 4. This is a typical result of the application of the DLP method to the estimation of kinetic constants. This behavior can be observed in previous publications [1,3,9]. Many points were left out of the layers just for illustration purposes; however, the vast majority of data are inside of layers corresponding to 98%, 99% and 99% for V_{max}, K_m and K_p, respectively. In terms of dispersion, 56% of calculated V_{max} values are in the range 0.95–1.05; 22% of calculated Km values are in the range 0.95–1.05 and 24% of calculated K_p values are in the range 9.5–10.5. The amount of data plotted in Figures 3 and 4 was obtained to estimate the kinetic constants V_{max}, K_m, and K_p by calculating the median for each of them. This is the procedure needed in routine experiments to determine the inhibition constant K_p. The statistical evaluation of DLP requires the repetition of this procedure to avoid bias and erroneous conclusions. A total of 1000 experimental runs were carried out to evaluate and compare the quality of the estimations between the DLP and the LS methods. A comparison of the lower values of the sum of squared residuals (*SSR*) between DLP and LS is plotted in Figure 5.

The DLP method obtained lower values of *SSR* for all the values of variance in the range from 0.001 to 0.020 when both procedures—calculation of the median of K_m/K_p and the median of K_p—were carried out. The calculation of the median of K_m/K_p was slightly superior for almost all variances explored. For example, at variance 0.010, the calculation of the median of K_m/K_p obtained a frequency of 0.8 compared to 0.78 for calculation of the median of K_p. This was not significant to conclude that one procedure was better than the other. However, when both calculations were compared regarding the *SSR* between them, the result depended on the variance, as shown in Figure 6a. The calculation of the median of K_m/K_p obtained a lower *SSR* at variance values lower than 0.010. The calculation of the median of K_p offered lower values of *SSR* at variances higher than 0.010. The effect of the ratio K_p/K_m was slightly superior to the calculation of the median of K_p, as shown in Figure 6b. However, this was not significantly different. In consequence, which procedure is better, in terms of the *SSR*, depends on the experimental error.

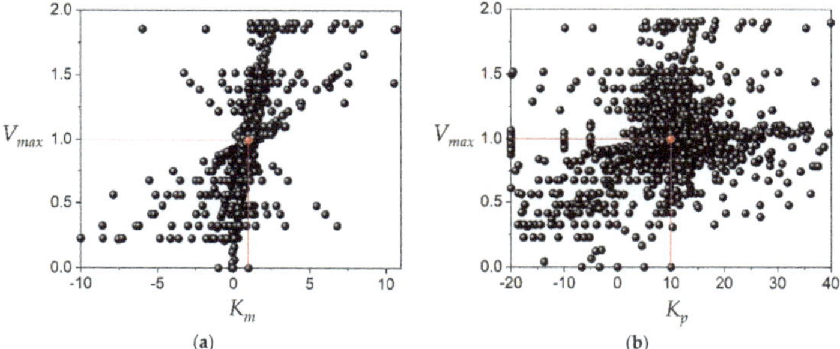

Figure 4. DLP for the competitive inhibition equation to estimate V_{max}, K_m, and K_p with real values of 1, 1, and 10, respectively, considering a normal distribution of error with variance 0.01. (**a**) Plot of V_{max} vs. K_m; (**b**) plot of V_{max} vs. K_p. The median values (red dots) are projected on each axis. Some data are outside of layers for illustration purposes only. Data inside of layers is 98%, 99% and 99% for V_{max}, K_m and K_p, respectively.

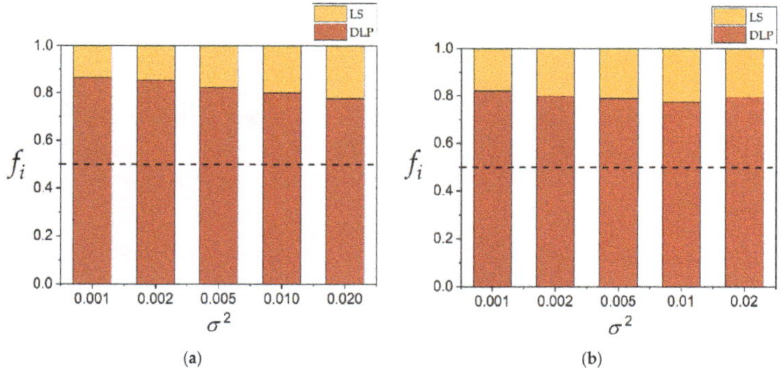

Figure 5. Frequency (f_i) of the lower sum of squared residuals (*SSR*) obtained between DLP and least-squares (LS) methods as a function of the variance (σ^2) of the normal distribution of error (ε_i). The total number of experiments was 1000. (**a**) DLP applied to the median of K_m/K_p; (**b**) DLP applied to the median of K_p. The dotted line represents the frequency 0.5, where both methods, LS and DLP, obtained the same statistical quality. The results showed that DLP obtained a lower *SSR* than LS approximately in 80% of the estimations.

The relative error of the estimated kinetic constants by LS and DLP methods was plotted for 1000 experimental runs in Figure 7. The DLP always resulted in a lower accumulation of relative error during estimations of the three kinetic constants. The product inhibition constant K_p accumulated the highest relative error when estimated by both LS and DLP methods. However, the accumulated relative error was less than half with DLP compared to LS. This result shows that DLP is a very accurate method to estimate the competitive inhibition constant. The product inhibition constant K_p estimated with DLP method accumulated a lower relative error than the LS method even at higher variance σ^2 (experimental error), as shown in Figure 8. The behavior patterns of both the median of K_m/K_p and the median of K_p were repeated. The median of K_m/K_p was slightly more accurate at variances lower than 0.010. At higher values, the median of K_p was more accurate.

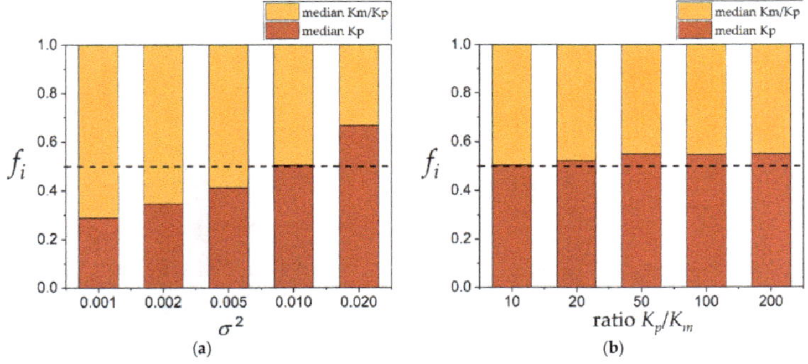

Figure 6. Frequency (f_i) of the lower *SSR* obtained between both the calculation of the median of K_m/K_p and the median of K_p. (**a**) Results of frequency as a function of the variance (σ^2) of the normal distribution of error (ε_i). (**b**) Results of frequency as a function of the ratio K_m/K_p. The interpretation is the same as Figure 5.

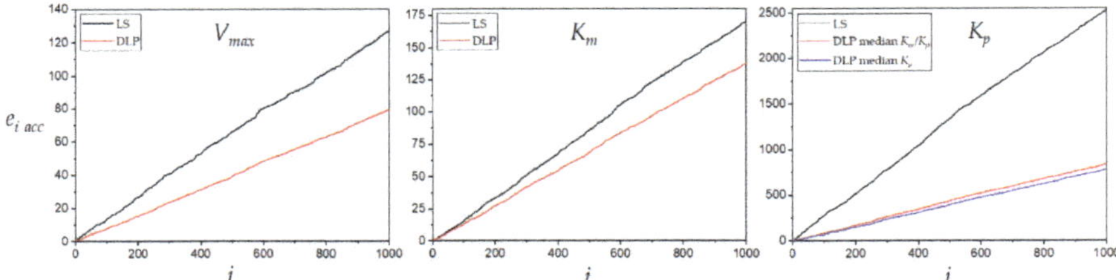

Figure 7. Relative error (e_i) for the estimations of kinetic constants V_{max}, K_m, and K_p by LS and DLP methods accumulated during iterations (i) of 1000 experimental runs with a normal distribution of error (ε_i) and variance (σ^2) 0.010. The K_p was estimated by DLP using the median of K_m/K_p (red line) and the median of K_p (blue line).

The effect of outliers was analyzed by adding fixed error values to the initial rate v_{34} ($S = 1$; $P = 10$). The added error ranged from −0.01 to +0.01. The results are shown in Figure 9. The estimated values of V_{max} by LS were constant due to an artifact of the estimation method. The estimated V_{max} was obtained by choosing the highest value from the five non-linear regressions corresponding to the five datasets of product concentrations. The error was added to the initial rate value corresponding to the third dataset of product concentration; in consequence, the highest value of V_{max} was not perturbed, and it was always the same value. The DLP method estimated K_m and the K_p with more accuracy and with less perturbation than the LS method. Although the LS method estimated K_p with more accuracy at some high positive values of fixed error, the results clearly showed that DLP was more robust than the LS method. The perturbation generated by the added error is plotted in Figure 10. The perturbation of Vmax was omitted due to the artifact during its estimation by the LS method. The results indicated that the estimation of kinetic constants K_m and K_p was much less perturbed when the DLP method was used. Almost all the perturbations were maintained below 1%, compared to the LS method, where perturbations were in the range from 4% to 8%.

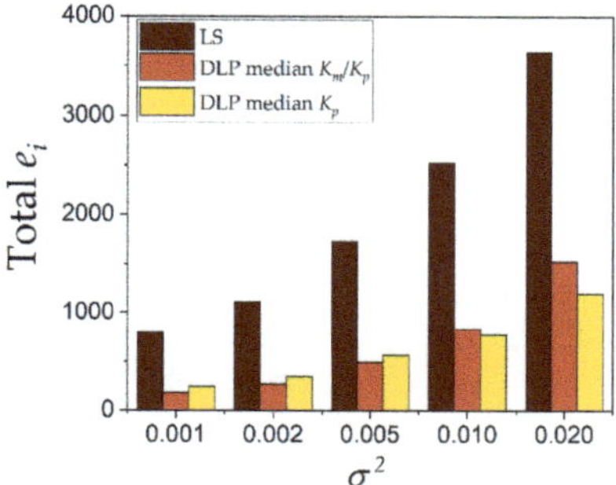

Figure 8. Total relative error (e_i) accumulated for estimations of the product inhibition constants K_p by LS and DLP methods after 1000 experimental runs plotted against variance (σ^2) of the normal distribution of error (ε_i). The K_p was estimated by DLP using the median of K_m/K_p (red bars) and the median of K_p (yellow bars).

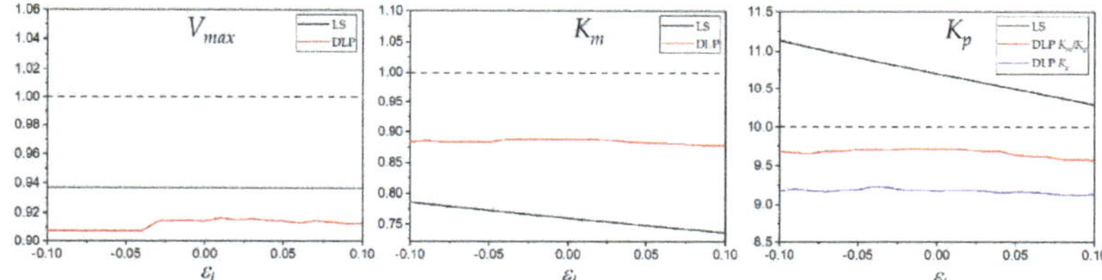

Figure 9. Effect of the fixed error added to the initial rate v_{34} on the estimated values of the kinetic constants V_{max}, K_m, and K_p by LS and DLP methods. Real kinetic constant value (dotted line), estimated values by LS (black line) and DLP (red and blue lines). The K_p was estimated by DLP using the median of K_m/K_p (red line) and the median of K_p (blue line).

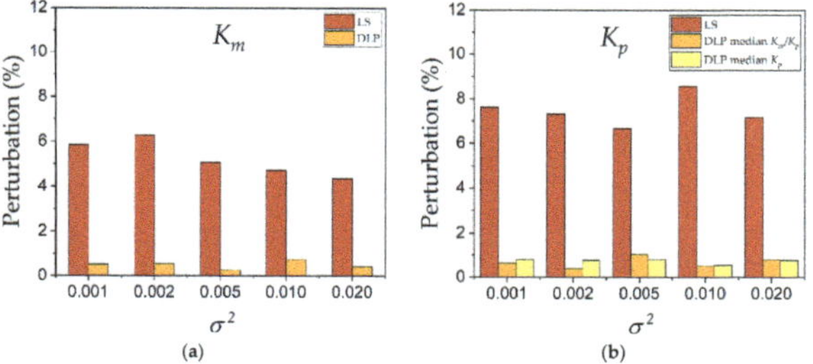

Figure 10. Perturbation percentage on the estimations of K_m and K_p by LS and DLP methods caused by adding fixed error to the initial rate v_{34} at different values of variance (σ^2). (**a**) K_m. (**b**) K_p. The perturbation corresponds to the difference between the minimum and the maximum value obtained respect to the original value of the kinetic constant (Equation (12)).

The DLP method exhibited the same pattern of behavior than that observed during application to the uncompetitive inhibition equation [3]. Lower *SSR* and estimation errors compared to the LS method were observed during estimations of the competitive inhibition constant. Again, the DLP method demonstrated accuracy and robustness during the estimation of kinetic constants. The possibility of using the DLP method to estimate inhibition constants in pharmaceutical and biotechnological studies was demonstrated. The performance of the DLP method applied to the competitive inhibition equation was different compared to the application of the substrate-uncompetitive inhibition equation previously studied. The addition of a variable product concentration caused mathematical indetermination during the calculations. The problem was overcome by avoiding using the same three product concentrations in Equations (4) and (5). This obligated a reduction in the number of combinations from 6545 to 6370, where 175 combinations resulted in the indetermination of the matrix (Equation (4)). Another difference is the dependence of K_p on the value of K_m, which can be observed in Equation (5), where the calculation consisted of the K_m/K_p value, and K_p was calculated after obtaining K_m. A significant implication of this is the transmission of the K_m error to K_p. Interestingly, based on the observations, this effect was not important. In general, the estimation of K_p by DLP was more accurate and robust than estimation by the LS method. The application of DLP to the uncompetitive inhibition equation exhibited a higher accuracy just in some cases compared to the LS method. In terms of robustness, the DLP was highly superior to the LS method. Presently, in this study, a higher superiority in terms of accuracy and robustness was observed. Despite the higher complexity of the competitive inhibition equation, requiring one more independent variable than the substrate-uncompetitive inhibition equation, the results were better. The transmission of error from K_m to K_p could be the key. This error is surely amplified when transmitted from one constant to another, and based on the observations, the effect was better softened by DLP compared to the LS method.

As Cornish-Bowden and Endrenyi commented [10], the median method cannot readily be generalized to equations of more than two parameters. This sentence became true when we found a mathematical issue during the application of the DLP to the product's inhibition equation, where an indeterminate system of equations (Equation (5)) is generated when the same value for the product concentration was used. However, this problem can be solved by selecting the proper combination of data. We, again, demonstrated that DLP can be applied to equations with more than two parameters. Presently, we also demonstrated the application to equations with more than two variables.

We conclude that the DLP method is better than the LS method to estimate the product's competitive inhibition constant in terms of accuracy and robustness. We now count on a new tool to estimate the product-competitive inhibition constant with reliable results.

3. Method

3.1. Experimental Design

The set of substrate and product concentrations was designed by first setting the real values of kinetic constants V_{max}, K_m, and K_p. The chosen values were 1 for V_{max} and K_m, and five different K_p values, including 10, 20, 50, 100 and 200, to study the effect on the estimation of the kinetic constants. These values correspond to the dimensionless kinetic constants in the case of V_{max} and K_m. Dimensionless V_{max} is the limit of $S/(S + K_m)$ when S tends to infinity. Dimensionless K_m is obtained when the kinetic equation is written in the form $S/K_m/(S/K_m + 1)$. In the case of K_p, the dimensionless value corresponds to K_p/K_m. As indicated by Cornish-Bowden et al. [2,10], the unitary values of V_{max} and K_m involved no loss of generality because they finally depend on the units of measurement. The values of substrate and product concentrations are listed in Table 1, corresponding to $n = 7$ and $m = 5$, according to the nomenclature in the scheme of Figure 1.

The chosen values of substrate concentrations were based on their distribution around the K_m value from 0.1 K_m to 10 K_m, with three values below K_m and three values above K_m, as recommended by Cornish-Bowden [11]. In the case of product concentrations, the choice was based on a K_p value-centered distribution. The experimental design consisted of 35 reaction rates (v_{ij}), which were calculated using the product-competitive inhibition equation (Equation (1)) assigning relative errors from a normal population of error ε_i with variance σ^2 (Equation (7)) to simulate the experimental error.

$$v_{ij} = \frac{V_{max}S_{ij}}{K_m + S_{ij} + \frac{K_m}{K_p}P_j}(1 + \varepsilon_i) \tag{7}$$

The calculated reaction rates (v_{ij}) were used as experimental values to estimate the kinetic constants V_{max}, K_m, and K_p. The estimations were carried out by DLP, using Equation (4), and by the LS method. The number of possible combinations of three initial rate values (v_{ij}) to calculate the kinetic constants V_{max}, K_m, and K_p was 6370, according to Equations (4) and (5). The median for each kinetic constant was obtained from this list. This procedure corresponded to one experiment to calculate the estimators of the kinetic constants, which is equivalent to an empirical procedure based on the initial rates method. A total of 1000 experiments were run to compare the accuracy of both methods and the quality of parameter estimations. An algorithm was written in Python software to perform all the calculations presented in this article. The algorithm can be downloaded from the repository indicated in the Supplementary Material. Estimation of K_p was carried out by two optional procedures: (i) calculating K_m/K_p for each of the 1000 experiments, calculating the median for K_m/K_p, and finally obtaining the estimated K_p using the corresponding K_m value, and (ii) calculating the K_p for each of the 1000 experiments and calculating the median of K_p (see Equation (5)). In the case of the LS method, a non-linear regression was applied to Equation (8), where the apparent K_m (K_m^{app}) was estimated. According to the experimental design, five values of V_{max} and K_m^{app} were obtained, one for each product concentration. As V_{max} is not affected by competitive inhibition, a higher value was chosen as the estimated V_{max} through the LS method. The K_m^{app} values were plotted against the product concentrations in a secondary plot, and the values of K_m and K_p were obtained by linear regression of Equation (9).

$$v = \frac{V_{max}S}{K_m^{app} + S} \tag{8}$$

$$K_m^{app} = K_m + \frac{K_m}{K_p}P \tag{9}$$

The effect of the K_p/K_m ratio was also evaluated. The values of K_p considered were 10, 20, 50, 100, and 200. The K_p/K_m ratios corresponded to the same values, considering that K_m was 1. As the values of K_p were modified, the experimental values of product concentration (P) were changed to maintain the same values of apparent K_m, according to Equation (9), among the different datasets to avoid negative effects in the experimental design.

3.2. Accuracy Test

The sum of squared residuals (SSR) was calculated for each of the 1000 experiments for both methods, DLP and LS, to evaluate their accuracies, according to Equation (10).

$$SSR = \sum_{i=1}^{n}(v_i - \hat{v}_i) \tag{10}$$

where v_i is the ith value of the experimental rate (Equation (7)), and $\hat{v}_i$ is the ith value of the predicted rate calculated according to Equation (1). The relative error (e_i) from the difference between the real (θ) and the estimated value ($\hat{\theta}$) was calculated for all kinetic constants according to Equation (11).

$$e_i = \left|\frac{\hat{\theta} - \theta}{\theta}\right| \tag{11}$$

The number of experiments (out of 1000) in which the LS resulted in a lower SSR and e_i was calculated and compared with DLP for different values of the variance σ^2 for error ε_i.

3.3. Effect of Outliers

The sensitivity of both methods to outliers was tested by changing the initial rate value v_{34} from Table 1. This value corresponds to the substrate concentration equal to K_m and the product concentration equal to K_p. The initial rate v_{34} was changed by varying fixed values of ε_i from −0.1 to +0.1. The rest of the experimental initial rates were calculated as mentioned above using Equation (6) with random ε_i with $\sigma^2 = 0.01$. The perturbation generated on the estimated constant was calculated according to Equation (12).

$$\%Perturbation = \frac{\hat{\theta}_u - \hat{\theta}_l}{\theta} \cdot 100 \tag{12}$$

where $\hat{\theta}_u$ and $\hat{\theta}_l$ are the upper and lower estimated values of the real constant θ in the range of the fixed error.

Supplementary Materials: The following are available online at https://github.com/bastiansepulveda/Inhibation-Equation-by-Product-Algorithms.git, algorithms for the analysis of DLP applied to the product's competitive inhibition equation.

Author Contributions: Conceptualization, P.L.V. and B.S.; methodology, P.L.V., B.S. and D.G.; software, D.G. and B.S.; validation, P.L.V., D.G., B.S. and C.A.-C.; formal analysis, P.L.V. and C.A.-C.; writing—original draft preparation, P.L.V. and C.A.-C. All authors have read and agreed to the published version of the manuscript.

Funding: This research received no external funding.

Acknowledgments: Pedro Valencia would like to thank Pedro Gajardo for the opportunity of presenting the direct linear plot problem in his class Laboratory of Modeling.

Conflicts of Interest: The authors declare no conflict of interest.

References

1. Eisenthal, R.; Cornish-Bowden, A. The direct linear plot: A new graphical procedure for estimating enzyme kinetic parameters. *Biochem. J.* **1974**, *139*, 715–720. [CrossRef] [PubMed]
2. Cornish-Bowden, A.; Eisenthal, R. Statistical considerations in the estimation of enzyme kinetic parameters by the direct plot and other methods. *Biochem. J.* **1974**, *139*, 721–730. [CrossRef] [PubMed]
3. Valencia, P.L.; Astudillo-Castro, C.; Gajardo, D.; Flores, S. Application of the median method to estimate the kinetic constants of the substrate uncompetitive inhibition equation. *J. Theor. Biol.* **2017**, *418*, 122–128. [CrossRef] [PubMed]
4. Airas, R.K. The Computation of Hyperbolic Dependences in Enzyme Kinetics. *Biochem. J.* **1976**, *155*, 449–452. [CrossRef] [PubMed]
5. Porter, W.R.; Trager, W.F. Improved Non-Parametric Statistical Methods for the Estimation of Michaelis-Menten Kinetic Parameters by the Direct Linear Plot. *Biochem. J.* **1977**, *161*, 293–302. [CrossRef] [PubMed]
6. Reading, C.; Hepburn, P. The Inhibition of Staphylococcal Beta-Lactamase by Clavulanic Acid. *Biochem. J.* **1979**, *179*, 67–76. [CrossRef] [PubMed]
7. Bialojan, C.; Takai, A. Inhibitory Effect of a Marine-Sponge Toxin, Okadaic Acid, on Protein Phosphatases. *Specificity and Kinetics. Biochem. J.* **1988**, *256*, 283–290. [CrossRef] [PubMed]
8. Takai, A.; Mieskes, G. Inhibitory Effect of Okadaic Acid on the P-Nitrophenyl Phosphate Phosphatase Activity of Protein Phosphatases. *Biochem. J.* **1991**, *275*, 233–239. [CrossRef] [PubMed]
9. Cornish-Bowden, A.; Eisenthal, R. Estimation of Michaelis constant and maximum velocity from the direct linear plot. *Biochim. Biophys. Acta* **1978**, *523*, 268–272. [CrossRef]
10. Cornish-Bowden, A.; Endrenyi, L. Robust regression of enzyme kinetic data. *Biochem. J.* **1986**, *234*, 21–29. [CrossRef] [PubMed]
11. Cornish-Bowden, A. *Fundamentals of Enzyme Kinetics*, 4th ed.; Wiley-VCH Verlag & Co.: Weinheim, Germany, 2012.

Article

Polymer-Assisted Biocatalysis: Polyamide 4 Microparticles as Promising Carriers of Enzymatic Function

Nadya Dencheva [1], Joana Braz [1], Dieter Scheibel [2], Marc Malfois [3], Zlatan Denchev [1,*] and Ivan Gitsov [2,4,*]

1. IPC-Institute for Polymers and Composites, University of Minho, 4800-056 Guimarães, Portugal; nadiad@dep.uminho.pt (N.D.); joanabraz@dep.uminho.pt (J.B.)
2. Department of Chemistry, State University of New York-ESF, Syracuse, NY 13210, USA; dmscheib@syr.edu
3. ALBA Synchrotron Facility, Cerdanyola del Valés, 0890 Barcelona, Spain; mmalfois@cells.es
4. The Michael M. Szwarc Polymer Research Institute, Syracuse, NY 13210, USA

* Correspondence: denchev@dep.uminho.pt (Z.D.); igivanov@syr.edu (I.G.)

Received: 15 June 2020; Accepted: 6 July 2020; Published: 9 July 2020

Abstract: This study reports a new strategy for enzyme immobilization based on passive immobilization in neat and magnetically responsive polyamide 4 (PA4) highly porous particles. The microsized particulate supports were synthesized by low-temperature activated anionic ring-opening polymerization. The enzyme of choice was laccase from *Trametes versicolor* and was immobilized by either adsorption on prefabricated PA4 microparticles (PA4@iL) or by physical in situ entrapment during the PA4 synthesis (PA4@eL). The surface topography of all PA4 particulate supports and laccase conjugates, as well as their chemical and physical structure, were studied by microscopic, spectral, thermal, and synchrotron WAXS/SAXS methods. The laccase content and activity in each conjugate were determined by complementary spectral and enzyme activity measurements. PA4@eL samples displayed >93% enzyme retention after five incubation cycles in an aqueous medium, and the PA4@iL series retained ca. 60% of the laccase. The newly synthesized PA4-laccase complexes were successfully used in dyestuff decolorization aiming at potential applications in effluent remediation. All of them displayed excellent decolorization of positively charged dyestuffs reaching ~100% in 15 min. With negative dyes after 24 h the decolorization reached 55% for PA4@iL and 85% for PA4@eL. A second consecutive decolorization test revealed only a 5–10% decrease in effectiveness indicating the reusability potential of the laccase-PA4 conjugates.

Keywords: laccase; polyamide 4; enzyme immobilization; magnetic enzyme supports; enzyme decolorization

---

1. Introduction

Biocatalysts (extracellular enzymes or whole-cells) constitute an alternative to traditional catalytic systems, being able to cause transformation of natural or synthetic substrates under mild reaction conditions, with high substance-, stereo-, and regiospecificity, and lack of toxic byproducts [1,2]. With the advances in the production of relatively inexpensive free enzymes, the steady increase of their extensive implementation as effective catalysts in many industrial and biomedical applications is beyond any doubt [3].

The wide use of industrial-scale processes with enzymatic catalysts is restricted by the free enzymes' inactivation by different denaturant agents and by their difficult or even impossible recovery from the reaction medium after use. These problems can be resolved by immobilization (conjugation) of the enzymes to prefabricated matrices of various nature, geometry, and topography. Immobilized enzymes, apart from their application as reusable heterogeneous biocatalysts, can serve as appropriate platforms for the development of biosensors, biofuel cells, or microscale devices for controlled release of protein drugs [4].

The most frequently used immobilization techniques can be sorted into five main groups, each with its associated advantages and disadvantages [5]: (i) non-covalent adsorption or deposition; (ii) immobilization via ionic interactions; (iii) entrapment in a polymeric gel or capsule (encapsulation); (iv) covalent attachment to the support (tethering); and (v) cross-linking of the enzyme. In the first three cases (i–iii), the enzyme immobilization is related to physical phenomena, i.e., entropy effects, electrostatic interactions, or hydrogen bonding. The distinct advantage of these methods is that neither the enzyme, nor the support, needs to be chemically modified or pretreated, which contributes to a cleaner and cost-effective biocatalyst. Moreover, the enzyme retains mobility, thus avoiding unfavorable screening of the active site. However, conjugates based on physical interactions often suffer from enzyme leaching, especially if used in aqueous media. To avoid leakage, immobilization protocols from groups (iv–v) are used. In them, covalent bonds across the enzyme–support interface are produced or cross-linking of the polypeptide moieties of the enzyme is caused by appropriate chemical reagents. These chemical modifications should be carefully directed so as not to cause deactivation of the biomacromolecule. As a rule, the immobilization method is selected on a case-by-case basis, by trial and error, although attempts to optimize this process by computer-aided analysis and molecular modeling have also been communicated [3,6].

General requirements for polymer supports are that they should not deactivate the immobilized enzyme, and for many important applications they have to be biocompatible. Therefore, traces of solvents, emulsifiers, unreacted monomers, or other accompanying substances that could be toxic or potentially deactivate the immobilized enzyme must be totally excluded. More specifically, immobilization by enzyme entrapment requires formation of the polymer support in situ, e.g., by low-temperature processes in the presence of the respective enzyme, thus avoiding denaturation. Furthermore, the monomer, the polymer support, and the enzyme should possess suitable functional groups in order to enable covalent or non-covalent interactions before, during, or after the polymerization. Finally, the polymer support should have appropriate physical structure and be formable into controllable shapes and textures at various length scales [7].

A large number of supports have been employed for enzyme immobilization in the form of beads, porous particles, membranes, and micro- or nanosized fibers made of inorganic materials [8] or organic polymers such as polyacrylates, cross-linked styrene-based copolymers, smart polymers, and modified polysaccharides [3]. Smart biocatalysts for industrial and biomedical applications produced by conjugation of enzymes to stimuli-responsive polymer supports have attracted a growing interest [9]. The activity of the enzyme can be favorably modified by complexation with supramolecular linear-dendritic block copolymers [10] or amphiphilic block copolymers self-assembling into micelles or physical networks [11].

Among the big variety of industrially relevant enzymes, the group of laccases (EC 1.10.3.2, benzenediol:oxygen oxidoreductases) have generated significant biotechnological interest since they require molecular oxygen as the final electron acceptor and release only water as a by-product [12]. Due to their broad specificity laccases are being used in different industrial fields for very diverse purposes, from food additives and beverage processing to biomedical diagnosis, pulp delignification, wood fiber modification, chemical or medicinal synthesis, or in the production of biofuels [13]. They show great promise as 'green' synthesis and polymerization mediators [14–16] and catalysts for Bisphenol A removal [17]. Also, laccases have great potential to biotechnological applications of industrial effluent treatment including textile dye decolorization and degradation [18,19]. Enzymes from the laccase group have also been studied intensely in nanobiotechnology for the development of implantable biosensors and biofuel cells, as well as wastewater and soil remediation [20].

Synthetic polyamides are among the known supports for laccase immobilization since they are cheap and can be easily produced in different forms. Thus, Fatarella et al. [21] explored the preparation of polyamide 6 (PA6) film or electrospun nanofiber carriers for covalently bonded laccase from *Trametes versicolor*, reaching enzyme loadings of ca. 60% and 71%, respectively. As reported, the resultant PA6/laccase conjugates are useful in industrial biocatalyst applications. Furthermore, a nanofibrous membrane for 3,3′-dimethoxybenzidine detoxification was developed by Jasni et al. [22] on the basis of PA66/laccase/Fe^{3+} conjugate produced by covalent bonding of the enzyme to the support by glutaraldehyde crosslinking. The immobilized enzyme displayed enhanced storage stability and a good potential for emerging pollutant detoxification. In a study of Silva et al. [23], laccase from *Trametes hirsuta* was covalently immobilized on woven PA66 supports employing glutaraldehyde and a 1,6-hexanediamine spacer. The PA66-laccase conjugate obtained under optimized conditions displayed higher half-life time than the free enzyme and potential for application in membrane reactors for continuous decolorization of effluents.

To the best of our knowledge, at this point of time, no attempt has been reported on laccase immobilization by polyamide particulate supports. Very recently, Dencheva et al. [24] synthesized neat and magnetic responsive PA4 [(poly-2-pyrrolidone), PPD] microparticles in good yields, with controlled size, shape, and porosity using activated anionic ring-opening polymerization (AAROP) of 2-pyrrolidone (2PD) carried out at 40 °C. This low-temperature and solventless process produces magnetic PA4 microparticles with controllable size, shape, and structure. There are certain advantageous factors in this technique that can open promising routes towards novel recyclable and efficient biocatalysts: PA4 is biodegradable in various environments [25,26], its monomer 2-pyrrolidone can be produced by sustainable biosynthesis [27], and the polyamide support is synthesized directly in the form of porous microparticles without the need of any additional treatment. Furthermore, the chemical structure of PA4 with its dense amide linkages and final functional groups resembles that of the proteins, thus enhancing the H-bonding between the support and the immobilized enzyme. Evidently, the neat and magnetic responsive PA4 microparticles combine most of the requirements for effective supports of biomolecules. Moreover, they have already been found to be good supports for model proteins [24] or to be used in molecular imprinting [28].

The present work investigates the potential of neat and magnetic-responsive PA4 microparticles obtained by low-temperature AAROP to serve as supports for immobilization of laccase. Two immobilization strategies were employed: (i) physical adsorption of laccase upon preformed PA4 microparticles and (ii) in situ encapsulation (entrapment) of laccase during the AAROP of PA4 synthesis. The morphology and the crystalline structure of the PA4-laccase conjugates obtained by these two methods were analyzed by microscopy, spectral and X-ray scattering techniques, and the respective enzyme activities toward 2,2′-azino-bis (3-ethylbenzothiazoline-6-sulphonic acid (ABTS) were compared. The biocatalytic capability of the synthesized PA4-laccase conjugates, as well as their reusability, were tested in two reactions of environmental interest related to the treatment of effluents from the textile industry. The activity of the immobilized laccase was tested in decolorization reactions of two dyes, widely used as textile coloring agents (malachite green and bromophenol blue).

2. Results and Discussion

Three microparticulate PA4 supports, without and with magnetic susceptibility, designated as PA4, PA4-Fe, and PA4-Fe_3O_4, were synthesized by low-temperature AAROP. They are in the form of fine powders with white, grey or brownish color, respectively. The scheme for the AAROP representing the chemical structure of all substances involved is presented in Figure S1 of the Supporting Information. When the polymerization is carried out in the presence of magnetic micro/nanoparticles, the PA4 MP become susceptible to external magnetic fields [24].

The 2PD monomer is a cyclic analog of the amino acids and the chemical composition of PA4 is comparable to that of proteins and enzymes. This will expectedly enable intense H-bond formation between the laccase and the PA4 supports: neat PA4, PA4-Fe, and PA4-Fe_3O_4. The fact that AAROP was possible at 40 °C allowed the entrapment of active laccase into the shell of the forming PA4 MP during the polymerization, resulting in three PA4@eL samples. Separately, the enzyme was immobilized by physical adsorption on prefabricated PA4 supports, thus preparing the three PA4@iL samples. This work presents a comparative discussion on the morphology, structure, and the catalytic activity of the three conjugates immobilized by adsorption PA4@iL and the three conjugates immobilized by entrapment PA4@eL.

2.1. Synthesis and Morphology of Empty PA4, PA4-Fe, and PA4-Fe_3O_4 Microparticles

Some basic characteristics of the synthesized neat and magnetic PA4 microparticles (MP) are listed in Table 1. The yields of polymer were in the 59–63 wt % range meaning that the presence of iron micro- and iron (II, III) oxide nanosized particles in the reaction mixture did not affect the activity of DL/C20 catalytic system and the kinetics of the AAROP reaction. Intrinsic viscosity (η) (see the Supporting Information) of 0.926 dL/g was found for the neat PA4 MP, which is similar to the values of PA4 microspheres obtained by a different method involving AAROP of 2PD [29].

Table 1. Designation and some characteristics of the PA4 MP empty supports.

Sample	PA4 Yield, % [(a)]	% of Oligo-mers	Real Fe Content R_L, % [(b)]	η, dL·g^{-1}	d_{max}, μm	d_{max}/d_{min}
PA4	59.4	4.4	-	0.926	8-15	1.1–1.2
PA4-Fe	61.2	6.9	1.7	-	10-20	1.2–1.3 3.6–4.1
PA4-Fe_3O_4	62.7	4.7	1.2	-	10-15	1.2–1.3 3.4–4.0

[(a)] In relation to the 2PD monomer; [(b)] Determined by thermogravimetric analysis (TGA) according to Equation (2) (see Materials and Methods—Section 3.2).

The average maximum size of the particles d_{max} in all PA4 empty supports based on optical microscopy with image-processing was found to be between 8–20 μm (Table 1, Figure S2 of the Supporting Information). The d_{max}, of PA4 MP without magnetic load is the smallest, ranging between 8 and 15 μm. Similar values of 10–15 μm are registered for the support with nanosized Fe_3O_4 loads, while in the presence of Fe the upper limit of d_{max} grows above 20 μm. The Fe- and Fe_3O_4 containing PA4 microparticles become less spherical, as seen from the roundness parameter d_{max}/d_{min} (Table 1, Figure S2). This means that the PA4-Fe MP would contain in their core up to 4–5 iron microparticles whose proper size is 3–5 μm, while the PA4-Fe_3O_4 core would expectedly contain iron oxide particles in the nanometer length scale [24,28]. The d_{max}/d_{min} in the range of 3–4 found in both magnetic empty supports are attributable to self-assembly of magnetized Fe and Fe_3O_4 to higher aspect ratio aggregates with their subsequent coating with PA4 during the AAROP.

All PA4 supports have porous structure as demonstrated by BET analysis (Table S1 of the Supporting Information). The neat PA4 microparticles showed the largest specific surface area, S_{BET} = 2.766 m^2/g with a total pore volume V_{total} = 0.012 cm^3/g, followed by PA4-Fe_3O_4 (2.542 m^2/g; 0.008 cm^3/g) and PA4-Fe (2.052 m^2/g; 0.004 cm^3/g) samples. The largest average pore size σ_{ave} = 85 Å was found for the PA4 MP, the values of PA4-Fe_3O_4 and PA4-Fe being 42 and 32 Å, respectively. These changes in σ_{ave}, V_{total}, and S_{BET} should be attributed to the presence of dense magnetic loads in the microparticles core of the last two samples.

More details on the morphology of the empty supports' particles can be obtained by SEM (Figure 1).

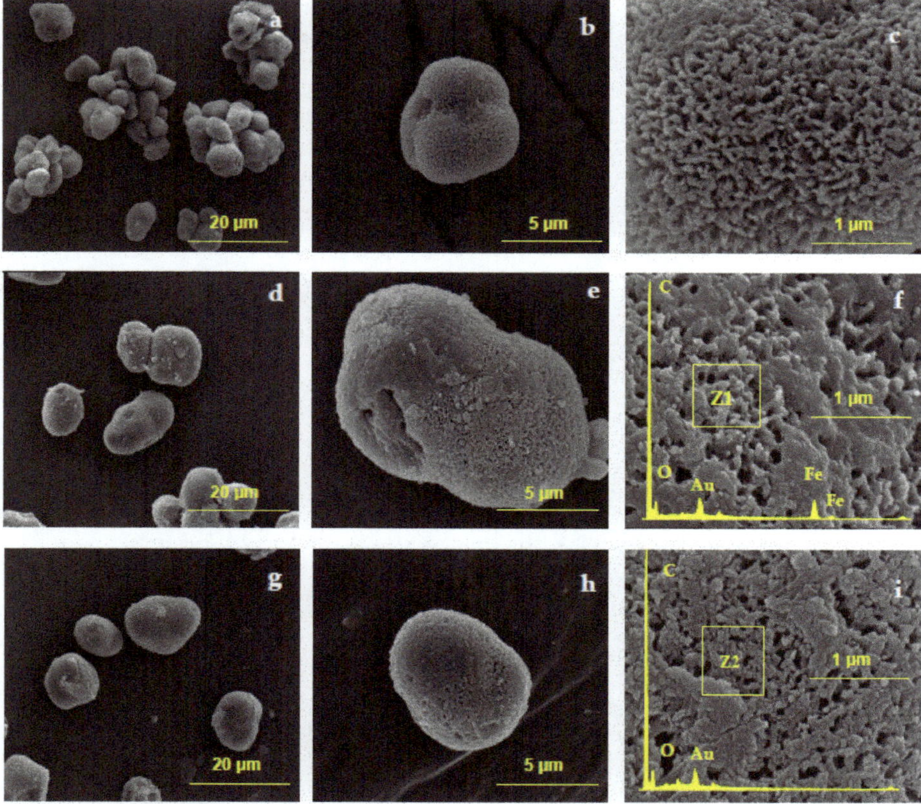

Figure 1. SEM images of empty PA4 particulate supports: (**a**–**c**) PA4; (**d**–**f**) PA4-Fe; (**g**–**i**) PA4-Fe_3O_4.

The micrographs of the neat PA4 support (Figure 1a–c) display spheroidal particles with sizes of the individual entity between 5–8 μm, the latter forming also aggregates with average sizes close to 20 μm. At larger magnifications, the PA4 particles display porous, scaffold-like topology with visible pore diameters in the range of 100–150 nm. The use of Fe and Fe_3O_4 particles (Figure 1d,e,g–i, respectively), seem to decrease the number of surface pores changing their distribution but with no significant change of the pore size. Moreover, as seen from some selected EDX traces in Figure 1f,i, the FeKα and FeKβ peaks only appear if the electron beam is directed into the pores (location Z1), rather than if it hits a non-porous spot of the particles surface (location Z2). This means that most of the magnetic particulate fillers (especially the much larger Fe particles) are embedded in the core of the respective MP, thus confirming the supposed precipitation–crystallization mechanism of the polyamide MP formation [24].

2.2. Immobilization of Laccase on PA4 MP Supports

2.2.1. Immobilization by Physical Adsorption

As previously mentioned, owing to the analogous chemical structure of PA4 and the peptide-containing biomolecules, a high capacity toward H-bonding formation is expected between the laccase and the PA4 supports during immobilization by adsorption. Since the laccase isoelectric point is in the range of 3–4, if the adsorption is carried out at $pH > 4$ the enzyme will be negatively charged. From the Z-potential measurements (see the Supporting Information, Table S2) it can be seen that under such conditions the three PA4 supports will also be negatively charged with values of −35 eV at pH 7 and between −10 ÷ −12 eV at pH 5. Nevertheless, preliminary adsorption tests confirmed that larger amounts of laccase are adsorbed in DDW at neutral pH than in PB, pH 5. This means that the immobilization by adsorption is not upset by possible electrostatic repulsion between the enzyme and the support, therefore all adsorption experiments were performed at pH 7.

Table 2 shows that, as expected, the particles of the three PA4@iL laccase-adsorbed samples do not change their average size and roundness. Judging from the SEM images in Figure 2, the topography of all particulate PA4@iL conjugates is noticeably smoother as compared to the starting PA4 support particles, which is better expressed in the case of the PA4-iL sample (Figure 2c).

Table 2. Designation and some characteristics of the PA4 supports with adsorbed laccase (PA4@iL samples).

Sample	Real Fe Content, R_L, % [(a)]	d_{max}, µm	d_{max}/d_{min}
PA4-iL	-	8–15	1.1–1.2
PA4-Fe-iL	1.4	12–20	1.2–1.3 3.6–4.1
PA4-Fe_3O_4 -iL	1.3	10–15	1.2–1.3 3.4–4.0

[(a)] Determined by TGA, according to Equation (2) (see Materials and Methods—Section 3.2).

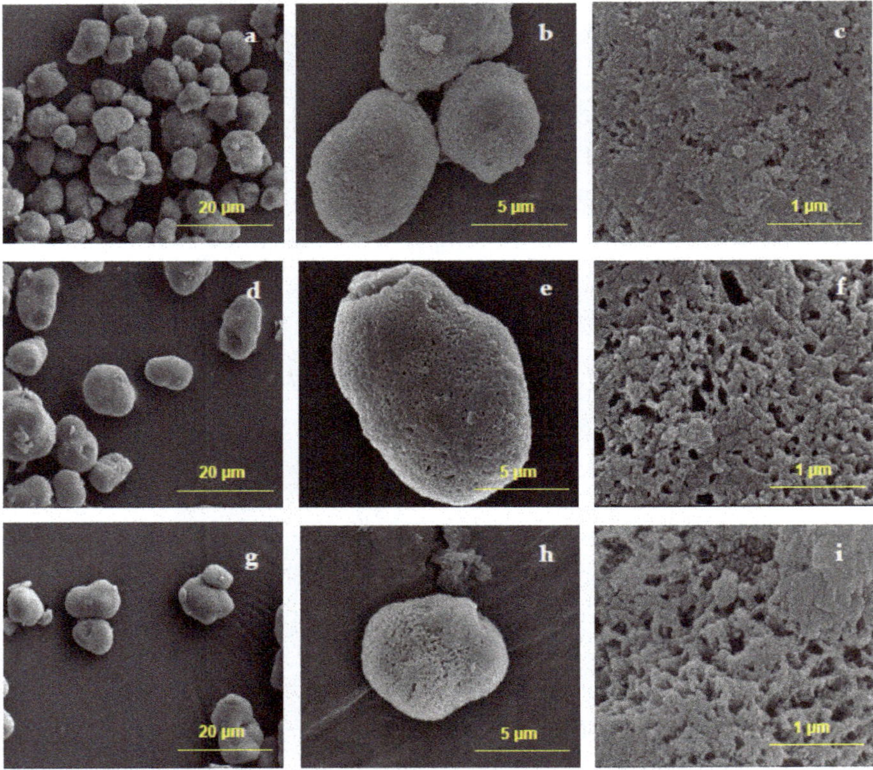

Figure 2. SEM images of laccase-adsorbed PA4@iL: (**a**–**c**) PA4@iL; (**d**–**f**) PA4-Fe@iL; (**g**–**i**) PA4-Fe_3O_4@iL.

The next step in the PA4@iL preparation was to quantify the amount of the adsorbed enzyme. Using the common method for determination of immobilized laccase, after adsorption completion, the residual supernatants were subjected to activity test toward ABTS. All three samples showed no activity, leading to the conclusion that there is no residual laccase in the supernatant, therefore almost 100% of the enzyme was adsorbed upon the PA4 supports (Table 3). This was a superior immobilization efficiency as compared to previous studies disclosing laccase immobilization by covalent bonding on PA6 [21] or carbon nanotube supports [30] with immobilization effectiveness (IE) values of 59–71% or 60–90%, respectively. However, the lack of activity of the residual supernatant means that it is free of active laccase but could contain some amounts of inactive enzyme. Therefore, a second method to verify the laccase content in the supernatant was used.

Table 3. Enzyme quantification in the PA4@iL samples.

Sample	Residual Laccase Activity, μkat·mL^{-1} [(a)]	Immobilized Laccase, mg·mL^{-1} [(b)]	Immobilized Laccase, mg·mL^{-1} [(c)]	Laccase on Support, mg	Laccase, mg·g^{-1} Support	Immobilization Efficiency, IE, % [(d)]
PA4-*iL*	0.0002	1.999	1.265	6.32	31.62	63.3
PA4-Fe-*iL*	0.0009	1.999	1.612	8.06	40.29	80.6
PA4-Fe_3O_4-*iL*	0.0002	1.999	1.072	5.36	26.81	53.6

[(a)] Activity of the residual laccase in the supernatant after immobilization; [(b)] Immobilized laccase calculated by standard curve based on the activation test toward ABTS; [(c)] Determined by UV analysis of the supernatant after immobilization using a standard curve based on the absorbance at λ = 286 nm; [(d)] Immobilization effectiveness (IE) is the ratio between the amounts of immobilized laccase and starting laccase used in the adsorption process (see Equation (3), Materials and Methods—Section 3.3.2).

An alternative method for laccase quantification can be based on its UV absorbance in the 280–290 nm range [31,32] due to the presence of aromatic amino acid residues: tryptophan, tyrosine, or phenylalanine. Thus, all supernatants were studied for residual laccase by UV spectroscopy measuring the intensity of the band appearing at 286 nm. As indicated in Table 3, 19–46% of residual laccase was found in the supernatants, which showed no activity in the previous test with ABTS. A possible reason for this behavior is that during the multiple adsorption/desorption processes in the PA4@iL sample preparation, a structural change occurs leading to lowering of the redox potential of the desorbed laccase, this effect depends on the PA4 support composition. Based on UV spectroscopy data, the maximum amount of laccase (40.3 mg/g or 81% from the enzyme amount) was adsorbed on the PA4-Fe MP support, whereas the PA4-Fe_3O_4 MP displayed the lowest IE of 54%. These IE values were considered more reliable and were therefore used in all further calculations.

Employing the approach of Qui et al. [33], the data in Table 3 can provide information about the mechanism of laccase adsorption on the PA4 supports of this study. Thus, since the size of laccase macromolecule is found to be 6.5 × 5.5 × 4.5 nm [34], its largest footprint on a surface will be ca. 35.8 nm^2. The surface areas of the three PA4@MP samples determined by BET (Table S1) were in the range of 2.05–2.78 m^2/g. Assuming that laccase molecules arrange in ideal monolayers when adsorbing and that the laccase molecular weight is ca. 85,000 Da, then 1.0 g of PA4 support should theoretically adsorb 8–11 mg enzyme. According to Table 3, however, the real values vary in the 27–40 mg range, i.e., about 3–4 times larger. Evidently, the laccase adsorption occurs in multilayers, implying lateral intra- and interlayer interactions between adsorption sites. This finding is in good agreement with our previous work on protein adsorption upon similar PA4 microparticles [28] proving that without special treatment of the PA4 support the adsorption data are consistent with the Freundlich isotherm, thus proving the multilayer adsorption model.

2.2.2. Immobilization by Entrapment

Entrapment is defined as physical retention of enzymes in a porous solid matrix [35]. In our case, the laccase entrapment occurs during AAROP. The enzyme is first suspended in the monomer solution, and a subsequent polymerization process keeps the biomolecule trapped, preventing direct contact with the environment. To obtain laccase-entrapped PA4 MP with or without magnetic particles, the same low temperature AAROP was used as for the preparation of the empty PA4 supports, however carried out in the presence of the enzyme. The final 2PD conversion to PA4 was in the 45–58 wt % range (Table 4), i.e., in average 5–10% lower than of the empty PA4 supports (Table 1). This is explained with the more rigorous washing/purification procedure to eliminate not entrapped enzyme and oligomers with lower conversion to PA4. It can therefore be concluded that the presence of laccase in the AAROP reaction mixture did not significantly upset the catalytic system permitting to use the same protocol for synthesis and purification as for the empty PA4 particulate supports. However, Table 4 indicates larger average particle sizes of 15–25 µm and lesser roundness values of 1.2–1.3 in the PA4@eL samples than in the respective empty and laccase-adsorbed supports (Table 2).

Table 4. Designation and some characteristics of PA4 supports with entrapped laccase (PA4@eL).

Sample	PA4-eL Yield, % [(a)]	Real Fe Content, R_L, % [(a)]	d_{max}, µm	d_{max}/d_{min}
PA4-eL	45.3	-	15–20	1.1–1.2
PA4-Fe-eL	48.4	1.7	15–25	1.2–1.3
PA4-Fe_3O_4 -eL	57.6	1.9	15–25	1.2–1.3

[(a)] In relation to the 2PD monomer; [(b)] Determined by TGA, according to Equation (2) (see Materials and Methods—Section 3.2).

The possibility to carry out AAROP to PA4 in the presence of laccase at 40 °C had two major advantages. First, no denaturation or other disruption of the enzyme's secondary or tertiary structure caused by temperature will occur during the polymerization process. Second, in AAROP of all lactams, the chain propagation is accompanied by PA4 crystallization that affixes the topology of the enzyme-loaded microparticles and results in their precipitation from the reaction medium, thus facilitating the entrapment.

As seen from Table 5 showing the laccase quantification in the PA4@eL series, the total amount of entrapped laccase was 45–58 mg, corresponding to 13–18 mg enzyme per gram of MP. These concentrations are up to 2.5 times lower than in the case of immobilization by adsorption (Table 3). As to the entrapment factor EF, it is dependent on the 2PD conversion to PA4. Thus, the highest degree of 2PD conversion of ca. 70% was achieved with the PA4-Fe_3O_4@eL sample, which accounted for the minimum amount of entrapped enzyme per gram support in this case.

Table 5. Enzyme quantification in PA4@eL samples.

Sample	Laccase in AAROP Mixture, mg	Yield of PA4 MP, % [(a)]	Laccase on Support, mg [(b)]	Laccase $mg{\cdot}g^{-1}$ Support	Entrapment Efficiency, EE, % [(c)]
PA4-e*L*	85.10	45.29	50.91	17.86	59.8
PA4-Fe-e*L*	85.10	48.30	52.00	16.72	61.1
PA4-Fe_3O_4-e*L*	85.10	57.58	59.25	13.47	69.6

[(a)] Based on the sample weight after removal of excessive laccase and oligomer extraction; [(b)] Calculated by UV–VIS detection of the laccase in the combined DDW supernatants of the threefold wash. For more details, see the Experimental part; [(c)] EE is the ratio between the quantity of the entrapped and the starting laccase used in the syntheses (see Equation (4), Materials and Methods—Section 3.3.3).

Table 4 indicates larger average particle sizes of 15–25 µm and lesser roundness values of 1.2–1.3 in the PA4@eL samples than in the respective empty and laccase-adsorbed supports. Figure 3 displays selected SEM images at different magnifications of the three PA4 supports carrying entrapped enzyme.

Figure 3. SEM images of laccase-entrapped PA4@eL: (**a–c**) PA4@eL; (**d–f**) PA4-Fe@eL; (**g–i**) PA4-Fe_3O_4@eL.

Unlike the three empty PA4 supports or the magnetic-loaded laccase conjugates PA4@iL, the particles' shapes in PA4@eL samples are far from being spherical or elliptical. Most probably, this is due to the fact that upon its dispersion in the 2PD monomer the lyophilized enzyme produced aggregates that were subsequently covered by PA4. The growing PA4 molecules wind up upon these aggregates and, after reaching a critical molecular weight, crystallize upon them interacting by H-bonds with the enzyme. As a result, the specific topography of the PA4@eL conjugates is produced with visible average pore diameters up to 300–350 nm, being almost twice as large as in the other samples and with quite distinct shape.

2.3. Structure Characterization of PA4 Empty Supports and Laccase Conjugates

The PA4 porous microparticles obtained by AAROP are used for the first time as enzyme supports. The laccase-entrapped PA4@eL samples are also synthesized for the first time. Therefore, some initial structural characterization of all samples of this study was necessary in order to be able to explain their catalytic activity and decide about their potential as enzyme supports.

2.3.1. FTIR Spectroscopy

A FT-IR spectra comparison between the empty PA4 microparticles and the laccase-adsorbed and entrapped samples are presented in Figure 4. In all samples, the bands at 3300 cm^{-1} were assigned to the valence stretching vibrations of hydrogen atoms in secondary NH groups. The shoulder with a maximum at 3450 cm^{-1} was attributed to stretching vibrations in primary amines corresponding to terminal amine groups. Also, the spectra show well defined peaks for Amide I at 1631.5 cm^{-1} and Amide II at 1535.0 cm^{-1} with almost identical intensities. This is a clear indication for fixation of the trans-conformation of the NH–CO group, being typical for high molecular weight polyamides and proteins.

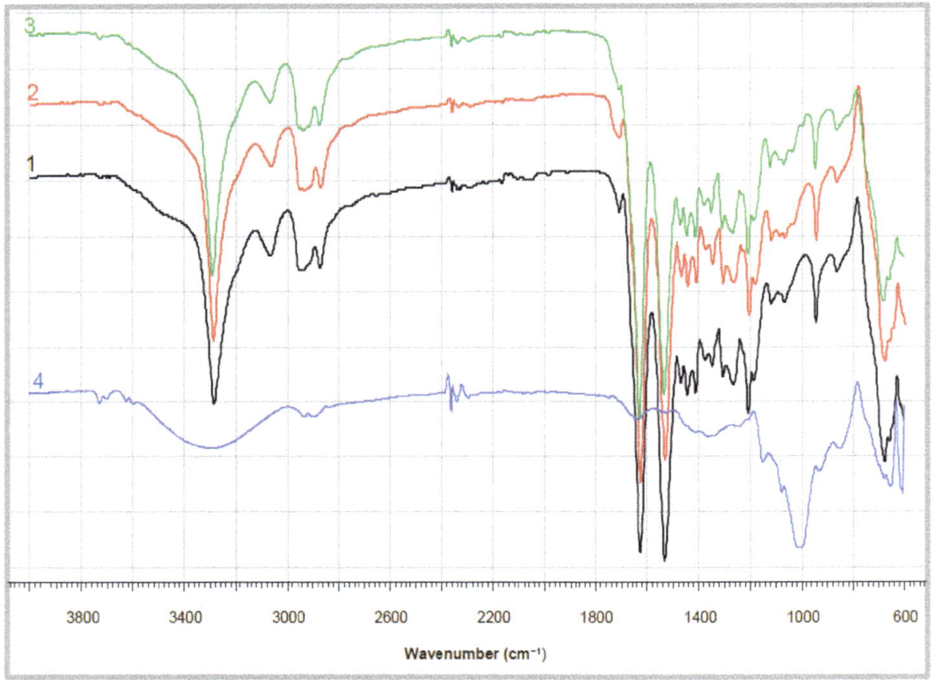

Figure 4. FTIR spectra with ATR of: 1—empty PA4 support; 2—PA4@iL; 3—PA4@eL; 4—native laccase.

The IR spectra in Figure 4 indicate undoubtedly that even at polymerization temperatures as low as 40 °C AAROP of 2DP led to high molecular weight polyamide in all samples studied. This was important to confirm for the PA4@eL and PA4@iL samples, in which the determination of (η) by viscosimetry is technically impossible. All curves display also a weak peak at 1710–1712 cm^{-1} attributable to terminal carboxyl groups. In the neat PA4 sample (curve 1) this narrow and well-resolved band belongs to the –COOH terminus of the polyamide macromolecule. In the enzyme-adsorbed and entrapped samples (curves 2 and 3) a broader composite peak appears due to superposition of the –COOH signal of PA4 with such belonging to laccase carboxyl groups. Curve 4 indicates that a weak and broad band centered at 1641.0 cm^{-1} really exists in the free laccase spectrum. However, FT-IR cannot provide more detailed quantitative or qualitative information about the entrapped or adsorbed laccase.

2.3.2. DSC and TGA

A typical shortcoming of PA4 is the intense degradation that precedes or accompanies the melting occurring normally in the broad range of 230–260 °C [24,36]. As shown previously, the presence of a model protein adsorbed onto PA4 microparticles can significantly increase the thermal resistance of the latter [28]. The TGA traces of all samples in this work displayed in Figure 5 confirm that the presence of laccase in the PA4@iL series can also considerably improve the thermal stability of the supports. Thus, the temperature of initial degradation T_d^{in} of PA4-iL is 255 °C, which is 38 °C higher than of the respective PA4 empty support particles. For the PA4-Fe-iL/PA4-Fe and $PA4\text{-}Fe_3O_4\text{-}iL/PA4\text{-}Fe_3O_4$ pairs $\Delta T_d^{in} = 20$ and 26 °C, respectively. Similar behavior was observed in PA4 obtained by AAROP of 2PD with subsequent conversion of the terminal –COOH into $–NH_2$ groups [37], which in our case may have occurred during the heating scan in the TGA.

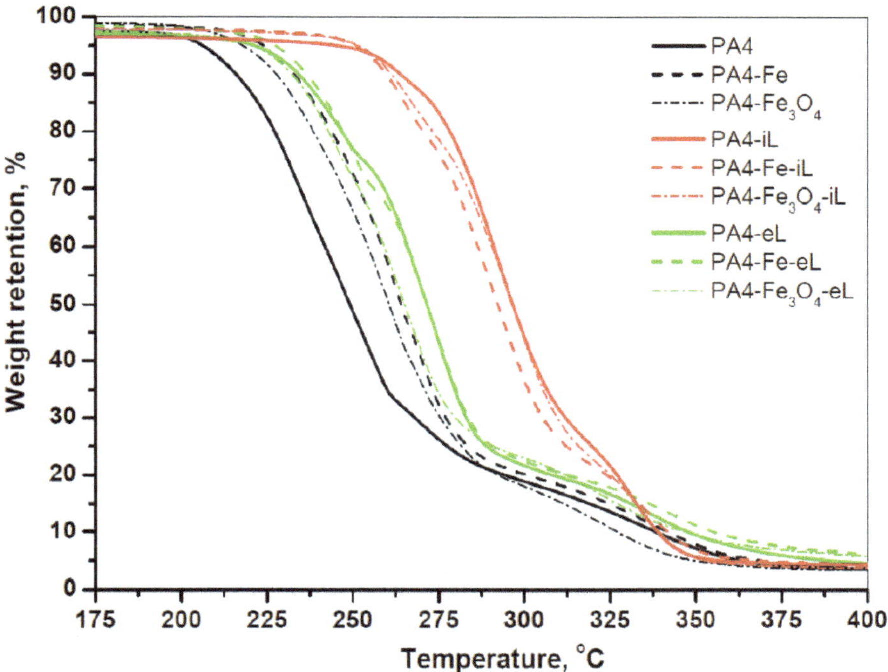

Figure 5. TGA curves at 10 deg/min of empty PA4 supports, laccase adsorbed (PA4@iL) conjugates, and laccase-entrapped (PA4@eL) conjugates.

It should be noted that the presence of magnetic particles alone in PA4 MP increases the T_d^{in} with 15–20 °C in comparison with the neat PA4 MP, most probably due to heat dissipation phenomena. At the same time, the ΔT_d^{in} for the entrapped PA4-eL/PA4 pair is only 12 °C. It can be hypothesized that in the case of the laccase-entrapped PA4@eL samples the said conversion of PA4 terminal groups is difficult or impossible. The TGA study permitted also to determine the carbonized residue at 600 °C of each sample of this study and to calculate on this basis the real content of Fe or Fe_3O_4 by means of Equation (2) (see Materials and Methods—Section 3.2).

The DSC curves of all samples presented in Figure 6 and the consolidated data extracted from them (Table S3 in the Supporting Information) display a trend of increasing the melting temperature T_m of PA4 in the presence of enzyme. After immobilization by adsorption in the PA4@iL series, the T_m of the three resulting conjugates goes over 260 °C irrespective of the support type. This is definitely higher than the T_m of the three empty supports, being in the range of 235–247 °C. As for the laccase-entrapped samples, the two of them i.e., PA4-eL and PA4-Fe-eL melt at slightly higher or similar T_m, whereas the $PA4\text{-}Fe_3O_4\text{-}eL$ sample is the only one with a T_m with 11 °C lower than the respective empty support. The fact that both TGA and DSC display similar degradation behavior and melting temperatures for the empty supports and the laccase-entrapped samples means that the molecular weight of PA4 in these two sets should be similar and relatively high.

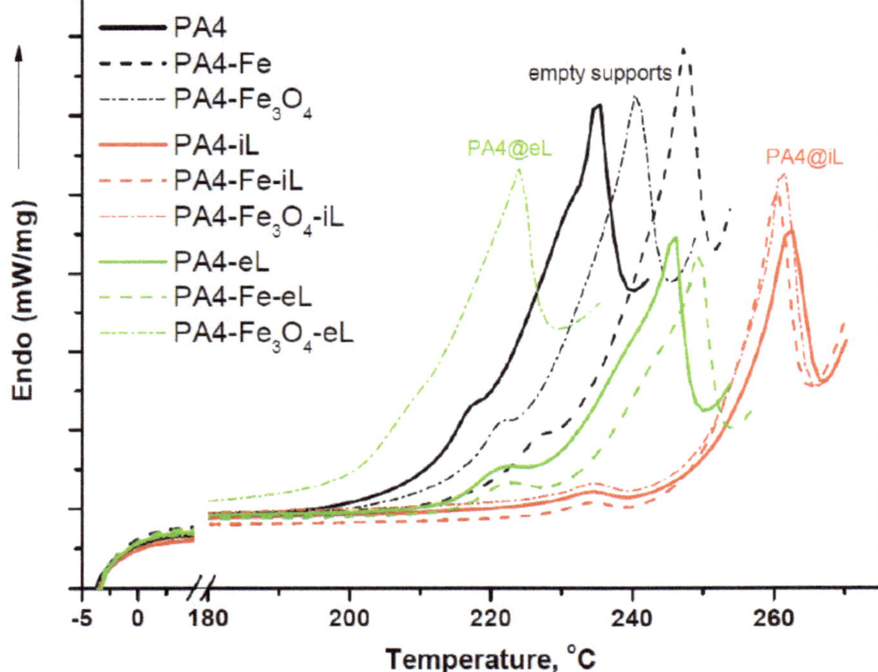

Figure 6. DSC (first scan, heating 10 deg/min) of empty PA4 supports, laccase adsorbed (PA4@iL) and laccase-entrapped (PA4@eL) conjugates.

2.3.3. Synchrotron WAXS

The activity of laccase physically adsorbed onto or entrapped into PA4 MP and the efficiency of the immobilization process in each case will directly depend on the nature and strength of the interactions at the enzyme/PA4 support interface. The postulated intense H-bond formation between the laccase and its structural analogue PA4 should have some influence on the enzyme configuration, or the crystalline structure of the polymeric support that can be probed by X-ray scattering techniques. It is important to know also whether or not the interior pores and channels of the PA4 microparticles that are impossible to access for direct SEM observation are filled with enzyme. In an attempt to evaluate these factors in the empty PA4 supports and in the respective adsorbed or entrapped PA4-laccase conjugates, synchrotron WAXS and SAXS were employed. Moreover, no structure studies about the PA4 crystalline structure by synchrotron X-ray were reported up to now.

Figure 7 shows a comparison between the linear WAXS patterns of samples representing the three empty supports, as well as the three PA4@iL immobilized and the three PA4@eL entrapped samples. All WAXS patterns display two strong reflections at $q \approx 14.5$ nm^{-1} and 17.0 nm^{-1} that, according to Bellinger et al. [38], should be ascribed to the monoclinic unit cell of the α-PA4 with $d_{\alpha[200]} = 4.33$ Å and $d_{\alpha[020]} = 3.69$ Å. Moreover, in accordance with the same study, it should be postulated that the PA4 chains are parallel to the lamellar normal and that an amide group is incorporated in the fold. Such incorporation does not take place in PA6 or PA66, however it is very typical for the β-bends in proteins [39]. Consequently, PA4 is really a closer structural analogue to all protein-containing biomolecules than other polyamides.

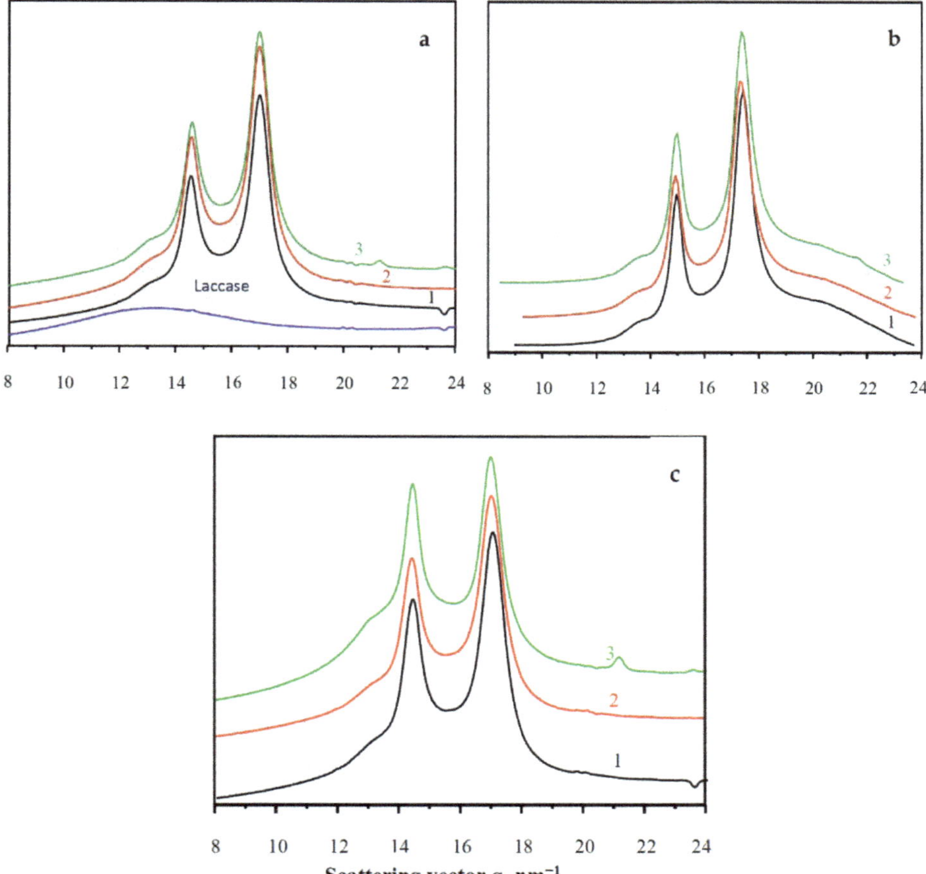

Figure 7. Background corrected linear WAXS patters of: (**a**) neat PA4 supports; (**b**) PA4@iL; (**c**) PA4@eL conjugates. 1—PA4; 2—PA4-Fe: 3—PA4 Fe_3O_4 supports. The curves are shifted along the vertical axis for better visibility.

The visual inspection of the WAXS patterns of the three PA4 supports in Figure 7a suggest that the presence of Fe or Fe_3O_4 fillers does not seem to change the PA4 crystalline structure leaving the angular position and the intensities of the two α-PA4 reflections unaffected. Figure 7a displays also the pattern of a free laccase producing a wide diffuse scattering peak (halo) typical of amorphous materials and centered at $q_a^{free} = 13$ nm^{-1}, i.e., the free laccase is amorphous at the length scale of various angstroms that is probed by WAXS.

The patterns of PA4@iL samples (Figure 7b) represent a superposition of the laccase and PA4 support scatterings. The two α-PA4 reflections apparently maintain their form and position, but the amorphous laccase halo appears centered at $q_a^{ads} \approx 20$ nm^{-1}. It is attributable to the adsorbed bulk laccase deposited on the surface and within the pores of the PA4 particulate support. The shift $\Delta q \approx 7$ nm^{-1} is significant and can be explained as follows. Since the amorphous halo in WAXS is related to intermolecular interactions [40], its position must be dependent on the degree of packing of the molecules (i.e., the density) of the respective amorphous phase. As pointed out by Alexander [41], the dependence of the amorphous halo angular position q_a on the intermolecular distance r_a is given as

$$r_a \sim \frac{\lambda}{2q_a}, \text{ nm} \tag{1}$$

which is the reciprocal dependence typical of all diffraction phenomena. Thus, the larger the scattering vector q_a, value is, the smaller the intermolecular distance and consequently the higher the density of this phase will be. This means that the adsorbed enzyme in the PA4@iL samples has a denser packing as compared to the free one, confirming the supposition for intensive interaction between the laccase and PA4 via multiple H-bonds.

The patterns of the entrapped PA4@eL samples (Figure 7c) do not show any difference in comparison to the empty PA4 supports in Figure 7a. This observation leads to two conclusions: (i) the enzyme arrested in the PA4 particles during AAROP cannot form a separate amorphous reflection and (ii) the entrapped enzyme does not upset the crystallization of the α-PA4 polymorph. It can be therefore hypothesized that, in the PA4@eL series, the enzyme macromolecules are distributed within the amorphous phase of the semi-crystalline PA4 support quite homogeneously with no significant interaction between one another.

Further information about the crystalline structure of the samples can be extracted after deconvolution of the WAXS patterns in Figure 8 by peak fitting. This procedure and the subsequent quantification of the α- and β-PA4 crystalline phases is made according to earlier publications, resolving the crystalline parameters and polymorph structure of PA4 [38] and PA6 [42]. All structural information from the fitted WAXS patterns is presented in Table 6.

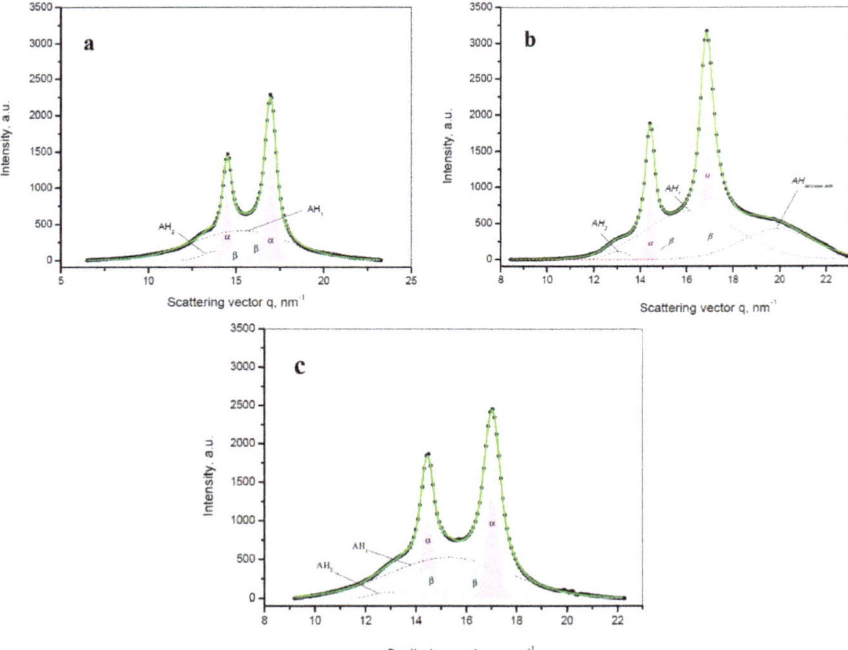

Figure 8. Examples for WAXS pattern deconvolution by peak fitting: (**a**) PA4 support; (**b**) PA4-iL; (**c**) PA4-eL.

Table 6. WAXS analysis of empty PA4 supports and PA4-laccase conjugates.

Sample	α, %	β, %	X_c^{WAXS}%	α/β	Δq_a, nm^{-1}	$d_{\alpha(200)}$ Å	$d_{\alpha(020)}$ Å	$d_{\beta(002)}$ Å	$d_{\beta(200)}$ Å
PA4	22.85	32.56	55.41	0.70	-	4.33	3.70	4.30	3.72
PA4-Fe	21.44	30.26	51.70	0.71	-	4.33	3.70	4.24	3.72
PA4-Fe_3O_4	23.71	28.12	51.83	0.84	-	4.33	3.70	4.27	3.73
PA4-iL	17.14 27.39	19.13 30.58	36.27 57.97	0.90 0.90	6.84	4.35	3.72	4.36	3.69
PA4-Fe-iL	18.43 21.84	20.96 24.84	39.39 46.68	0.88 0.88	7.21	4.36	3.73	4.34	3.71
PA4-Fe_3O_4-iL	17.09 20.77	20.80 25.28	37.90 46.06	0.82 0.82	6.50	4.34	3.72	4.35	3.69
PA4-eL	21.92	28.43	50.35	0.77	-	4.34	3.69	4.30	3.69
PA4-Fe-eL	21.54	26.62	48.16	0.81	-	4.36	3.69	4.15	3.71
PA4-Fe_3O_4-eL	21.91	30.60	45.25	0.72	-	4.36	3.68	4.30	3.70

Notes: For the d-spacings indexation presented the chain axis coincides with the b-axis [38]. The bolded values for the PA4@iL samples are determined excluding the adsorbed laccase amorphous reflection; $\Delta q_a = q_a^{ads} - q_a^{free}$; X_c^{WAXS} = WAXS crystallinity index. For more information, see the text.

All example deconvolutions in Figure 8 show that excellent fits with regression coefficients $R^2 \geq 0.99$ were only possible if along with the two peaks of α−PA4 and the two amorphous halos AH_1 and AH_2 two more crystalline peaks were considered that should be assigned to an additional monoclinic phase designated by β-PA4. It was first described in the early work of Frederiks et al. [43] and later on shown to co-exist with α-PA4 [24].

Table 6 shows that the values of the long spacings of these two PA4 phases almost coincide. Thus, the reflections related to $d_{\beta(002)}$ and $d_{\beta(200)}$ are quite away from each other, which is typical of a monoclinic unit cell, contrary to the two γ-PA6 peaks that are often described as belonging to a pseudo-hexagonal unit cell [42,44].

Table 6 displays the structural data of all studied samples.

The α/β ratio in the empty PA4 supports and in the laccase-entrapped samples PA4@eL (Table 6) seems to be relatively constant varying between 0.70–0.85. The α/β ratio rises to ca. 0.90 upon physical adsorption of laccase in the PA4@iL samples, meaning that the PA4-enzyme interaction via hydrogen bonds may cause some slight β−α transition.

According to Table 6, the WAXS crystallinity indices X_c^{WAXS} of the empty PA4 supports vary in the narrow range of 52–55%, i.e., values relatively high for polyamides. Entrapping laccase by in situ AAROP drops the crystallinity with about 5%, being within the margin of the experimental error of the deconvolution method. More fluctuations in X_c^{WAXS} appear in the laccase-adsorbed samples. We are inclined to explain this with an increased error in the deconvolution due to interactions between the laccase halo $AH_{laccase}$ and AH_2 diffuse peak related to the PA4 own amorphous fraction, rather than to alteration of the crystallinity index.

As regards the Δq_a values displayed in Table 6 for all samples with laccase adsorption, it should be noted that this difference is larger with the PA4-Fe-iL, followed by the PA4-iL and the PA4-Fe_3O_4-iL samples. In accordance with Equation (1), in this sequence the density of the adsorbed bulk laccase is expected to decrease as a function of the polymer support composition.

2.3.4. Synchrotron SAXS

The use of synchrotron SAXS allows further clarification of the structure of the PA4 MP before and after laccase immobilization or entrapment. This method probes density periodicities with dimensions in the 20–250 angstroms range, which includes the sizes of the crystalline lamellae typically found in semi-crystalline polymers.

Figure 9 presents the background-subtracted and Lorentz-corrected SAXS linear profiles of the three empty PA4 supports (neat PA4 MP and such containing Fe or Fe_3O_4 fillers) and the respective laccase adsorbed (PA4@iL) and entrapped (PA4@eL) samples. To enable comparison, Figure 9b also contains the SAXS curve of the free laccase.

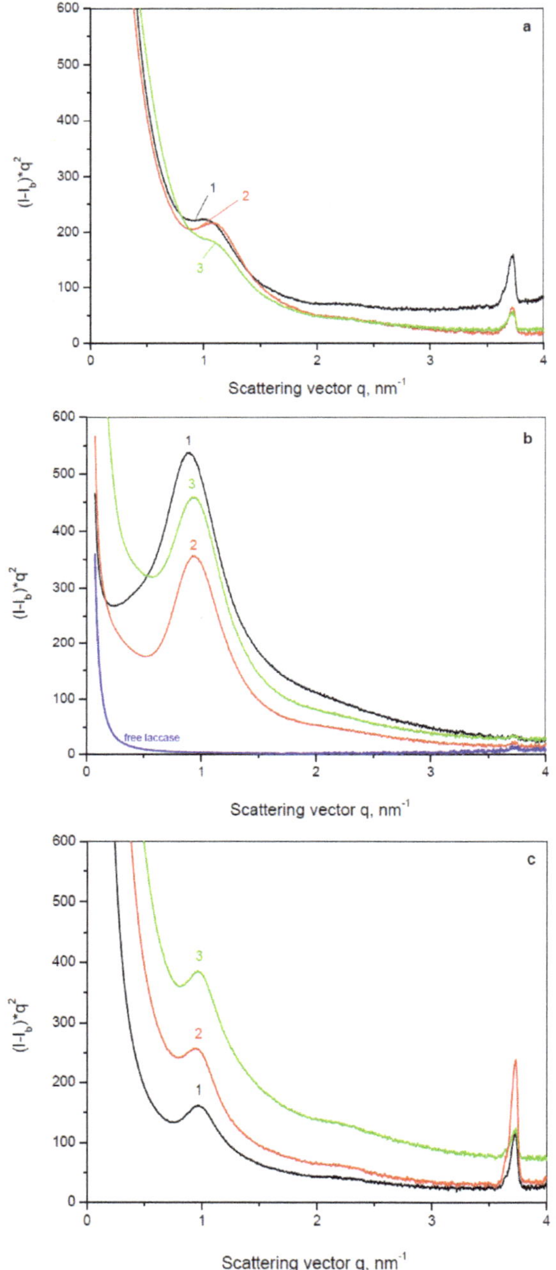

Figure 9. Linear SAXS patterns of: (**a**) empty PA4 supports; (**b**) PA4@iL and (**c**) PA4@eL samples. 1—PA4; 2—PA4-FE; 3—PA4-Fe_3O_4 supports. The laccase pattern is presented in Figure 9b for comparison.

It can be seen that the three empty PA4 supports (Figure 9a) show relatively well resolved Bragg peaks, indicating lamellar stack morphology. The neat PA4 displays a long spacing value L_B = 62 Å, i.e., close to the 63–66 Å established previously in PA4 bulk samples [38]. The presence of Fe or

Fe_3O_4 slightly reduces L_B to 57 Å. Having in mind that the total WAXS crystallinity index of the three empty supports is above 50% (Table 6), it should be concluded that it is the crystalline phase which is predominant in the lamellae. All the three PA4 empty supports display well-expressed and narrow SAXS peaks at $q_{max} \approx 3.75$ nm^{-1} corresponding to $L_B = 1.5$–1.7 nm that should be related to nanoporosity in the PA4 microparticles not observable directly by SEM but undoubtedly established by BET (Table S1). It is important to note that, as expected, this nanoporosity-related SAXS peak disappears if the empty support particles undergo melting and recrystallization (Figure S3 of the Supporting Information, the curves at 270 °C and at 30 °C after heating up to 270 °C).

As seen from Figure 9b, the free laccase does not show any periodicity in the SAXS q-range, which agrees with the WAXS results. Nevertheless, the laccase-adsorbed PA4@iL samples display very well resolved SAXS peaks with L_B = 67–70 Å, suggesting a much better phase contrast between the densities of the amorphous and crystalline regions, as compared to the respective neat PA4 supports (Figure 9a).

It should be noted that no significant changes in the crystalline structure of the PA4 supports could possibly occur during the physical adsorption of laccase. Therefore, the SAXS curves in Figure 9b allow the conclusion that after laccase immobilization most of the pores and channels of the empty supports get filled with enzyme whose density should be comparable to that of the amorphous PA4. This creates a clearer density gradient between the amorphous and crystalline fractions of the lamellar periodicity resulting in better resolved SAXS peaks of the PA4@iL samples. Notably, the narrow SAXS peaks at high q-values of the empty supports disappear completely after laccase adsorption, confirming its relation to the PA4 nanoporosity.

Considering the SAXS patterns in Figure 9c with their clear peaks with L_B = 65–67Å leads to the conclusion that a lamellar stack system similar to that of the empty PA4 supports is built after AAROP of 2PD in the presence of laccase that produces the PA4@eL samples. Let us note that the SAXS peaks with $q_{ma} \approx 3.75$ nm^{-1} remain present in all samples with enzyme entrapment, being the most intense in the PA4-Fe-eL sample. This finding can be logically explained if the enzyme is arrested within the amorphous phase of the PA4 support, leaving its nanoporosity unobstructed. At the same time, in the laccase adsorbed samples, all enzyme seems to be located within the cavities of the particulate support, completely obstructing its nanoporosity. This is an important structural conclusion that will be used in the explanation of the enzyme activities of the two types PA4-laccase conjugates.

2.4. Specific Activity of Laccase Immobilized by Adsorption and Entrapment

The catalytic activity of all PA4-laccase conjugates was studied using ABTS as a substrate and calculating its rate of oxidation in each case. As shown in preliminary studies, the free laccase displays highest activity at pH 5, so this condition was assumed for all activity tests (0.0125 M PB, 25 °C). Table 7 contains data on the total, specific, and relative activities of each PA4@iL and PA4@eL samples, taking the free laccase activity as 100%. As expected, the laccases immobilized by either adsorption or entrapment, was less active compared to the free laccase, the reduction of specific activity being different for the two immobilization strategies. The laccase-adsorbed samples (PA4@iL) were about 1.7–2.6 times less active, whereas the entrapped laccases displayed 2.4–6.2 times lower specific activity, as compared to the free laccase. Notably, the surface-immobilized laccase samples were from 1.4 to 3 times more active than the corresponding entrapped counterparts.

These results were expected having in mind the SEM and SAXS studies of the PA4@iL samples. After the adsorption, there is a large amount of 'exposed' laccase covering the surface of the PA4@iL and entering into their pores/cavities located near the surface. As seen directly from SEM and indirectly from the SAXS data, the topography of the entrapped samples (PA4@eL) is completely different. During the AAROP the laccase macromolecules are covered by a PA4 shell so there should be much less (or no) 'exposed' laccase in these samples. The porous shell makes more difficult the access of the ABTS substrate molecules to the active site of the entrapped enzyme. Moreover, the resulting oxidized cation-radical $ABTS^{+\cdot}$ should return to the aqueous medium where it is quantified by UV–VIS, which is also hampered by the PA4 MP shell.

Table 7. Specific activity of laccase immobilized by adsorption or entrapment.

Sample	Laccase Activity, μkat·mL^{-1}	Laccase Activity, μkat·mL^{-1}·g^{-1} Support	Specific Laccase Activity, μkat·mL^{-1}·mg^{-1} Laccase	Relative Laccase Activity, %
Free laccase	0.1472	-	0.1472	100
PA4-iL	0.0508	2.5417	0.0804	54.60
PA4-Fe-iL	0.0714	3.5694	0.0886	60.18
PA4-Fe_3O_4-iL	0.0303	1.5139	0.0565	38.36
PA4-eL	0.0085	0.4246	0.0238	16.15
PA4-Fe-eL	0.0203	1.0139	0.0606	41.19
PA4-Fe_3O_4-eL	0.0075	0.3750	0.0278	18.91

Note: The substrate is 0.1 mL ABTS (5 mM in DDW). For more details, see the Materials and Methods, Section 3.3.4.

As seen from Table 7, the conjugate with the highest laccase activity is the PA4-Fe-iL sample, in which the laccase immobilization effectiveness is the highest (ca. 81%, Table 4). Logically, the enzyme densification measured by WAXS data (Table 6, Δq values) is the largest in this sample. Among the entrapped samples, it is also the Fe-containing PA4-Fe-eL that displays the highest activity. These data suggest some synergism between the laccase activity and the presence of Fe^0 in the PA4 support.

2.5. Laccase Retention Studies

Since both of our immobilization strategies count on H-bond formation between the PA4 support and laccase, leaching of enzyme will be an inevitable feature of PA4@iL and PA4@eL samples, as it is in all conjugates with no covalent bonding [33]. Figure 10 displays the laccase retention in each conjugate type as a function of the support composition, in five consecutive application cycles.

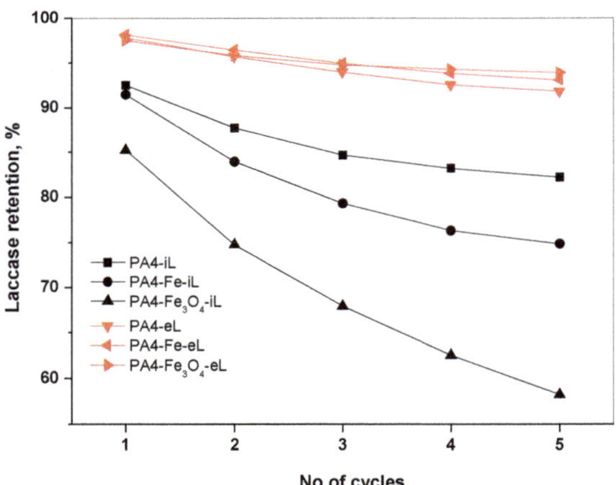

Figure 10. Laccase retention as a function of the immobilization strategy and cycles of application. Conditions: 50 mg of the respective PA4@iL or PA4@eL sample were dispersed in 5 mL PB, pH 0.0125M, pH 5, and shaken at 37 °C for 15 min. For more details, see the Experimental Part Section 3.3.4.

The amount of laccase leached was below the detection limit of the spectroscopy method measuring the absorbance at λ = 286 nm. Therefore, the supernatants after every application cycle were subjected to activity testing, in which the higher the rate of laccase-catalyzed oxidation of ABTS substrate, the higher the percentage of leached laccase (Section 2.3.4). The enzyme content in the starting conjugates (Tables 4 and 5) was considered 100%.

Figure 10 shows that the leaching from the PA4@eL samples is quite low and even after the fifth cycle the enzyme retention is close to 95%, not depending on the PA4 support. This can be explained with the steric hindrance of the enzyme entrapped deep in the PA4 shell. The laccase release from the PA4@iL samples is notably higher, whereby the strongest drop is observed during cycles 1–3. Apparently, the laccase that is closer to the surface leaches first at higher rates, and after that a stabilization is attained. Moreover, the leaching rate in this series clearly depends on the PA4 support composition. The leaching is the strongest in the PA4-Fe_3O_4-iL sample ranging from 15% (cycle 1) to a total 42% (after cycle 5). The other two samples of this series display leaching percentages of 15–22% after the last cycle. In summary, the entrapment strategy affords better enzyme retention than immobilization by absorption.

2.6. Decolorization of Dyestuff Employing PA4-Laccase Conjugates

Malachite green (MG) is a cationic (i.e., positively charged) dyestuff widely used in the pigment industry and in agriculture. Bromophenol blue (BPB) is negatively charged with major applications as industrial or laboratory dyestuff and color marker. Both dyes contain polyaromatic hydrocarbon moieties (Figure S4 of Supporting Information) that have carcinogenic and mutagenic health effects. MG and BPB are frequently found in effluents from the textile industry and agriculture, so their neutralization is an important problem. As was proven in a previous report [45], MG and BPB can be successfully degraded by laccase oxidation. Therefore, the two dyes were selected in this study as substrates for decolorization with both PA4@iL and PA4@eL samples, produced by two different immobilization strategies. In both cases, the oxidizing activity of the conjugates was evaluated without any mediator.

Figure 11 illustrates the decolorization kinetics of MG and BPB by the PA4@iL and PA4@eL samples. For comparison, the action of the free laccase was also studied at the same conditions. All PA4-laccase conjugates showed 90–95% effectiveness of the MG decolorization after 15 min only, irrespective of the immobilization strategy, whereas about 300 min were necessary for the free laccase to reach similar values (Figure 11a,b). These yields are worth comparing to the data of Bagewadi et al. [46] who, in a similar experiment, reported 95–100% MG decolorization, but after 960 min using laccase from *Trichoderma harzianum* immobilized in a sol–gel matrix and 1-hydroxybenzotriazole (HBT) mediator.

The decolorization of BPB occurred quite differently with a clear dependence on the way of enzyme incorporation into the PA4 support microparticles. The adsorption-immobilized PA4@iL samples were less active than the free laccase, showing a lower decolorization effectiveness within the whole 24 h period that peaked only around 25–50% (Figure 11c,d). At the same time, the two laccase-entrapped samples PA4-eL and PA4-Fe-eL displayed values of 70–80% in 120 min, the free laccase value being 45% at this time. A final decolorization of 80–90% was achieved after 24 h against 100% of the free enzyme. The PA4-Fe_3O_4-eL sample maintains an effectiveness of ca. 50% after 150 min that almost coincides with that of the free laccase. From this point on, the decolorization rate of the free enzyme continues to increase linearly, whereas that of the Fe_3O_4-containing conjugate saturates. Our results on the BPB decolorization can be favorably compared to the data of Forootanfar et al. [47] who performed decolorization studies of BPB with free laccase from *Trametes versicolor* and reached effectiveness values of 20% after 180 min without mediator and 32% with HBT mediator

Comparing the rates of decolorization of the two dyes by free laccase (Figure 11) suggests that the MG substrate is more susceptible to oxidative degradation than BPB. This experimental fact can be related to their different chemical structure. Thus, BPB contains bulky and heavy groups (four Br atoms and one SO_3 group) and its molecular weight is almost twice as high as that of MG, making BPB more difficult to degrade [46].

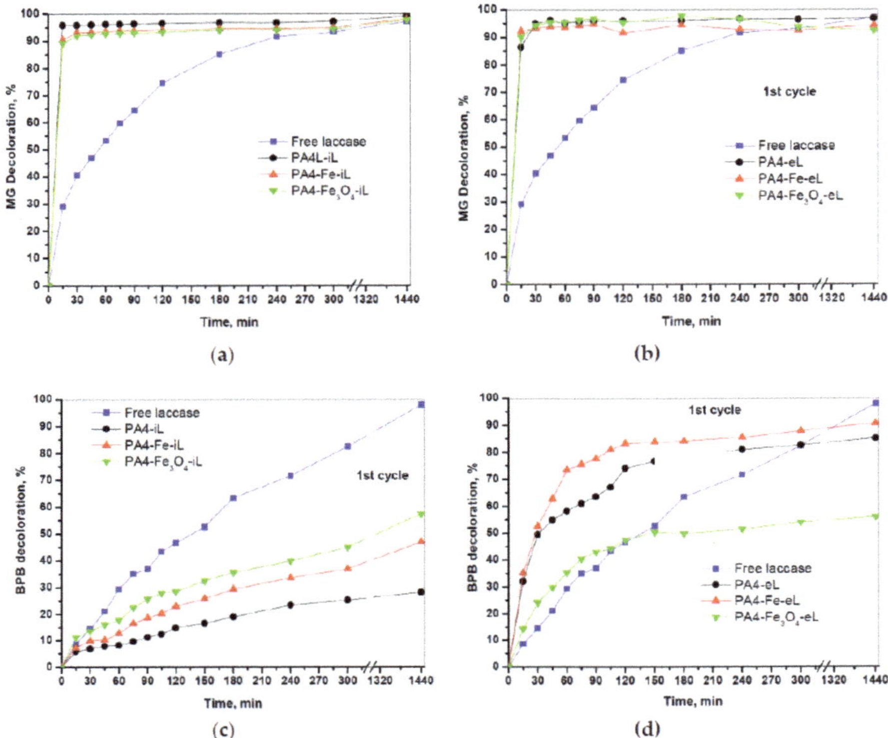

Figure 11. Decolorization of: (**a**,**b**)—malachite green (MG) and (**c**,**d**)—bromophenol blue (BPB) by PA4@iL or PA4@eL samples. The decolorization percentage was calculated according to Equation (5) (see Materials and Methods, Section 3.3.5).

When immobilized enzymes are used, theoretically, the disappearance of the color of the initial dye solution can be caused both by enzymatic action and by adsorption in the support, whereby the contribution of the latter could be quite significant and must be taken into account [48]. To assess the contribution of the physical adsorption in the above decolorization experiments, the empty PA4 supports were incubated in DDW MG and BPB solutions. Further treatment was exactly the same as indicated in Section 2.4. The results for MG and BPB are presented in Figure 12.

In the case of MG (Figure 12a), a ramp to 95% decolorization effectiveness was achieved within 15 min time with all three empty supports. This effect can only be due to the extremely high adsorption capacity of the PA4 microparticles (Z-potential of −36 eV at pH7) toward the MG cation. Clearly, this vigorous adsorption process is much faster than the enzymatic oxidative degradation of MG by free laccase. In the case of BPB, however (Figure 12b), the PA4 particles of the support adsorb only 2–3% dye (PA4 MP) or ca. 15% (Fe and Fe_3O_4-containing PA4 MP), these amounts being constant after 30 min exposure. The lower adsorption of BPB should be attributed to the electrostatic repulsion between the negatively charged dye molecules and PA4 microparticles. Evidently, the presence of Fe^0 or $Fe^{2+,3+}$ in the PA4 support slightly enhances the BPB removal by physical adsorption.

The results shown in Figure 12 help assess the contribution of the adsorption by the supports in the experiments in Figure 11. The instant decolorization of the MG substrate by PA4@iL or PA4@eL conjugates in Figure 11a,b should be explained by strong adsorption in the support particles, which influences the MG removal more than the oxidative degradation by laccase. In the case of BPB, the results in Figure 11c,d could be corrected by subtraction of the adsorption-caused dye removal thus allowing the evaluation of the sole enzymatic decolorization. The corrected data is presented in Figure 13.

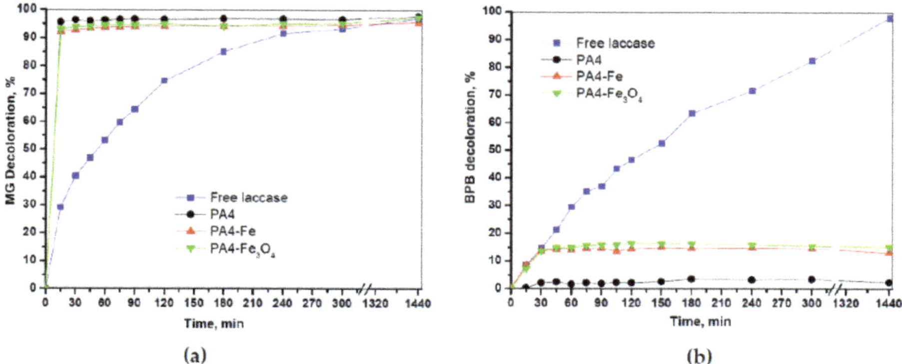

Figure 12. Decolorization of MG (**a**) and BPB (**b**) solutions by empty PA4 MP supports due to physical adsorption. The decolorization kinetics by free laccase are also given for comparison. The decolorization percentage was calculated according to Equation (5) (see Materials and Methods, Section 3.3.5). The structures of MG and BPB are presented in Figure S4 of the Supporting Information.

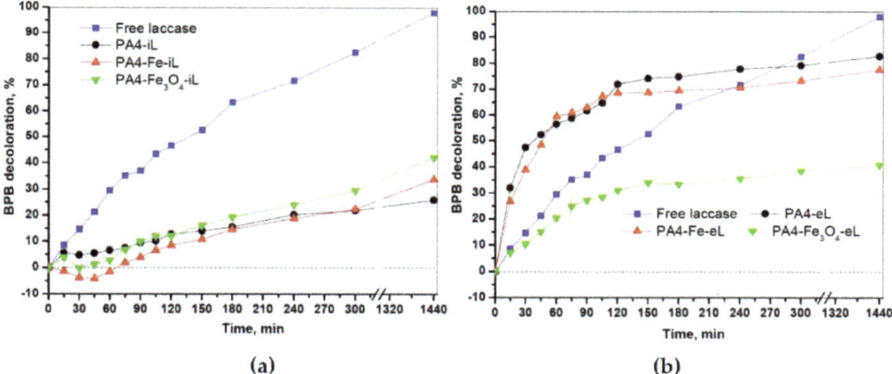

Figure 13. Resultant decolorization (RD) of BPB solution by: (**a**) PA4@iL samples and (**b**) PA4@eL samples. RD is obtained by subtracting the adsorption-caused decolorization (Figure 13) from the total decolorization (Figure 12).

The PA4@iL samples display slower decolorization kinetics than the free laccase in the whole 24 h interval studied. The Fe^0 and $Fe^{2+,3+}$ containing samples of this series display negative values of ΔA in the beginning of the experiment due to predominant adsorption-caused dye removal. As to the laccase entrapped samples, the PA4-eL and the PA4-Fe-eL samples show excellent decolorization kinetics, being faster than that of the free laccase during the first 3 h of the experiment. The maximum decolorization effectiveness of BPB due only to the enzyme action was between 65–75%. The $Fe^{2+,3+}$-containing support displayed a clear negative deviation from this behavior. Since the specific and relative laccase activities in the PA4-Fe_3O_4-eL sample are even slightly higher than those of the PA4-eL sample, the significantly lower decolorization kinetics in the former case cannot be attributed to lesser amount or lesser starting activity of the enzyme. Most probably, some laccase inactivation occurs by BPB functional groups during the decolorization process.

On the other hand, as seen from Table 7, the samples with entrapped laccase displayed lower activity toward the ABTS substrate, as compared to both free and PA4-immobilized laccases. The results in Figure 12 can be explained with some deactivation of the free laccase and that in the iL samples, while the catalytic activity of the eL samples remained constant. It should be noted that all samples were stored at exactly the same conditions and for the same time duration.

The potential of practical application of the PA4@iL and PA4@eL conjugates will be directly related to the possibility to remove them rapidly and completely from the reaction mixture and use them in several consecutive cycles. Figure S5 in the Supporting Information visualizes the removal of the PA4-Fe-eL sample from its suspension in DDW by means of a constant magnet. Figure 14 displays the second decolorization cycle of MG and BPB dyes by PA4@iL and PA4@eL conjugates.

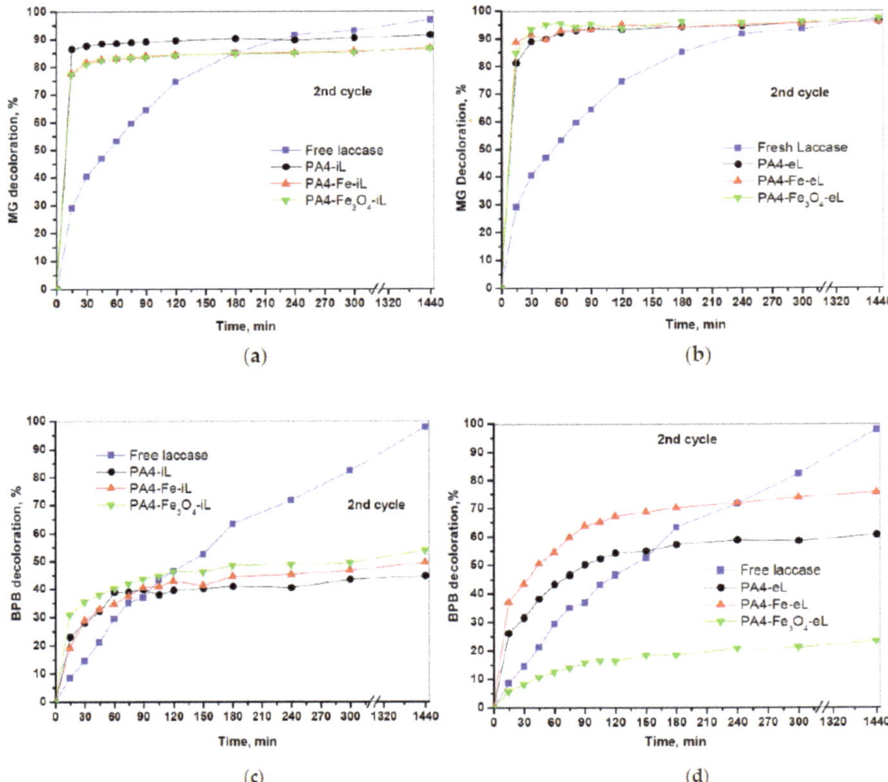

Figure 14. Second cycle of decolorization of: (**a**,**b**)—malachite green (MG) and (**c**,**d**)—bromophenol blue (BPB) by PA4@iL and PA4@eL samples. The decolorization percentage was calculated according to Equation (5) (see Materials and Methods, Section 3.3.5).

The comparison of Figures 11 and 14 show that in the case of MG both the PA4@iL and PA4@eL conjugates lose less than 5% of their decolorization capacity during the second cycle, which means they can be reused in many additional decolorizing cycles. However, in this case, the contribution of the enzymatic reaction cannot be evaluated due to the very rapid adsorption process. In regard to the BPB decolorization, where the contribution of the adsorption is negligible, the laccase-entrapped samples lose ~10% of their decolorization effectiveness. Supposing that this rate of loss is maintained in subsequent cycles, at least 3–5 more uses will be possible for PA4@iL and PA4@eL series, respectively, without adding free laccase.

3. Materials and Methods

3.1. Materials

The 2PD functional monomer for the PA4 synthesis and all solvents used in this work are of analytical grade and were supplied by Merck Life Science, Algés, Portugal. The laccase from *Trametes versicolor* (≥0.5 U/mg) was also purchased from the same supplied and used without further purification. As activator of the anionic polymerization of 2PD the commercial product Brüggolen C20 from Brüggemann Chemical GmbH, Heilbronn, Germany was employed containing, according to the manufacturer, 80 wt % of aliphatic diisocyanate blocked in ε-caprolactam. The polymerization initiator sodium dicaprolactamato-bis-(2-methoxyethoxo)-aluminate (Dilactamate, DL), which is also a commercial product was supplied by Katchem, Prague, Czech Republic and used without further treatment. Soft, non-insulated iron particles (Fe content >99.8%), with average diameters of 3–5 µm were kindly donated by the manufacturer BASF, Ludwigshafen, Germany. The Fe_3O_4 magnetic particles are a product of Merck, Algés, Portugal with >99% purity and grain sizes of 50–80 nm. Diammonium 2,2′-azino-bis(3-ethylbenzothiazoline-6-sulfonate) (ABTS) with a purity of ≥98% (HPLC), as well as malachite green and bromophenol blue dyes, were purchased from the same supplier. All buffer solutions in this work were prepared with double-distilled water (DDW).

3.2. Instrumentation and Methods

Fourier-transform infra-red spectroscopy with attenuated total reflection (FTIR-ATR): The FTIR-ATR spectra were collected in a Perkin-Elmer Spectrum 100 apparatus (Waltham, MA, USA) using a horizontal ATR attachment with ZnSe crystal. Spectra were acquired between 4000 and 600 cm^{-1} accumulating up to 16 spectra with a resolution of 2 cm^{-1}. The PA4 samples were studied in the form of fine powders.

Ultraviolet–visible spectroscopy (UV–VIS): The UV–VIS analysis was performed on a Shimadzu UV-2501 PC spectrometer (Kyoto, Japan). The absorbance at λ = 286 nm of aqueous solutions placed in quartz cuvettes was measured to determine the protein content and calculate the immobilization efficiency (IE). The assessment of the laccase activity was performed using ABTS as substrate, measuring the absorbance at λ = 414 nm as previously indicated by Claus et al. [19]. One unit of laccase activity expressed in µkatals corresponds to the amount of enzyme transforming 1 µmol ABTS per second at pH 5.0 and 25 °C. The absorbance in decolorization experiments was measured at the respective wavelength of each dye that was λ = 616 nm for malachite green and λ = 591 nm for bromophenol blue.

Scanning electron microscopy (SEM): The SEM studies were performed in a NanoSEM-200 apparatus of FEI Nova (Hillsboro, OR, USA) using mixed secondary electron/back-scattered electron in-lens detection. The pulverulent samples were observed after sputter-coating with Au/Pd alloy in a 208 HR equipment of Cressington Scientific Instruments (Watford, UK) with high-resolution thickness control.

Thermo-gravimetric analysis (TGA): The real iron load, R_L, and the thermal stability of all neat PA4 micro-particulate supports and laccase@PA4 MP samples were established by means of thermo-gravimetric analysis (TGA) in a Q500 gravimetric balance (TA Instruments, New Castle, DE, USA), heating the samples in the 40–600 °C range at 20 °C/min in a nitrogen atmosphere. The R_L was calculated according to Equation (2)

$$R_L = (R_i - R_{PA4}) \times 100, \tag{2}$$

where R_{PA4} is the carbonized residue at 600 °C of the neat PA4 particles and R_i represents the carbonized residue of the respective Fe- or Fe_3O_4—containing PA4 MP or laccase@PA4 MP samples.

Differential scanning calorimetry (DSC): The DSC measurements were carried out in a 200 F3 equipment of Netzsch (Selb, Germany) at a heating/cooling rate of 10 °C/min under nitrogen purge.

The samples were heated to 290 °C, cooled down to 0 °C, and then heated back to 290 °C. The typical sample weights were in the 10–15 mg range.

Synchrotron X-ray studies: Synchrotron wide- (WAXS) and small-angle X-ray scattering (SAXS) measurements were performed in the NCD-SWEET beamline of the ALBA synchrotron facility in Barcelona, Spain [49]. Two-dimensional detectors were used, namely LH255-HS (Rayonix, Evanston, IL, USA) and Pilatus 1M (Dectris, Baden Daettwil, Switzerland) for registering the WAXS and SAXS patterns, respectively. The sample-to-detector distance was set to 111.7 mm for WAXS and 2700 mm for SAXS measurements, the λ of the incident beam being 0.1 nm and the beam size 0.35 × 0.38 mm (h × v). The 2D data were reduced to 1D data using pyFAI software [50]. For processing of the WAXS and SAXS patterns, the commercial package Peakfit 4.12 (2016) by SYSTAT (San Jose, CA, USA) was implemented.

3.3. Sample Preparation and Activity Testing

3.3.1. Synthesis of Empty or Magnetic PA4 MP

The low-temperature AAROP of 2PD to empty PA4 MP or to PA4 MP with magnetic response was described in detail in a previous publication [24]. Generally, 0.2 mol of 2PD were stirred with 1 wt % of Fe micro- or Fe_3O_4 nanoparticles at 25 °C for 30 min. Subsequently, the C20 activator and the DL initiator were added under stirring, in an inert atmosphere. Then, the temperature of the reaction mixture was set to 40 °C and the pressure to 50 mbar for the next 6 h. After AAROP completion, the resulting reaction mixture was dispersed in acetone and filtered, followed by a two-fold wash with methanol, thus removing most of the unreacted monomer. In order to eliminate the PA4 oligomers, the resulting fine powders were extracted with methanol in a Soxhlet for 4 h, dried in vacuum, and stored in a desiccator. For the synthesis of the empty PA4 MP, the procedure was the same but without addition of any magnetic nanoparticles. Altogether, three types of PA4 MP supports were synthesized: PA4, PA4-Fe, and PA4-Fe_3O_4. Some basic characteristics of these samples are presented in Table 1.

3.3.2. Immobilization of Laccase by Physical Adsorption

A typical immobilization was carried out by first preparing the laccase solution in DDW with concentration of 2 mg/mL. 200 mg of each PA4 MP sample were introduced into 5 mL of this laccase solution and the three sample tubes were incubated at room temperature for 24 h using a laboratory orbital shaker. Thereafter, the samples were centrifuged, the supernatant was decanted, and the laccase-immobilized PA4 MP samples were washed two times with distilled water to remove the non-adsorbed laccase. UV analysis was performed to determine the residual laccase and calculate the immobilization efficiency, IE, expressed as

$$IE = \frac{C_0 - C_s}{C_0} \times 100,\ \% \tag{3}$$

where C_0 is the starting laccase concentration and C_s is the laccase content in the resultant supernatant after completion of the immobilization process. Two methods were applied for the determination of C_s. In the first one, the absorbance at λ = 286 nm characteristic for the proteic part of the enzyme was measured and the laccase concentration was determined using a standard calibration plot. In the second method, the laccase activity toward ABTS was tested and C_s was assessed using an activity standard calibration plot. Altogether, three samples of laccase immobilized by adsorption on PA4 MP were prepared and designated as PA4-iL, PA4-Fe-iL, and PA4-Fe_3O_4-iL. They were stored in semi-dried conditions at 4 °C with basic characteristics listed in Table 2.

3.3.3. Immobilization of Laccase by Entrapment

The entrapment of laccase into the PA4 MP was carried out in situ during their synthesis by AAROP. Thus, 1.0 wt % of the enzyme in respect to the monomer was used, stirring their mixture for 30 min in an inert atmosphere at room temperature. Thereafter, the respective magnetic micro/nano particles (if necessary) and the DL/C20 catalytic system were introduced and the AAROP was conducted under the conditions described in Section 2.3.1. After 6 h of polymerization, the raw product was dispersed in acetone and filtered. The resulting laccase-entrapped PA4 MP samples were washed three times with DDW and stored at 4 °C. The water from each washing was stored separately and further subjected to UV analysis to determine the efficiency of the laccase entrapment (EE) using the equation

$$EE = \frac{(C_0 - \Sigma\ C_i)}{C_0} \times 100,\ \% \quad (4)$$

where C_0 is the starting laccase concentration in the reaction mixture and C_i (i = 1–3) is the laccase content in each DDW washing. To assess the content of the non-entrapped laccase, the two methods already described in Section 2.3.2. were applied. Thus, three PA4 MP samples with entrapped laccase were synthesized, namely: PA4-eL, PA4-Fe-eL, and PA4-Fe3O4-eL, with basic characteristics given in Table 3.

3.3.4. Laccase Activity Assay

Native laccase and all conjugates of laccase-immobilized PA4@-iL and entrapped PA4@-eL were assayed for their activities using ABTS as a color-generating substrate. The rate of the formation of ABTS-cation radical ($ABTS^{+\cdot}$) due to the catalytic action of laccase was proportional to the enzyme activity. In a typical assay of free laccase, 0.9 mL of the enzyme solution (1.0 mg/mL in 0.0125 M phosphate buffer, PB, pH 5) were introduced into the spectrometer cell, followed by 0.1 mL of freshly prepared 5 mM solution of ABTS in DDW. During the next 2 min the absorbance at 414 nm (ε_{414} = 36,000 $M^{-1}{\cdot}cm^{-1}$) was measured every 5 s. To determine the activity of the PA4@iL and PA4@eL conjugates, 20 mg of each sample were added to 1.0 mL of 0.0125 M PB, pH 5. The mixtures were stirred for 5 min followed by addition of 0.1 mL of ABTS solution. Then, the samples were centrifuged and 1 mL from the decanted supernatant was subjected to UV analysis to measure the absorbance at 414 nm for 10 min. In order to assay the amount and the activity of the laccase that possibly leached from both adsorption-immobilized and entrapped laccase samples, they were subjected to the following procedure: 50 mg of the respective PA4@iL or PA4@eL sample were dispersed in 5 mL PB, pH 0.0125M, pH 5, and shaken at 37 °C for 15 min. Thereafter, the system was centrifuged and the supernatant decanted and subjected to UV–VIS to determine the laccase concentration based on activity test calibration curve. Then, new portion of fresh PB was added and the above procedure was repeated up to five consecutive cycles.

In all tests the enzyme activities were expressed in microkatals. One microkatal of enzyme activity (μkat) was defined as the amount of enzyme that converted 1.0 μmol of ABTS to $ABTS^{+\cdot}$ per second, at pH 5 and 25 °C. The enzyme activity assay was always performed in duplicate, and the standard deviations in measurements were consistently below 3%. Throughout this study, enzyme activities are expressed in μkat per milliliter ($\mu kat{\cdot}mL^{-1}$).

3.3.5. Application of Free, Adsorbed, and Entrapped Laccase

The enzyme-catalyzed decolorization of two structurally different dyes -malachite green (λ_{max} = 616 nm) and bromophenol blue (λ_{max} = 591 nm) were studied in the presence of free, immobilized and entrapped laccases without any mediator. The process was followed using UV–VIS spectroscopy by measuring the decrease in the λ_{max} absorbance of each dye. The reaction mixture for the decolorization contained a final concentration of 0.0153 mg/mL of individual dye in 0.0125 M PB, pH 5, and 1.0 mg free laccase (0.5 U) or 10–30 mg of adsorption immobilized or entrapped PA4@laccase conjugates (~0.5 U) in a total volume of 1.3 mL. All reaction mixtures were incubated at 30 °C for 24 h under shaking. Aliquots of 1 mL were withdrawn from each sample at certain intervals to measure the residual dye concentration. After the UV–VIS measurement the analyte was returned to the reaction mixture. After 24 h the immobilized or entrapped samples were reused for a second decolorization cycle. For that purpose, the liquid phase was drained, the immobilized or entrapped laccase conjugates were washed consecutively with 2 × 1 mL ethanol, DDW and 0.0125M PB, pH 5 to remove the adsorbed dyes, and then introduced into a fresh dye solution. All UV–VIS measurements were performed twice to determine the standard deviation. The percentage of decolorization ΔA was calculated as

$$\Delta A = \frac{A_i - A_t}{A_i} \times 100,\ \% \tag{5}$$

where, A_i is the initial absorbance of the dye at the respective wavelength, and A_t is the absorbance of the test sample.

4. Conclusions

This is the first study reporting on PA4 porous microparticles as effective supports for enzyme immobilization and describes PA4-laccase conjugates with all aspects of their preparation and characterization. The main objective was to investigate three new micron-sized particulate porous supports based on PA4 obtained by low-temperature AAROP, with and without magnetic susceptibility, establishing whether or not they are suitable for laccase immobilization. The structure and morphology were investigated by microscopic, spectral, thermal, and synchrotron WAXS/SAXS technics.

Two immobilization strategies—i.e., by adsorption or entrapment—were applied. Comparing the adsorption-immobilized (PA4@iL) and entrapped (PA4@eL) samples, an important difference in the synthetic procedure of the PA4@eL should be noted. There, the polymerization was carried out in the presence of the enzyme, whereby the PA4 is formed around the laccase molecules leading to their distribution within the shell of the porous microparticles. In the case of PA4@iL, the enzyme is adsorbed onto the surface and fills the pores of prefabricated PA4 microparticles. The SEM studies confirm these morphological differences. On the other hand, the WAXS/SAXS studies provide evidence for the different way of immobilization of the enzyme in PA4@eL and PA4@iL.

The adsorption strategy resulted in high content of immobilized laccase upon the PA4 supports (27–40 mg/g), the immobilization efficiency being in the range of 54–81%. The entrapment strategy produced 60–70% immobilization efficiency with enzyme content of 14–18 mg/g support. The laccase-adsorbed samples PA4@iL showed up to three times higher specific activity as compared to their entrapped laccase counterparts PA4@eL, whereby the highest specific activity of 0.09 μkat·mL^{-1} was registered with PA4-Fe-iL sample. The relative laccase activity of the PA4-based conjugates determined in relation to the free enzyme was between 16–41% for the PA4@eL samples and 38–60% for the PA4@iL series. The PA4@eL samples displayed >93% enzyme retention after five cycles of incubation, whereas for the PA4@iL series this value was ~60%. The potential application of the adsorption- and entrapment-immobilized PA4-laccase conjugates was tested in the enzymatic decolorization of two synthetic dyes. All laccase conjugates displayed excellent decolorization of malachite green dyestuff reaching ~100% in 15 min, which was mostly due to dye adsorption upon the PA4 support. The decolorization after 24 h of the bromophenol reached 55% by PA4@iL and 85% by PA4@eL samples. The reuse of the laccase-PA4 conjugates in a second consecutive decolorization test resulted in only up to 10% decrease in their effectiveness. Supposing that this will be the decrease after each cycle of utilization, it seems that the laccase-PA4 conjugates will be sufficiently active to use for 3–5 more consecutive cycles.

The present study justifies further investigations on PA4 microparticles as effective enzyme supports for biotechnological applications. Compared to the common polyamides as PA6, PA66, and PA12, PA4 has the advantage of biodegradability and possesses a crystalline structure, which is a closer structural analogue of proteic biomolecules. The possibility to apply different immobilization strategies and to use magnetic field as external stimulus could open the way to the use of PA4 microparticulate supports of this study in smart green catalytic systems.

Supplementary Materials: The following are available online at http://www.mdpi.com/2073-4344/10/7/767/s1. Figure S1: Chemical reactions occurring during AAROP of 2PD to PA4 microparticles; Figure S2: Histograms of the average size (d_{max}) and roundness (d_{max}/d_{min}) distributions for PA4 MP supports based on optical microscopy; Figure S3: Disappearance of the scattering peak at q = 3.73 nm^{-1} in the SAXS patterns after melting and recrystallization of PA4 MP support; Figure S4. Structures of the dyes used in the discoloration studies with PA4@iL and PA4@eL laccase conjugates; Figure S5. PA4-Fe-eL sample fast removal from DDW suspension by means of a constant magnet; Table S1: Data from BET for the empty PA4 supports; Table S2: Z-potential values of empty particulate PA4-based supports; Table S3: Consolidated DSC data from an initial heating scan, subsequent cooling scan, and a second heating scan at 10 deg/min.

Author Contributions: Conceptualization, N.D., Z.D., and I.G.; Data curation, N.D., J.B., and D.S.; Formal analysis, N.D., J.B., and Z.D.; Funding acquisition, Z.D., N.D., and I.G.; Investigation, N.D., J.B., D.S., and M.M.; Methodology, N.D., Z.D., and I.G.; Software, M.M.; Supervision, Z.D. and I.G.; Validation, N.D., M.M., and Z.D.; Writing—original draft preparation, N.D. and Z.D.; Writing—review and editing, I.G. and Z.D. All authors have read and agreed to the published version of the manuscript.

Funding: The authors gratefully acknowledge the financial support of the project TSSiPRO NORTE-01-0145-FEDER-000015, supported by the regional operation program NORTE2020, under the Portugal 2020 Partnership Agreement, through the European Regional Development Fund, as well as the support by National Funds through *Fundação para a Ciência e Tecnologia* (FCT), project UID/CTM/50025/2019. D.S. and I.G. wish to thank the Research Foundation of the State of New York for partial funding through Materials Network of Excellence. N.D. is also grateful for the personal program-contract CTTI-51/18-IPC.

Conflicts of Interest: The authors declare no conflict of interest. The funders had no role in the design of the study, in the collection, analyses, or interpretation of data, in the writing of the manuscript, or in the decision to publish the results.

References

1. Bull, A.T.; Bunch, A.W.; Robinson, G.K. Biocatalysts for clean industrial products and processes. *Curr. Opin. Microbiol.* **1999**, *2*, 246–251. [CrossRef]
2. Sheldon, R.A.; Rantwijk, F.V. Biocatalysis for sustainable organic synthesis. *Aust. J. Chem.* **2004**, *57*, 281–289. [CrossRef]
3. Torres-Salas, P.; Monte-Martinez, A.; Cutiño-Avila, B.; Rodriguez-Colinas, B.; Alcalde, M.; Ballesteros, A.O.; Plou, F.J. Immobilized Biocatalysts–novel approaches and tools for binding enzymes to supports. *Adv. Mater.* **2011**, *23*, 5275–5282. [CrossRef]
4. Ge, J.; Lu, D.; Liu, Z.; Liu, Z. Recent advances in nanostructured biocatalysts. *Biochem. Eng. J.* **2009**, *44*, 53–59. [CrossRef]
5. Hanefeld, U.; Gardossi, L.; Magner, E. Understanding enzyme immobilization. *Chem. Soc. Rev.* **2009**, *38*, 453–468. [CrossRef] [PubMed]
6. Hudson, S.; Magner, E.; Cooney, J.; Hodnett, B. Methodology for the immobilization of enzymes onto mesoporous materials. *J. Phys. Chem. B* **2005**, *109*, 19496–19506. [CrossRef]
7. Dwevedi, A. Basics of enzyme immobilization. In *Enzyme Immobilization*; Springer: Cham, Switzerland, 2016; Chapter 2; pp. 21–44, ISBN 978-3-319-41416-4. [CrossRef]
8. Sigurdardóttir, S.B.; Lehmann, J.; Ovtar, S.; Grivel, J.C.; Negra, M.; Kaiser, A.; Pinelo, M. Enzyme immobilization on inorganic surfaces for membrane reactor applications: Mass transfer challenges, enzyme leakage and reuse of materials. *Adv. Synth. Catal.* **2018**, *360*, 2578–2607. [CrossRef]
9. Shakya, A.K.; Nandakumar, K.S. An update on smart biocatalysts for industrial and biomedical applications. *J. R. Soc. Interface* **2018**, *15*, 20180062. [CrossRef] [PubMed]
10. Gitsov, I.; Hamzik, J.; Ryan, J.; Simonyan, A.; Nakas, J.P.; Omori, S.; Krastanov, A.; Cohen, T.; Tanenbaum, S.W. Enzymatic nanoreactors for environmentally benign biotransformations – 1. Formation and catalytic activity of supramolecular complexes of laccase and linear-dendritic blockcopolymers. *Biomacromolecules* **2008**, *9*, 804–811. [CrossRef]
11. Scheibel, D.M.; Gitsov, I. Polymer-assisted biocatalysis: Effects of macromolecular architectures on the stability and catalytic activity of immobilized enzymes toward water-soluble and water-insoluble substrates. *ACS Omega* **2018**, *3*, 1700–1709. [CrossRef]
12. Mate, D.M.; Alcalde, M. Laccase: A multi-purpose biocatalyst at the forefront of biotechnology. *Microb. Biotechnol.* **2017**, *10*, 1457–1467. [CrossRef] [PubMed]

13. Couto, S.L.; Herrera, J.L. Lacasses in the textile industry. *Biotechnol. Mol. Biol. Rev.* **2006**, *1*, 115–120. Available online: https://academicjournals.org/journal/BMBR/article-full-text-pdf/1C6FD3040214 (accessed on 5 March 2020).
14. Gitsov, I.; Simonyan, A.; Wang, L.; Krastanov, A.; Tanenbaum, S.W.; Kiemle, D. Polymer-assisted biocatalysis: Unprecedented enzymatic oxidation of fullerene in aqueous medium. *J. Polym. Sci. Part A Polym. Chem.* **2012**, *50*, 119–126. [CrossRef]
15. Gitsov, I.; Wang, L.; Vladimirov, N.; Simonyan, A.; Kiemle, D.J.; Schütz, A. "Green" synthesis of unnatural poly(amino acid)s with zwitterionic character and pH-responsive solution behavior, mediated by linear-dendritic laccase complexes. *Biomacromolecules* **2014**, *15*, 4082–4095. [CrossRef] [PubMed]
16. Scheibel, D.M.; Gitsov, I. Unprecedented enzymatic synthesis of perfectly structured alternating copolymers via "green" reaction cocatalyzed by laccase and lipase compartmentalized within supramolecular complexes. *Biomacromolecules* **2019**, *20*, 927–936. [CrossRef] [PubMed]
17. Gitsov, I.; Simonyan, A. "Green" Synthesis of Bisphenol Polymers and Copolymers, Mediated by Supramolecular Complexes of Laccase and Linear-Dendritic Block Copolymers. In *Green Polymer Chemistry*; Cheng, R.A., Gross, H.N., Eds.; American Chemical Society: Washington, DC, USA, 2013.
18. Bilal, M.; Rasheed, T.; Nabeel, F.; Iqbal, H.M.N.; Zhao, Y. Hazardous contaminants in the environment and their laccase-assisted degradation–A review. *J. Environ. Manag.* **2019**, *234*, 253–264. [CrossRef] [PubMed]
19. Claus, H.; Faber, G.; König, H. Redox-mediated decolorization of synthetic dyes by fungal laccases. *Appl. Microbiol. Biotechnol.* **2002**, *59*, 672–678. [CrossRef] [PubMed]
20. Maloney, J.; Dong, C.; Campbell, A.S.; Dinu, C.Z. Emerging enzyme-based technologies for wastewater treatment. In Green Polymer Chemistry: Biobased Materials and Biocatalysis. *ACS Symp. Ser.* **2015**, *1192*, 73–75.
21. Fatarella, E.; Spinelli, D.; Ruzzante, M.; Pogni, R. Nylon 6 film and nanofiber carriers: Preparation and laccase immobilization performance. *J. Mol. Catal. B Enzym.* **2014**, *102*, 41–47. [CrossRef]
22. Jasni, M.J.F.; Sathishkumar, P.; Sornambikai, S.; Yusoff, A.R.M.; Ameen, F.; Buang, N.A.; Kadir, M.R.A.; Yusop, Z. Fabrication, characterization and application of laccase–nylon 6,6/Fe^{3+} composite nanofibrous membrane for 3,30-dimethoxybenzidine detoxification. *Bioprocess Biosyst. Eng.* **2017**, *40*, 191–200. Available online: https://link.springer.com/article/10.1007%2Fs00449-016-1686-6 (accessed on 20 February 2020). [CrossRef]
23. Silva, C.; Silva, C.J.; Zille, A.; Guebitz, G.M.; Cavaco-Paulo, A. Laccase immobilization on enzymatically functionalized polyamide 6,6 fibers. *Enzym. Microb. Technol.* **2007**, *41*, 867–875. [CrossRef]
24. Dencheva, N.; Braz, J.; Nunes, T.G.; Oliveira, F.D.; Denchev, Z. One-pot low temperature synthesis and characterization of hybrid poly(2-pyrrolidone) microparticles suitable for protein immobilization. *Polymer* **2018**, *145*, 402–4015. [CrossRef]
25. Yamano, N.; Nakayama, A.; Kawasaki, N.; Yamamoto, N.; Aiba, S. Mechanism and characterization of polyamide 4 degradation by Pseudomonas sp. *J. Polym. Environ.* **2008**, *16*, 141–146. Available online: https://link.springer.com/article/10.1007/s10924-008-0090-y (accessed on 10 March 2020). [CrossRef]
26. Tachibana, K.; Urano, Y.; Numata, K. Biodegradability of nylon 4 film in a marine environment. *Polym. Degrad. Stabil.* **2013**, *98*, 1847–1851. [CrossRef]
27. Park, S.J.; Kim, E.Y.; Noh, W.; Oh, Y.H.; Kim, H.Y.; Song, B.K.; Cho, K.M.; Hong, S.H.; Lee, S.H.; Jegal, J. Synthesis of nylon 4 from gamma-aminobutyrate (GABA) produced by recombinant Escherichia coli. *Bioproc. Biosyst. Eng.* **2013**, *36*, 885–892. Available online: https://link.springer.com/article/10.1007/s00449-012-0821-2 (accessed on 10 March 2020). [CrossRef] [PubMed]
28. Dencheva, N.V.; Oliveira, F.D.; Braz, J.F.; Denchev, Z. Bovine serum albumin-imprinted magnetic poly(2-pyrrolidone) microparticles for protein recognition. *Eur. Polym. J.* **2020**, *122*, 109375. [CrossRef]
29. Kim, N.; Kim, J.H.; Nam, S.W.; Jeon, B.S.; Kim, Y.J. Preparation of nylon 4 microspheres via heterogeneous polymerization of 2-pyrrolidone in a paraffin oil continuous phase. *J. Ind. Eng. Chem.* **2015**, *28*, 236–240. [CrossRef]
30. Costa, J.; Lima, M.J.; Sampaio, M.J.; Neves, M.C.; Faria, J.L.; Morales-Torres, S.; Tavares, A.P.M.; Silva, C.G. Enhanced biocatalytic sustainability of laccase by immobilization on functionalized carbon nanotubes/polysulfone membranes. *Chem. Eng. J.* **2019**, *355*, 974–985. [CrossRef]

31. Liers, C.; Ullrich, R.; Pecyna, M.; Schlosser, D.; Hofrichter, M. Production, purification and partial enzymatic and molecular characterization of a laccase from the wood-rotting ascomycete Xylaria polymorpha. *Enz. Microb. Technol.* **2007**, *41*, 785–793. [CrossRef]
32. Atalla, M.; Zeinab, H.; Eman, R.; Amani, A.; Abeer, A. Characterization and kinetic properties of the purified Trematosphaeria mangrovei laccase enzyme. *Saudi J. Biol. Sci.* **2013**, *20*, 373–381. [CrossRef]
33. Qiu, H.; Xu, C.; Huang, X.; Ding, Y.; Qu, Y.; Gao, P. Immobilization of laccase on nanoporous gold: Comparative studies on the immobilization strategies and the particle size effects. *J. Phys. Chem. C* **2009**, *113*, 2521–2525. [CrossRef]
34. Piontek, K.; Antorini, M.; Choinowski, T. Crystal structure of a laccase from Trametes versicolor at 1.90 Å resolution containing a full complement of coppers. *J. Biol. Chem.* **2002**, *277*, 37663–37669. [CrossRef] [PubMed]
35. Fernández-Fernández, M.; Sanromán, M.Á.; Moldes, D. Recent developments and applications of immobilized laccase. *Biotechnol. Adv.* **2013**, *31*, 1808–1825. [CrossRef] [PubMed]
36. Kawasaki, N.; Nakayama, A.; Yamano, N.; Takeda, S.; Kawata, Y.; Yamamoto, N.; Aiba, S. Synthesis, thermal and mechanical properties and biodegradation of branched PA4. *Polymer* **2005**, *46*, 9987–9993. [CrossRef]
37. Tachibana, K.; Hashimoto, K.; Tansho, N.; Okawa, H. Chemical modification of chain end in nylon 4 and improvement of its thermal stability. *J. Polym. Sci. Part A Polym. Chem.* **2011**, *49*, 2495–2503. [CrossRef]
38. Bellinger, M.A.; Waddon, A.J.; Atkins, E.D.T.; MacKnight, W.J. Structure and morphology of nylon 4 chain-folded lamellar crystals. *Macromolecules* **1994**, *27*, 2130–2135. [CrossRef]
39. Schulz, G.E.; Schirmer, R.H. *Principles of Protein Structure*; Springer Verlag: New York, NY, USA, 1987. [CrossRef]
40. Bartczak, Z.; Galeski, A.; Argon, A.S.; Cohen, R.E. On the plastic deformation of the amorphous component in semicrystalline polymers. *Polymer* **1996**, *37*, 2113–2123. [CrossRef]
41. Alexander, L.E. *X-ray Diffraction Methods in Polymer Science*; Wiley-Interscience: New York, NY, USA, 1969.
42. Dencheva, N.; Nunes, T.; Oliveira, M.J.; Denchev, Z. Microfibrillar composites based on polyamide/polyethylene blends. 1. Structure investigations in oriented and isotropic PA6. *Polymer* **2005**, *46*, 887–901. [CrossRef]
43. Fredericks, J.; Doyne, T.H.; Spague, R.S. Crystallographic studies of nylon 4. II. On the β and δ polymorphs of Nylon 4. *J. Polym. Sci. Polym. Phys.* **1966**, *4*, 913–922. [CrossRef]
44. Fornes, T.D.; Paul, D.R. Crystallization behavior of Nylon nanocomposites. *Polymer* **2003**, *44*, 3945–3961. [CrossRef]
45. Teerapatsakul, C.; Parra, C.R.; Keshavarz, T.; Chitradon, L. Repeated batch for dye degradation in an airlift bioreactor by laccase entrapped in copper alginate. *Int. Biodeterior. Biodegrad.* **2017**, *120*, 52–57. [CrossRef]
46. Bagewadi, Z.K.; Mulla, S.I.; Ninnekar, H.Z. Purification and immobilization of laccase from Trichoderma harzianum strain HZN10 and its application in dye decolorization. *J. Genet. Eng. Biotechnol.* **2017**, *15*, 139–150. [CrossRef] [PubMed]
47. Forootanfar, H.; Moezzi, A.; Khozani, M.; Mahmoudjanlou, Y.; Ameri, A.; Niknejad, F.; Faramarzi, M.A. Synthetic dye decolorization by three sources of fungal laccase. *Iran. J. Environ. Health Sci. Eng.* **2012**, *9*, 27–37. Available online: https://link.springer.com/article/10.1186%2F1735-2746-9-27 (accessed on 18 January 2020). [CrossRef]
48. Peralta-Zamora, P.; Pereira, C.M.; Tiburtius, E.; Moraes, S.G.; Rosa, M.A.; Minussi, R.C.; Durán, N. Decolorization of reactive dyes by immobilized laccase. *Appl. Catal. B Environ.* **2003**, *42*, 131–144. [CrossRef]
49. González, J.B.; González, N.; Colldelram, C.; Ribó, L.; Fontserè, A.; Manas, G.J.; Villanueva, J.; Llonch, M.; Peña, G.; Gevorgyan, A.; et al. NCD-SWEET Beamline Upgrade. In Proceedings of the 10th Mechanical Engineering Design Synchrotron Radiation Equipment Instruments, Paris, Franch, 25–29 June 2018; pp. 374–376. [CrossRef]
50. Ashiotis, G.; Deschildre, A.; Nawaz, Z.; Wright, J.P.; Karkoulis, D.; Picca, F.E.; Kieffer, J. The fast azimuthal integration Python library: pyFAI. *J. Appl. Crystallogr.* **2015**, *48*, 510–519. [CrossRef] [PubMed]

Article

Engineering of Bifunctional Enzymes with Uricase and Peroxidase Activities for Simple and Rapid Quantification of Uric Acid in Biological Samples

Thanawat Phuadraksa [1], Jurairat Chittrakanwong [1,2], Kittitouch Tullayaprayouch [1], Naruthai Onsirisakul [3], Sineewanlaya Wichit [1] and Sakda Yainoy [1,*]

1 Department of Clinical Microbiology and Applied Technology, Faculty of Medical Technology, Mahidol University, Bangkok 10700, Thailand; thanawat.phd@student.mahidol.ac.th (T.P.); jurairat.chi@gmail.com (J.C.); kittitouch.tul@student.mahidol.ac.th (K.T.); sineewanlaya.wic@mahidol.ac.th (S.W.)

2 Program in Applied Biological Sciences: Environmental Health, Chulabhorn Graduate Institute, Chulabhorn Royal Academy, Bangkok 10210, Thailand

3 Center for Research and Innovation, Faculty of Medical Technology, Mahidol University, Bangkok 10700, Thailand; naruthai.ons@mahidol.ac.th

* Correspondence: sakda.yai@mahidol.ac.th; Tel.: +66-2-441-4370

Received: 25 March 2020; Accepted: 13 April 2020; Published: 14 April 2020

Abstract: Serum uric acid (SUA) is an important biomarker for prognosis and management of gout and other diseases. The development of a low-cost, simple, rapid and reliable assay for SUA detection is of great importance. In the present study, to save the cost of enzyme production and to shorten the reaction time for uric acid quantification, bifunctional proteins with uricase and peroxidase activities were engineered. In-frame fusion of *Candida utilis* uricase (CUOX) and *Vitreoscilla* hemoglobin (VHb) resulted in two versions of the bifunctional protein, CUOX-VHb (CV) and VHb-CUOX (VC). To our knowledge, this is the first report to describe the production of proteins with uricase and peroxidase activities. Based on the measurement of the initial rates of the coupled reaction (between uricase and peroxidase), CV was proven to be the most efficient enzyme followed by VC and native enzymes (CUOX+VHb), respectively. CV was further applied for the development of an assay for colorimetric detection of SUA, which was based on VHb-catalyzed oxidation of Amplex Red in the presence of hydrogen peroxide (H_2O_2). Under the optimized conditions, the assay exhibited a linear relationship between the absorbance and UA concentration over the range of 2.5 to 50 μM, with a detection limit of 1 μM. In addition, the assay can be performed at a single pH (8.0) so adjustment of the pH for peroxidase activity was not required. This advantage helped to further reduce costs and time. The developed assay was also successfully applied to detect UA in pooled human serum with the recoveries over 94.8%. These results suggest that the proposed assay holds great potential for clinical application.

Keywords: uricase; *Vitreoscilla* hemoglobin; bifunctional enzyme; uric acid; colorimetric detection

---

1. Introduction

Uric acid (UA) is the end product of purine nucleotide catabolism in the human body. The normal ranges of serum uric acid (SUA) in females and males are 2.4–6.0 mg/dL (143–357 μM) and 3.4–7.0 mg/dL (202–416 μM), respectively. Elevation of SUA greater than 6.8 mg/dL (405 μM) is recognized as hyperuricemia, which is known to associate with various diseases, e.g., gout, nephrolithiasis, kidney failure, hypertension, and cardiovascular diseases [1]. Therefore, monitoring of SUA levels is essential for the prevention, diagnosis and treatment of these diseases. To measure the SUA level, various analytical methods such as spectral (UV, colorimetry, fluorescence), electrochemical (voltammetry, electrochemiluminescence, surface plasmon resonance), chromatographic (liquid and gas phase) and capillary electrophoresis methods have been described. The details of these techniques have been extensively reviewed elsewhere [2]. Although various modern techniques have been proposed, the most widely employed technique for UA determination by automated systems in clinical laboratories is classical spectrophotometry [3]. This is because the technique is designed with an automation-friendly protocol and the cost is more favorable. The principle of this technique is based on the cascade reactions of uricase and peroxidase enzymes [4,5].

In the presence of oxygen, uricase oxidizes uric acid to allantoin, carbon dioxide and hydrogen peroxide (H_2O_2). The produced H_2O_2 is then consumed by peroxidase for the oxidation of a chromogenic substrate to a color product, which can be measured spectrophotometrically. Different types of chromogenic substrate (e.g., 3-methyl-2-benzothiazolinone hydrazine (MBTH) plus N,N-dimethylaniline (DMA) [6], 2,4-dichlorophenol (2, 4-DCP) plus 4-aminophenazone [7], p-hydroxybenzoate (PHBA) plus 4-aminoantipyrine (4-APP) [8], 3,5-dichloro-2-hydroxybenzene sulfonic acid plus 4-aminophenazone [9], o-dianisidine [4], 2,2′-azino-bis (3-ethylbenzthiazoline-6-sulphonic acid) (ABTS) [10], and 3,3′,5,5′-tetramethylbenzidine (TMB) [11]) and different sources of uricase have been explored. Among the tested enzymes, microbial uricase (e.g., *Aspergillus flavus*, *Candida utilis*, *Bacillus fastidiosus* and *Arthrobacter globiformis*) has been shown to be superior to uricase from other higher organisms; microbial enzymes exhibited more stability and higher enzymatic activity [12–15]. Generally, uricase is more stable and exhibits the highest enzymatic activity at pH ≥ 8.0 [16], thus the measurement of UA using uricase is performed at alkaline pH. Besides uricase, the other key enzyme of the cascade reactions is a peroxidase. As opposed to the wide variety of uricase sources, the only source of peroxidase is the horseradish plant (horseradish peroxidase; HRP). This may be due to the fact that HRP has high activity, high stability and can be produced from the farmed plant (*Armoracia rusticana*). Nevertheless, the activity of HRP is strictly dependent on pH. The optimal pH range for the HRP reaction is 5.5–6.0 [17]. As a consequence, to assay SUA using uricase and HRP, the measurement must be done with a two-step procedure. Uricase reaction is performed first at pH ≥ 8.0 then the reaction pH is changed to <6 to facilitate HRP action [6]. This makes the uricase/HRP-based method more complicated and several reaction buffers with different pH must be prepared. In fact, one could also perform the test at a more neutral pH (7.0–7.5) [18]. However, at this pH range, HRP activity is reduced to less than 80% [17]. Besides the pH incompatibility issue, classical production of HRP from *Armoracia rusticana* roots is undesirable, since sufficient production of high-quality enzyme is hampered by seasonal availability, long cultivation times, low production yields and low product homogeneity [19,20]. To produce a uniform enzyme with defined characteristics, the production of recombinant HRP (rHRP) using *Saccharomyces cerevisiae* [21] and *Pichia pastoris* [20–22] has been attempted. The advantage of using yeast cells is that the recombinant protein can obtain glycosylation as produced in plant cells. Nonetheless, glycosylation of rHRP in yeast cells has been reported to occur non-specifically, resulting in more complicated downstream processes [20]. Besides yeast cells, production of rHRP in *E. coli* have also been explored [23–27]. The enzyme can be produced up to 1 g/1-Liter culture. However, the expressed protein formed inclusion bodies (IBs), which need to be recovered through cumbersome protein-refolding processes. Although up to 10% of rHRP can be successfully recovered from IBs, the enzyme specific activity was markedly reduced [26]. Besides direct expression in cytoplasm, rHRP was also targeted to express in the periplasmic space of *E. coli*. Nevertheless, the production yields and enzymatic activity of the enzyme were even lower than that of the enzyme recovered from IBs [26].

Previously, the peroxidase activity of *Vitreoscilla* hemoglobin (VHb) has been reported by our group [28–30]. The protein can catalyze a wide variety of reducing substrates such as ferrocene carboxylic acid, dopamine, L-dopa, ABTS [28] and Amplex Red [29]. Although the peroxidase activity of VHb is much lower than that of HRP, soluble and active protein can be massively produced in *E. coli*. Homogeneous protein with well-defined characteristics can be obtained by simple purification using immobilized metal affinity chromatography (IMAC). The protein can also be engineered as a fusion protein while maintaining high solubility and sufficient peroxidase activity. More importantly, VHb exhibits maximum peroxidase activity at pH 7.0 [28,31], which is closely related to the optimal pH of uricase (8.0–9.0). These observations imply that uricase and VHb may be combined in a single reaction for the detection of UA without the need for changing the reaction pH.

Therefore, in this study, quantification of UA using uricase and VHb was investigated. To reduce the cost for enzyme production and to improve the reaction velocity of uricase-peroxidase cascade reactions, which, in turn, can reduce the reaction time, bifunctional proteins with uricase and peroxidase activities were engineered. The proteins were created by direct fusion of *Candida utilis* uricase (CUOX) and VHb. The details for production, characterization and successful application of the bifunctional protein for detection of UA in human serum are reported here.

2. Results and Discussion

2.1. Production of Native and Bifunctional Proteins with Uricase and Peroxidase Activities

Recombinant CUOX, VHb, CUOX-VHb (CV) fusion protein and VHb-CUOX-VHb (VC) fusion proteins were successfully expressed in *E. coli* BL21 (DE3) and purified to homogeneity. Genes encoding CUOX and VHb were fused in-frame, resulting in direct joining of the C-terminus of the first protein to the N-terminus of the second protein. Schematic representation for the production of a fusion protein (CV) is shown in Figure 1. To facilitate protein purification by IMAC, all proteins were designed to express with a hexahistidine tag at the N-terminus. After IMAC-purification, SDS-PAGE analysis revealed that CUOX, VHb and CV fusion protein had a molecular weight (MW) under denaturing conditions of ~35, ~17 and ~50 kDa, respectively. However, VC fusion protein presented with two protein bands with MW of ~50 kDa and ~25 kDa (data not shown). This result indicated that VC was susceptible to protein degradation, which could be due to steric effects presented within the engineered polypeptide chain. The degrading point is believed to be in the CUOX region because the purified fraction had a MW larger than that of the VHb subunit. Nevertheless, as in other IMAC-purified proteins, the mixture of full-length and degraded VC was subjected to further purification by gel-filtration chromatography. As expected, CUOX, VHb and CV showed a single protein peak in the chromatogram while VC showed two separated protein peaks (data not shown). This result suggested that the full-length and the degraded fractions of VC can be successfully separated from each other by gel-filtration chromatography.

All purified proteins were again subjected to SDS-PAGE analysis, which revealed that they were purified to more than 95% homogeneity and the MW under denaturing condition of CUOX, VHb, CV and VC were ~35, ~17, ~50 and ~50 kDa, respectively (Figure 2a). These MW values were closely related to the values predicted by the ExPASy Compute pI/Mw tool (https://web.expasy.org/compute_pi/), which were ~35.7, ~17.4, ~51.7 and ~51.7 kDa, respectively. A 1-L culture of *E. coli* yielded purified CUOX, VHb, CV and VC up to 413, 118, 51 and 23 mg, respectively. In addition, based on gel-filtration chromatography, MW under the natural condition of all proteins was calculated. The result revealed that CUOX, VHb, CV and VC had natural MW of ~143 kDa, ~35 kDa, ~207 kDa and ~207 kDa, respectively (Figure 2b). This result showed that recombinant CUOX and VHb were produced in their natural forms, which are tetrameric and dimeric forms, respectively. More importantly, this result confirmed that the fusion proteins assembled into the tetrameric complex as proposed in Figure 1. One molecule of the tetrameric complex was composed of four CUOX subunits, arranged in its naturally occurring tetrameric structure, and four VHb subunits, arranged in monomeric rather than dimeric form.

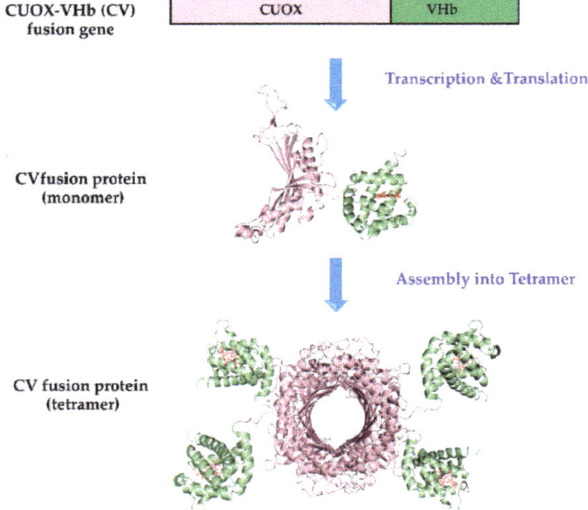

Figure 1. Schematic representation of the strategy for the engineering of bifunctional protein with uricase and peroxidase activities. In-frame fusion of *Candida utilis* uricase (CUOX) and *Vitreoscilla* hemoglobin (VHb) genes results in CUOX-VHb fusion protein, then, four monomers of the fusion protein assemble into a tetrameric complex. Construction of VHb-CUOX fusion protein was also performed by the same technique except that the VHb gene was put in front of the CUOX gene.

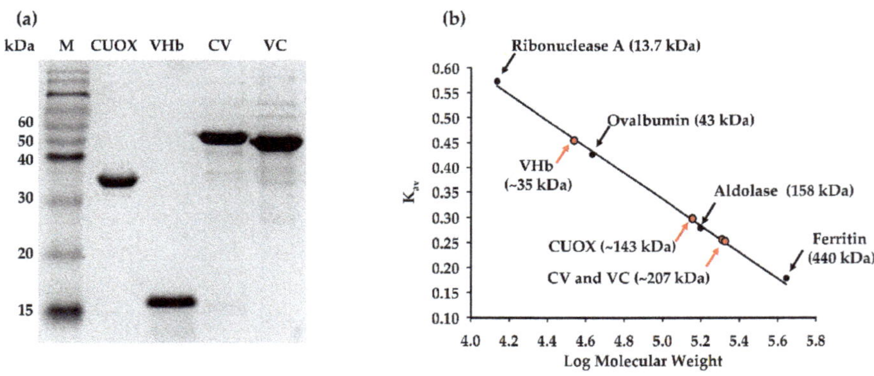

Figure 2. (**a**) SDS-PAGE analysis showing purity after protein purifications and molecular weight under the denaturing condition of native and fusion proteins. Lane 1, protein molecular weight marker; lane 2, purified CUOX; lane 3, purified VHb; lane 4, purified CUOX-VHb fusion protein; and lane 5, purified VHb-CUOX fusion protein. (**b**) The gel-filtration chromatography calibration curve was performed according to the manufacturer's protocol (GE Healthcare Life Sciences, PA, USA). The linear curve representing K_{av} values versus molecular weights of the standard proteins including ribonuclease A, ovalbumin, aldolase and ferritin. K_{av} of CUOX, VHb and fusion proteins are shown in red circles.

Nevertheless, VHb has been known to possess monomer-dimer equilibrium [32] and has been reported to confer maximum peroxidase activity when present in monomeric form [31]. Therefore, the absence of dimeric VHb seemed to have no effect on the peroxidase activity of the fusion proteins. Besides the abovementioned experiments, three-dimensional (3D) structure of the fusion proteins was predicted by the I-TASSER server [33]. The model with the highest C-score was evaluated by various tools [34]. The secondary structure of the predicted model was evaluated by Ramachandran Plot through PROCHECK, which showed that the model had 71.9%, 24.3%, 2.1% and 1.7% of the amino acid residues in most favored regions, additional allowed regions, generously allowed regions and disallowed regions, respectively. The residue-wise local quality of the model structure was evaluated by Pro-Q, which revealed LG-score and MaxSub-score of 2.965 and 0.126, respectively. Evaluation of the predicted structure by ProSA showed that the Z-score value was −6.59, which was within the range observed for a native set of proteins of the same size. The quality of the predicted 3D structure was also evaluated by Verify3D. The result showed that the model possessed 79.3% of amino acid with a 3D-1D profile score ≥0.2. Although the obtained value was not that high, this value was very close to the expected value (80%) of a satisfactory predicted model. Overall, the predicted model had fairly good and sufficient quality for further studies. The predicted model was then used to construct the 3D structure of the tetrameric complex by PyMol 2.3.3 software. The constructed model consisted of four CV-monomers, resulting in four active sites of CUOX and four active sites of VHb (Figure 3). Conversely, the construction of the 3D structure of the tetrameric complex of VC was not possible. The I-TAS SER-predicted model showed an infeasible structure, VHb and CUOX portions leaned closely toward each other, preventing tetramerization of this fusion protein (data not shown). This finding supported the abovementioned steric effect hypothesis on the degradation of VC. Nevertheless, although VC was indicated to be subject to steric effects, more than 20 mg of the protein in the tetrameric complex was successfully purified from a 1-L culture of *E. coli*. The apparent colors of the purified proteins are shown in Figure 4a. CUOX is colorless while the other proteins are red, which is characteristic of heme-containing VHb. The UV–vis absorption spectra of VHb, CV and VC exhibited the Soret peak at 412 nm (Figure 4b), indicating that all proteins contain a mixture of ferrous and ferric hemoglobin [35]. The ratios between absorbance at 412 nm and 280 nm of VHb, CV and VC were 4.3, 0.89 and 0.97, respectively. This result indicated that, as compared to that of VHb, heme content of CV and VC were 4.8- and 4.4-fold decreased, respectively. However, this is not surprising since some part of the fusion protein (CUOX part) did not contain heme prosthetic group. More importantly, both fusion proteins exhibited similar absorption spectra, including the Soret peak (412 nm) and α-bands (540 and 577 nm), to that of native VHb, implying that protein engineering in this study did not alter heme-binding characteristics of VHb.

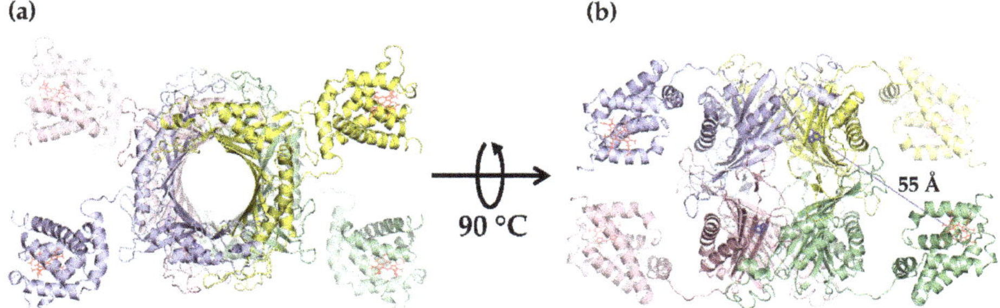

Figure 3. Tetrameric structure of CUOX-VHb fusion protein constructed by assembly of 3D structure of CUOX-VHb predicted by I-TASSER. One color represents one fusion protein chain. Heme prosthetic groups in VHb active sites are represented by red sticks while 5-amino 6-nitro uracil molecules sitting in uricase active sites are represented by blue sticks. (**a**) Top view of the tetramer. (**b**) Side view of the tetramer. The distance between the adjacent CUOX and VHb active sites is 55 Å.

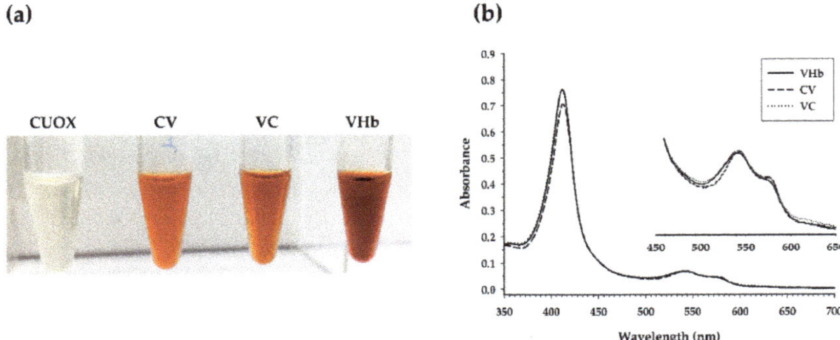

Figure 4. (**a**) The apparent colors of the purified proteins. From left to right: native CUOX, CUOX-VHb fusion protein, VHb-CUOX fusion protein and native VHb. (**b**) UV–vis absorption spectra of the purified proteins (250–700 nm). The Soret peak is represented at 412 nm while the α-bands are represented by the peak between 540 and 577 nm.

2.2. Characterization of Enzymatic Activity and Kinetic Parameters

The engineered proteins exhibited both uricase and peroxidase activities. Optimum pH of catalysis for uricase and peroxidase was 9.0 (Figure 5a) and 7.0 (Figure 5b), respectively. Specific enzymatic activities are shown in Table 1. CV and VC retained uricase activity 69% and 60% of that of CUOX, and retained peroxidase activity 3.4% and 2.7% of that of VHb, respectively. However, when protein molecular weight was taken into account CV and VC revealed relative uricase activity up to 99% and 88% of that of CUOX, and revealed peroxidase activity up to 20% and 15% of that of VHb, respectively (Table 1). These results indicated that the engineering strategy hardly disturbs structure and, hence, the function of uricase in the fusion proteins. Conversely, peroxidase activity of VHb in the fusion proteins was markedly decreased. This is due to the following facts: firstly, the molecular weights of the fusion proteins were ~3 times higher while the numbers of VHb active sites were the same. Thus U/μmol should be used for activity comparison rather than U/mg (Table 1). Secondly, as compared to HRP, the distal histidine and arginine are missing from the VHb active site, which has been suggested to be problematic for substrate binding and account for low activity [30]. This evidence implied that peroxidase activity of VHb is highly sensitive to structural change, which can occur with fusion proteins. To improve peroxidase activity, introduction of distal histidine into VHb active site using site-directed mutagenesis may be attempted [30]. Nevertheless, the obtained peroxidase activities of the fusion proteins were sufficient for further experiments. Notably, uricase and peroxidase activities of CV were higher than those of VC. These results suggested that the structure of CV may be more energetically favorable than that of VC. This assumption was supported by the presence of protein truncation in VC but not in CV. Besides specific activity, kinetic parameters were also studied (Table 2 and Figure 6). As compared to that of CUOX, the apparent K_m of CV and VC were reduced to 65.82% and 58.22% while the k_{cat} increased 1.33-fold and 1.61-fold, respectively. These characteristics resulted in 2-fold and 2.72-fold increase in catalytic efficiency of CV and VC, respectively.

(a)

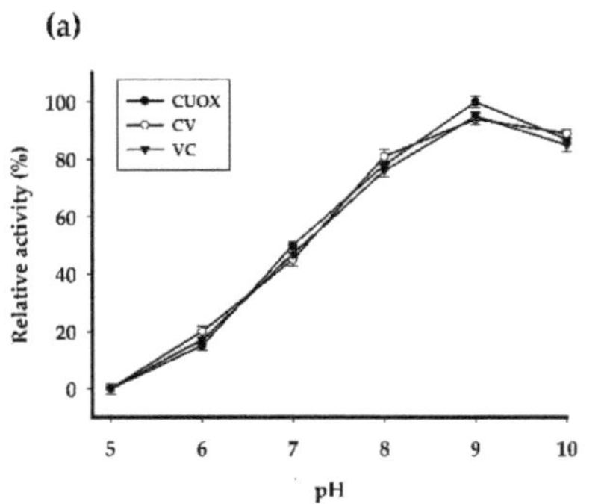

(b)

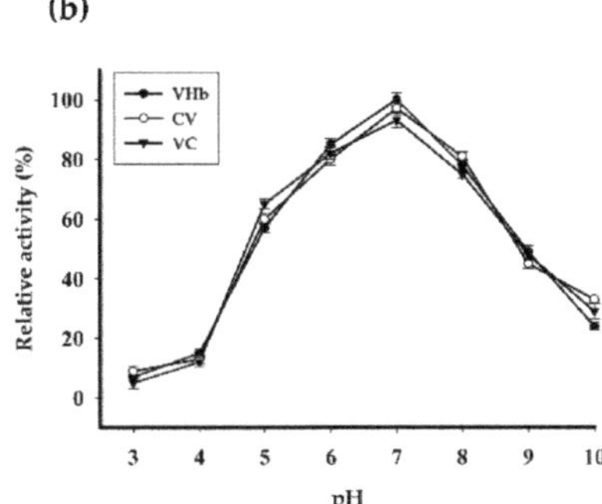

Figure 5. Effect of pH on (**a**) uricase and (**b**) peroxidase activities of native and fusion proteins. Enzymatic activities were assayed at 25 °C, under different pH conditions.

Table 1. Molecular weight, specific and relative activity of the produced proteins.

Protein	Molecular Weight (kDa)	Uricase Activity				Peroxidase Activity			
		U/mg	U/µmol	U/mg (%)	U/µmol (%)	U/mg	U/µmol	U/mg (%)	U/µmol (%)
CUOX	143	9.7 ± 0.2	1387.1 ± 28.6	100	100	-	-	-	-
VHb	35	-	-	-	-	20.0 ± 0.5	700.00 ± 17.5	100	100
CV	207	6.7 ± 0.3 *	1386.9 ± 71.7	69.07	99.98	0.68 ± 0.1 **	140.76 ± 22.4	3.40	20.10
VC	207	5.9 ± 0.1 *	1221.3 ± 31.1	60.82	88.04	0.54 ± 0.1 **	111.78 ± 16.4	2.70	15.96

Uricase activities (*) and peroxidase activities (**) of CV and VC were not significantly different. The statistical analysis was performed by Mann-Whitney U test.

Table 2. Steady-state kinetics of the produced proteins.

Protein	Uricase			Peroxidase		
	K_m (mM)	k_{cat} (s^{-1})	k_{cat}/K_m ($mM^{-1}s^{-1}$)	$K_m^{H_2O_2}$ (mM)	k_{cat} (s^{-1})	$k_{cat}/K_m^{H_2O_2}$ ($mM^{-1}s^{-1}$)
CUOX	0.079	29.08	368.04	-	-	-
VHb	-	-	-	1.84	4.72	2.56
CV	0.052	38.77	735.01	2.12	2.15	1.02
VC	0.046	46.93	1003.83	4.04	2.02	0.50

(a)

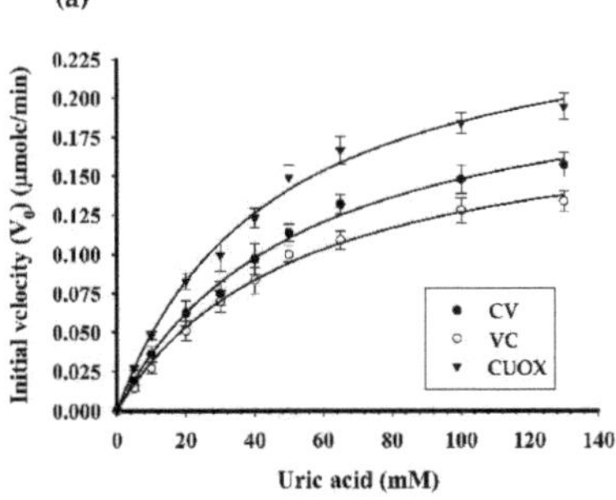

(b)

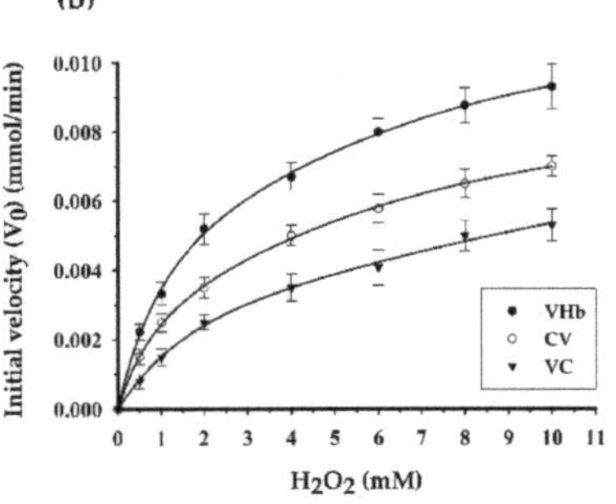

Figure 6. Comparison of uricase activity and peroxidase activity of native and fusion proteins. In the uricase assay (**a**), the rate of substrate oxidation was calculated based on a linear decrease in the absorbance of uric acid. The plots show the dependence of the initial rate of oxidation on uric acid concentration. Whereas in peroxidase assay (**b**), the rate of substrate oxidation was calculated based on a linear increase in the absorbance of resorufin. The plots show the dependence of the initial rate of oxidation on hydrogen peroxide concentration. All plots were fitted to the Michaelis-Menten equation.

For peroxidase kinetics, as compared to that of VHb, the apparent K_m of CV and VC were increased 1.15 and 2.20-fold, while the k_{cat} were decreased 2.2 and 2.3-fold, respectively. These kinetic behaviors resulted in 2.52-fold and 5.12-fold decreases in catalytic efficiency of CV and VC, respectively. In the present study, Amplex Red was used as a reducing substrate for the measurement of peroxidase activity and kinetic parameters. In fact, both ABTS and Amplex Red were employed but only the data obtained from using Amplex Red was presented, since only Amplex red was used as a chromogenic substrate in the cascade reaction.

2.3. Optimization of Two-Enzymatic Cascade Reactions for Quantification of Uric Acid

A schematic representation of the cascade reactions is shown in Figure 7. Uricase catalyzes the oxidation of uric acid to allantoin, carbon dioxide and hydrogen peroxide (H_2O_2). Then H_2O_2 and Amplex Red are stoichiometrically converted to the color product, resorufin. This final product can be followed by measuring either absorbance at 571 nm or fluorescence emission at 585 nm. Factors influencing catalysis of these cascade reactions including pH, reaction time, range of substrate concentration and enzyme (either natives or fusions) were investigated. In the previous section, we demonstrated that the optimal pH for catalysis of CUOX and VHb were 9.0 and 7.0, respectively (Figure 5a,b). However, at pH 7.0, CUOX retained uricase activity less than 50% while at pH 9.0, VHb retained peroxidase activity less than 50%. Thus, these two pH conditions are not appropriate for the cascade reactions. Nevertheless, at pH 8.0, both enzymes retained activity up to 75%–80%, and resorufin has been reported to be unstable at pH above 8.5 [36]. Therefore, pH 8.0 was chosen to study the cascade reactions. The concentrations of uric acid were varied from 5–40 µM and reaction times were varied from 5–40 min.

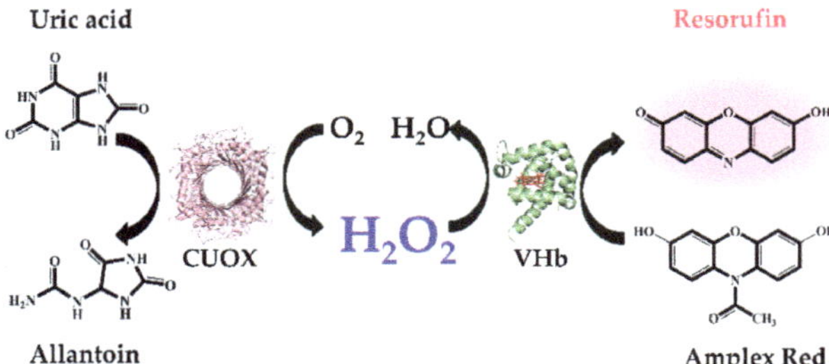

Figure 7. Schematic illustration of the colorimetric detection of uric acid using uricase and VHb (peroxidase)-catalyzed oxidation of Amplex Red.

Three enzymatic systems including; (i). native CUOX + native VHb, (ii). CV fusion protein, and (iii). VC fusion protein, were compared. In each system, the amount of uricase was fixed at 0.5 units/reaction. Based on the specific activities shown in Table 1, the ratio of uricase: peroxidase of CV was approximately 10:1 units while that of VC was approximately 11:1 units. Thus, the ratio of uricase: peroxidase in the system with native proteins was fixed at 10:1 units. However, at this ratio, the initial rates of the reactions were very slow. Therefore, the uricase: peroxidase ratio in the system with native proteins was changed to 10:2 units. The absorbance of the reactions was recorded immediately after all components were mixed. The absorbance over time of reactions with different concentrations of uric acid, catalyzed by native enzymes, CV and VC is shown in Figure 8a,c,e, respectively. At the maximum concentration of uric acid (40 µM), the initial rates of cascaded reactions catalyzed by native enzymes, CV and VC were 41.37, 258.62 and 82.75 nmol/min (resorufin), respectively, and at maximum reaction time (40 min), these reactions gave maximum absorbances at 0.1, 0.3 and 0.15, respectively. These results indicated that CV had higher catalytic efficiency for cascade reactions than the other protein systems. This high efficiency could be due to the direct transfer of an intermediate (H_2O_2) produced by uricase to the adjacent VHb active site without completely mixing with the bulk phase [37,38]. The schematic demonstration of direct substrate transfer is shown in Figure 9. Besides the absorbance over time of reactions with different concentrations of uric acid, the graphs of absorbance over certain uric acid concentrations at different time points were also plotted (Figure 8b,d,f). Reactions with native enzymes showed linearity over the tested range of uric acid concentrations only when measured at 5 min, but the signals were very weak (Figure 8b). Interestingly, CV gave linearity over the tested range of uric acid concentrations at all measured time points, 5–40 min (Figure 8d). However, the graphs of measurements made at 35 and 40 min were not different from that measured at 30 min, indicating that reaction time at 30 min is enough for the cascaded reactions to reach saturation. For VC, the graph with linearity over the tested range of uric acid concentrations was not obtained (Figure 8f).

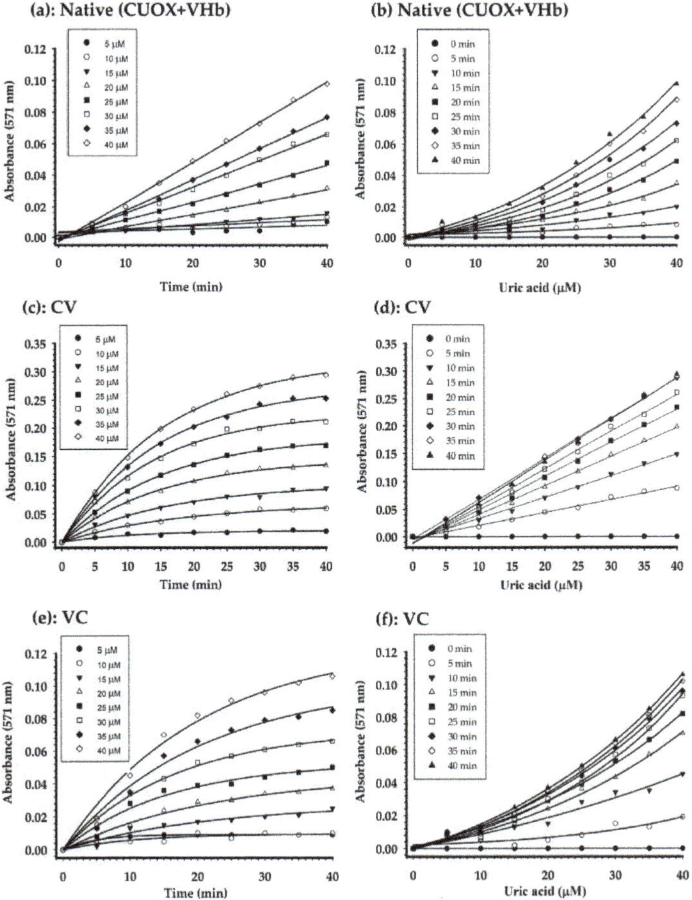

Figure 8. Quantification of uric acid using native enzyme (CUOX+VHb) (**a**,**b**), CUOX-VHb fusion protein (**c**,**d**) and VHb-CUOX fusion proteins (**e**,**f**). The graphs representing absorbance over time of reactions with different concentrations of uric acid (**a**,**c**,**e**) were transformed to the graphs showing calibration curves at different time points (**b**,**d**,**f**).

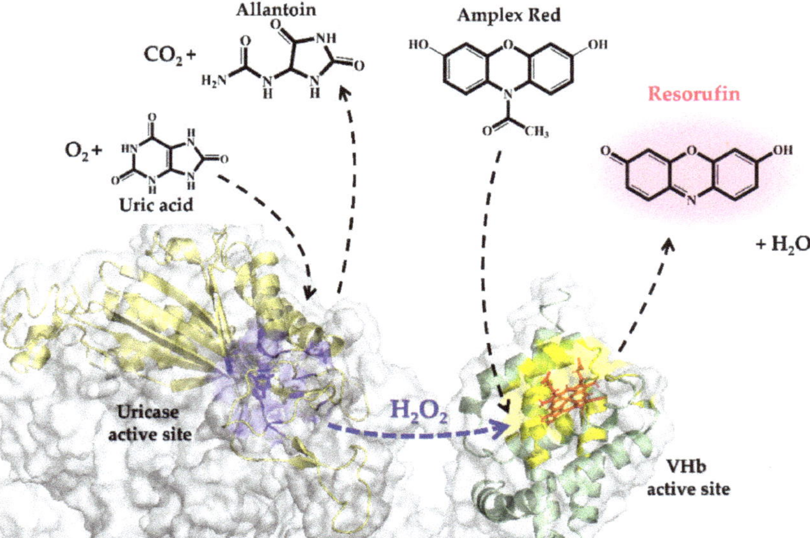

Figure 9. Schematic demonstration of hydrogen peroxide channeling between adjacent CUOX and VHb active sites.

2.4. Uric Acid Calibration Curve

Based on the results from the previous Section 2.3, CV was shown to produce the strongest resorufin signal, which should give the best sensitivity for the assay. Therefore, only CV was chosen for the construction of a linear calibration curve. The experiment was performed using final concentrations of uric acid ranging from 1–60 μM (1, 2.5, 5, 10, 15, 20, 25, 30, 35, 40, 45, 50, 55, 60 μM). The absorbance was measured after initiation of the reaction for 10 min. As shown in Figure 10a, the absorbance increases gradually when the UA concentration increases from 0–60 μM. The color variation of the reactions, from light-pink to deep-pink, is shown in the inset. The limit of detection (LOD) as observed by eyes is as low as 10 μM. The absorbances at 571 nm from all reactions were plotted and shown in Figure 10b. The linearity of the graph was obtained when UA concentration was within the range 2.5–50 μM. The linear regression equation was $A = 0.0038C - 0.0041$, where A is absorbance and C is concentration of UA (μM). The correlation of coefficient (R^2) was 0.9981. Comparison of the proposed assay with the other uricase-based colorimetric methods for detection of UA is shown in Table 3.

The proposed method is sensitive enough for UA detection, and is comparable or even better than the other methods. When the sample dilution factor in final reaction volume (1:10) was considered, the theoretical linear range of the proposed method is 25–500 μM, which covers the normal range of human serum uric acid, 142.77–416.41 μM (2.4–7.0 mg/dL). Besides good sensitivity, the proposed method consumed much less time than that of the other methods. In addition, most of the published methods required a 2-phases reaction, the first phase with pH $\geq$ 8 for uricase reaction and the second phase with pH $\leq$ 7 for peroxidase reaction. Thus, several reagents must be prepared and the assay must be performed through multiple steps. However, our method was performed in one step using one reaction buffer. After all reagents were mixed and stood for 10 min, the reaction was ready to be measured. More importantly, as compared to the other methods, the cost for peroxidase/peroxidase-like enzyme can be omitted, because in our work, single production of the enzyme (CV) yielded both uricase and peroxidase activities. Notably, as shown in Figure 8d, higher resorufin signals were observed when the reaction is incubated for more than 10 min. Therefore, the sensitivity of our assay may be improved if the test is incubated for more than 10 min. However, this should not exceed 30 min since the signal was not increased after 30 min.

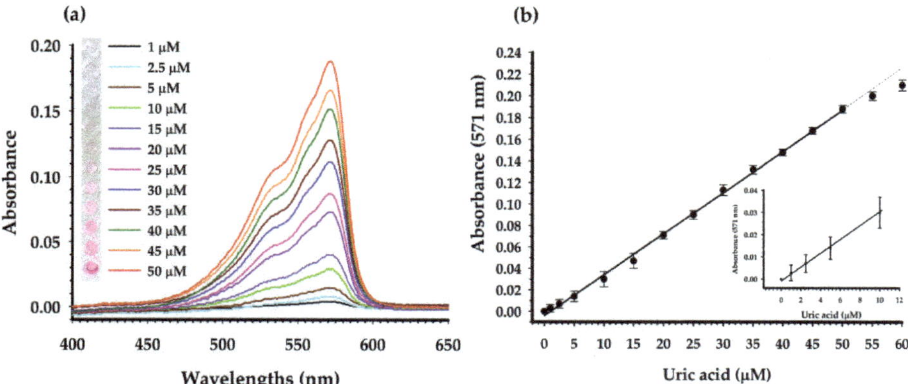

Figure 10. (**a**) Absorption spectra of the assay with different concentrations of uric acid (1–50 μM) at a reaction time of 10 min. Photograph of the corresponding reactions is shown in the inset. (**b**) A dose-response curve between uric acid concentration and absorbance at 571 nm was constructed.

Table 3. Comparison of different uricase-based colorimetric methods for detection of UA.

Analytical Method	Uricase/rxn	Peroxidase or Peroxidase-Like/rxn	Reaction Phase with Optimal Condition (pH/Temp./Time)	Total Reaction Time	Total Volume (μL)	Linear Range (μM)	LOD (μM)	Reference
1. Uricase/HRP/ oxPOD and GSH-MQDs	200 μg *	HRP 20 μg *	1. pH NR **, 37 °C, 30 min 2. pH 7.2–7.4, 37 °C, time NR **	>30 min	2000	1.2–100	0.2	[39]
2. Uricase/ MoS_2/TMB	2.5 μg *	MoS_2 0.18 μg	1. pH 8.5, 35 °C, 15 min 2. pH 4, 50 °C, 60 min	75 min	2000	0.5–100	0.3	[40]
3. Uricase/Cu^{2+}/ TMB	2.5 μg *	Cu^{2+} 8 μmol	1. pH 8.5, 35 °C, 15 min 2. pH 4, 45 °C, 60 min	75 min	2000	1–100	0.64	[41]
4. Uricase/ Co-Se-nano-crystalline/ 4-AAP/TOPS	1 U	Co-Se-nanocrystal 260 μg	1. pH 8.5, 40 °C, 10 min 2. pH 8.5, 40 °C, 30 min	40 min	4000	2–40	0.5	[42]
5. Uricase/ MIL53(Fe)/ TMB	0.1 U	MIL53(Fe) 20 μg	1. pH 9.0, 37 °C, 15 min 2. pH 4, 55 °C, 40 min	55 min	1000	4.5–60	1.3	[43]
6. Uricase/HRP/ 4-AAP/TOOS	0.4 U	HRP 1.2 U	1. pH 6.5, 25 °C, 5 min 2. pH 7.2–7.6, 25°C, 20 min	25 min	1020	Up to 34	12	[18]
7. UOX/VHb/ Amplex Red	0.05 U	VHb 0.005 U	1. pH 8.0, 30 °C, 10 min	10 min	100	2.5–50	1	This study

* Specific enzymatic activity (unit/mg) was not reported. ** NR = not reported. LOD = limit of detection.

2.5. Method Evaluation and Measurement of Uric Acid from Lyophilized Serum

Analysis of the standard curve showed that the assay had a good correlation between UA concentration and absorbance. The limit of detection (LOD) and limit of quantitation (LOQ) were 1 and 2.5 μM, respectively. The systematic error of the method was assessed by performing a recovery assay. The result showed that the recovery rate of UA obtained by the standard addition method was between 94.8% and 100.1%, with RSD (n = 5) less than 2.3% (Table 4). To investigate the feasibility of the proposed method for practical application, five samples of lyophilized human serum were assayed for UA by the proposed method and a clinical chemistry analyzer, then the results obtained from the two methods were compared. As shown in Table 5, the average relative error between the two methods was 3.06%. These results indicated that the proposed method has good potential for the detection of UA in human serum.

Table 4. Spiked recoveries of UA in pooled serum.

Detected (μM)	Added (μM)	Found (μM)	Recovery (%)	RSD (%, n = 5)
327 ± 5.2	50	374.4 ± 6.1	94.8	1.63
	75	399.4 ± 7.4	96.5	1.85
	100	424.0 ± 8.0	97.0	1.90
	125	452.6 ± 10.0	100.5	2.22
	150	477.2 ± 10.4	100.1	2.19

Table 5. Determination of UA in lyophilized human serum using the developed assay and a clinical chemistry analyzer.

Sample	Developed Assay	Automated Analyzer	Relative Error (%)	Average Relative Error (%)
Serum 1	285	298	4.5	3.06
Serum 2	349	360	3.1	
Serum 3	466	453	2.7	
Serum 4	539	525	2.5	
Serum 5	672	655	2.5	

3. Materials and Methods

3.1. Bacterial Strains, Plasmids and Chemicals

E. coli NovaBlue and BL21(DE3) strains, and pETDuet-1 plasmid were from Novagen (EMD Bioscience, Madison, WI, USA). Restriction enzymes and T4-DNA ligase were from Fermentas (Thermo Fisher Scientific, Waltham, MA, USA). KOD-Plus DNA polymerase was purchased from Toyobo (Osaka, Japan). Uric acid, lithium carbonate and Amplex Red were from Sigma–Aldrich (St. Louis, MO, USA). Hydrogen peroxide (30% v/v) was obtained from Merck. The concentration of uric acid, hydrogen peroxide and resorufin were standardized using the extinction coefficient of 12.6 $mM^{-1}cm^{-1}$ at 293 nm [44], 72.8 M^{-1} cm^{-1} at 230 nm [45] and 54,000 M^{-1} cm^{-1} at 571 nm [36], respectively.

3.2. DNA Manipulations

cDNA for *Candida utilis* uricase (XM_020212619.1) was synthesized by Integrated DNA Technologies (Skokie, Illinois). The gene was PCR-amplified using primers shown in Table 6. PCR product was then digested with *Bam*HI and *Hin*dIII, and cloned into pETDuet-1 plasmid pre-treated with the same enzymes. The constructed plasmid was designated as pETDuetCUOX. To construct the plasmid expressing CUOX-VHb fusion protein, endogenous *Xho*I restriction site and stop codon of the CUOX gene in pETDuetCUOX were removed by site-directed mutagenesis (SDM). *Hin*dIII restriction site was introduced prior to the start codon of the VHb gene in pET46VHb [29] by SDM. Then the VHb gene was excised by treating with *Hin*dIII and *Xho*I, and subcloned into pETDuetCUOX pretreated with the same enzymes. The resulting plasmid was designated as pETDuetCUOX-VHb. To construct the plasmid for expression of VHb-CUOX fusion protein, stop codon of VHb in pET46VHb was changed to *Bam*HI by SDM then the CUOX gene was excised from pETDuetCUOX by treating with *Bam*HI and *Xho*I then the gene was ligated to pET46VHb pretreated with the same enzymes. The resulting plasmid was designated as pET46VHb-CUOX. Sequences of all primers used for PCR and SDM reactions are shown in Table 6. The correctness of all recombinant plasmids was confirmed by DNA sequencing analysis.

Table 6. Primers for PCR amplification and site-directed mutagenesis. *Bam*HI and *Hind*III restriction sites were included in CUOX_FP (forward primer) and CUOX_RP (reverse primer), respectively. The restriction sites are indicated as underlined text.

Primers	Description	DNA sequence (5′ → 3′)
CUOX_FP	for construction of pETDuetCUOX	CGGGATCCCATGTCAACAACGCTCTCATC
CUOX_RP		CCCAAGCTTTTACAACTTGGTCTTCTCC
Remove *Xho*I CUOX_FP	for removal of *Xho*I site from CUOX gene	CCACCTTTGCTCTTGAGAACTCTCCATCTG
Remove *Xho*I CUOX_RP		CAGATGGAGAGTTCTCAAGAGCAAAGGTGG
Remove stop CUOX_FP	for removal of stop codon from CUOX gene	GAGAAGACCAAGTTGAAGCTTGCGGCCGC
Remove stop CUOX_RP		GCGGCCGCAAGCTTCAACTTGGTCTTCTC
Insert *Hind*III VHb_FP	for insertion of *Hind*III site to VHb gene	CATCACGTGGATGACGACAAGCTTATGTTAGACCAGCAAACC
Insert *Hind*III VHb_RP		GGTTTGCTGGTCTAACATAAGCTTGTCGTCATCCACGTGATG
Insert *Bam*HI VHb_FP	for insertion of *Bam*HI site to VHb gene	GCTCAAGCGGTTGAAGGATCCACCGGGCTTCTCCTC
Insert *Bam*HI VHb_RP		GAGGAGAAGCCCGGTGGATCCTTCAACCGCTTGAGC

3.3. Protein Expression and Purification

pETDuetCUOX, pET46VHb, pETDuetCUOX-VHb and pET46VHb-CUOX were individually transformed into *E. coli* BL21(DE3). Cells were grown in 1 L terrific broth supplemented with 100 μg/mL ampicillin at 37 °C with shaking at 180 rpm for 6–9 h. To induce protein expression and to improve the heme content of VHb, the cultures were supplemented with 1 mM isopropyl β-D-1-thiogalactopyranoside (IPTG) and 0.3 mM δ-aminolevulinic acid, respectively. Then the cultures were continued for 16 h at 30 °C with shaking at 125 rpm. Cells were collected by centrifugation, resuspended in 50 mM Tris-HCl buffer pH 8.0 supplemented with NaCl 500 and disrupted by ultrasonic disintegration. Then, cell lysate was cleared by centrifugation at 27,216× g for 10 min and the supernatant was filtered using a 0.45 μm syringe filter (Sartorius). Proteins were purified by IMAC using Ni-nitrilotriacetic acid (Ni-NTA) agarose column as previously described [34].

3.4. Molecular Weight Determination

To further purify and simultaneously determine the molecular weight of the proteins, each IMAC-purified protein was loaded onto a 16/60 Sephacryl S-300 HR gel-filtration chromatography column pre-equilibrated with 50 mM phosphate buffer at pH 7.2 containing 0.15 M EDTA. The elution was done with the same buffer at a flow rate of 1.0 mL/min. Standard protein markers ranging from 13.7 to 440 kDa were used for standard curve preparation. Experiments were run with an ÄKTA pure protein purification system (GE healthcare life sciences, Sweden) under the operation of the Unicorn 6.3 software.

3.5. Measurement of Enzymatic Activity and Kinetic Parameters

Uricase activity and kinetic parameters were determined as previously described [34]. In the kinetic assay, concentrations of uric acid ranging from 5–120 μM were used. One unit of uricase activity was defined as the amount of enzyme required to convert 1 μmol of uric acid to allantoin per minute at 25 °C, pH 8.0. Optimum pH for uricase reaction was studied over a pH range of 5.0–10.0 using 50 mM sodium citrate buffer (pH 5.0–6.0) and 50 mM Tris-HCl buffer (pH 7.0–10.0). VHb peroxidase activity and kinetic parameters were assayed as previously described by Kvist M, et al. [28]. In the kinetic assay, concentrations of H_2O_2 were varied from 0.5–10 μM. ABTS and Amplex Red were used as reducing substrates. One unit of peroxidase activity was defined as the amount of enzyme required for consumption/production of 1 μmol of substrate/product per minute at 25 °C. Optimum pH for peroxidase reaction was studied in buffers with pH 3.0–10.0 using 100 mM citrate–phosphate (pH 3.0–7.0) 100 mM Tris–HCl (pH 8.0–10.0).

3.6. Investigation of Sequential Reactions Catalyzed by Uricase (CUOX) and Peroxidase (VHb)

Catalysis of the consecutive reactions by uricase and peroxidase was studied. The principle of the reactions is shown in Figure 7. A volume of 50–400 μM of UA (50, 100, 150, 200, 250, 300, 350 and 400 μM), 1 mM of Amplex Red and 0.5 unit/mL of uricase (native CUOX, CV, and VC) were prepared in 50 mM sodium borate buffer, pH 8.0. The solution of native CUOX also contained 0.05 units of peroxidase (VHb). The assays were performed in a 96-well microtiter plate. The reactions were started by the addition of 10 μL of enzyme solution [46] to 90 μL of the reaction mixtures containing different concentrations of UA or serum samples. Please note that the standard UA or sample was diluted 10 times in the final reaction volume. The reaction system in a total volume of 100 μL consisted of 50 mM sodium borate buffer pH 8.0, 0.1 mM Amplex Red, 0.05 units of UOX, 0.005 units of peroxidase (VHb) and 5–40 μM of UA. Production of final reaction product (resorufin) was followed by measuring the absorbance at 571 nm using a microplate reader (Synergy HTX multi-mode, BioTek, Winooski, VT, USA), at 5-min intervals for 40 min.

3.7. Preparation of a UA Calibration Curve

All reagents were prepared as mentioned in the latest experiment (3.6), except that the range of UA concentrations was 10–600 μM (10, 25, 50, 100, 150, 200, 250, 300, 350, 400, 500, 550, 600 μM). The reactions were performed using only CV fusion protein and the absorbance was measured at 10 min after the addition of the enzyme. All assays were performed in triplicate.

3.8. Method Evaluation and Quantification of UA from Lyophilized Human Serum

The assay was evaluated in terms of linearity, sensitivity, the limit of detection (LOD), limit of quantitation (LOQ) and recovery. Recovery was evaluated by spiking known amounts of UA (50, 75, 100, 125, 150 μM) into pooled human serum samples (Sigma–Aldrich, St. Louis, MO, USA) then UA in each spiked sample was measured. To evaluate the reliability of the assay, five vials of lyophilized human serum with different concentrations of UA were purchased from The External Quality Assessment Schemes in Clinical Chemistry (EQAC), Faculty of Medical Technology, Mahidol University. UA value of each vial was measured using the developed assay and routine clinical method (measured by Roche Cobas C501 clinical chemistry analyzer).

3.9. Molecular Modeling

The 3D structures of single-chain CUOX-VHb and VHb-CUOX were modeled using I-TASSER server [33]. The amino acid sequences of CUOX-VHb and VHb-CUOX were uploaded in FASTA format then 3D structures were predicted in PDB format. The server generates five top models for each entry, the one with the highest confidence score (c-score) represents the best model. The models were evaluated for stereochemical quality by PROCHECK [47], Verify3D [48,49], ProSA [50] and Pro-Q [51]. The tetrameric structure of CUOX-VHb fusion protein was subsequently constructed using PyMol 2.3.3 software [52].

3.10. Statistical Analysis

Uricase activities and peroxidase activities between CV and VC were compared using the nonparametric Mann–Whitney U-test. The alpha value was set at 0.05.

4. Conclusions

Bifunctional proteins with uricase and peroxidase activities were constructed by direct fusion of CUOX and VHb. The engineered proteins offer the advantages of the low cost of enzyme production (single production yielded a protein with two enzymatic activities) and fast catalysis (due to substrate channeling between the two enzymes), which helps to shorten the reaction time. Based on VHb-catalyzed oxidation of Amplex Red, one of the engineered proteins, CV, was successfully applied for the development of a simple, rapid and reliable assay for colorimetric detection of UA. The assay had a good linear relationship (R^2 = 0.9981) between absorbance at 571 nm and uric acid concentration over the range of 2.5–50 μM, which covers the normal range of SUA. In addition, the assay can be performed at a single pH (8.0), thus, adjusting the pH for peroxidase activity is not required. This helps to save both cost and time. The proposed assay was successfully applied to detect UA in pooled human serum, suggesting a promising potential for clinical application.

Author Contributions: Conceptualization, T.P. and S.Y.; investigation, T.P., J.C., K.T., N.O. and S.Y.; writing original draft, T.P., S.W. and S.Y.; draft editing, T.P., N.O., S.W. and S.Y.; supervision, S.Y. All authors have read and agreed to the published version of the manuscript.

Funding: This work was supported by the Thailand Research Fund (MRG6080283 and RSA6280021). The APC was partially funded by Faculty of Medical Technology, Mahidol University.

Acknowledgments: The Thailand Research Fund financial support is acknowledged.

Conflicts of Interest: The authors declare no conflicts of interest.

References

1. Dalbeth, N.; Choi, H.K.; Joosten, L.A.B.; Khanna, P.P.; Matsuo, H.; Perez-Ruiz, F.; Stamp, L.K. Gout. *Nat. Rev. Dis. Primers* **2019**, *5*, 69. [CrossRef] [PubMed]
2. Wang, Q.; Wen, X.; Kong, J. Recent progress on uric acid detection: A review. *Crit. Rev. Anal. Chem.* **2019**, 1–17. [CrossRef] [PubMed]
3. Zhao, Y.; Yang, X.; Lu, W.; Liao, H.; Liao, F. Uricase based methods for determination of uric acid in serum. *Microchim. Acta* **2009**, *164*, 1–6. [CrossRef]
4. Domagk, G.F.; Schlicke, H.H. A colorimetric method using uricase and peroxidase for the determination of uric acid. *Anal. Biochem.* **1968**, *22*, 219–224. [CrossRef]
5. Price, C.P.; James, D.R. Analytical reviews in clinical biochemistry: The measurement of urate. *Ann. Clin. Biochem.* **1988**, *25*, 484–498. [CrossRef]
6. Gochman, N.; Schmitz, J.M. Automated determination of uric acid, with use of a uricase-peroxidase system. *Clin. Chem.* **1971**, *17*, 1154–1159. [CrossRef]
7. Klose, S.; Stoltz, M.; Munz, E.; Portenhauser, R. Determination of uric acid on continuous-flow (AutoAnalyzer II and SMA) systems with a uricase/phenol/4-aminophenazone color test. *Clin. Chem.* **1978**, *24*, 250–255. [CrossRef]
8. Trivedi, R.C.; Rebar, L.; Berta, E.; Stong, L. New enzymatic method for serum uric acid at 500 nm. *Clin. Chem.* **1978**, *24*, 1908–1911. [CrossRef]
9. Fossati, P.; Prencipe, L.; Berti, G. Use of 3,5-dichloro-2-hydroxybenzenesulfonic acid/4-aminophenazone chromogenic system in direct enzymic assay of uric acid in serum and urine. *Clin. Chem.* **1980**, *26*, 227–231. [CrossRef]
10. Majkić-Singh, N.; Stojanov, M.; Spasić, S.; Berkes, I. Spectrophotometric determination of serum uric acid by an enzymatic method with 2,2′-azino-di(3-ethylbenzthiazoline-6-sulfonate) (ABTS). *Clin. Chim. Acta* **1981**, *116*, 117–123. [CrossRef]
11. Zhao, H.; Wang, Z.; Jiao, X.; Zhang, L.; Lv, Y. Uricase-Based Highly Sensitive and Selective Spectrophotometric Determination of Uric Acid Using BSA-Stabilized Au Nanoclusters as Artificial Enzyme. *Spectrosc. Lett.* **2012**, *45*, 511–519. [CrossRef]
12. Legoux, R.; Delpech, B.; Dumont, X.; Guillemot, J.C.; Ramond, P.; Shire, D.; Caput, D.; Ferrara, P.; Loison, G. Cloning and expression in *Escherichia coli* of the gene encoding *Aspergillus flavus* urate oxidase. *J. Biol. Chem.* **1992**, *267*, 8565–8570. [PubMed]

13. Itaya, K.; Yamamoto, T.; Fukumoto, J. Studies on yeast uricase. *Agric. Biol. Chem.* **1967**, *31*, 1256–1264.
14. Zhao, Y.; Zhao, L.; Yang, G.; Tao, J.; Bu, Y.; Liao, F. Characterization of a uricase from *Bacillus fastidious* A.T.C.C. 26904 and its application to serum uric acid assay by a patented kinetic uricase method. *Biotechnol. Appl. Biochem.* **2006**, *45*, 75–80. [CrossRef] [PubMed]
15. Suzuki, K.; Sakasegawa, S.-I.; Misaki, H.; Sugiyama, M. Molecular cloning and expression of uricase gene from *Arthrobacter globiformis* in *Escherichia coli* and characterization of the gene product. *J. Biosci. Bioeng.* **2004**, *98*, 153–158. [CrossRef]
16. Bomalaski, J.S.; Holtsberg, F.W.; Ensor, C.M.; Clark, M.A. Uricase formulated with polyethylene glycol (uricase-PEG 20): Biochemical rationale and preclinical studies. *J. Rheumatol.* **2002**, *29*, 1942–1949.
17. Saud Al-Bagmi, M.; Shahnawaz Khan, M.; Alhasan Ismael, M.; Al-Senaidy, A.M.; Ben Bacha, A.; Mabood Husain, F.; Alamery, S.F. An efficient methodology for the purification of date palm peroxidase: Stability comparison with horseradish peroxidase (HRP). *Saudi J. Biol. Sci.* **2019**, *26*, 301–307. [CrossRef]
18. Huang, Y.; Chen, Y.; Yang, X.; Zhao, H.; Hu, X.; Pu, J.; Liao, J.; Long, G.; Liao, F. Optimization of pH values to formulate the bireagent kit for serum uric acid assay. *Biotechnol. Appl. Biochem.* **2015**, *62*, 137–144. [CrossRef]
19. Lavery, C.B.; Macinnis, M.C.; Macdonald, M.J.; Williams, J.B.; Spencer, C.A.; Burke, A.A.; Irwin, D.J.G.; D'Cunha, G.B. Purification of peroxidase from Horseradish (*Armoracia rusticana*) roots. *J. Agric. Food Chem.* **2010**, *58*, 8471–8476. [CrossRef] [PubMed]
20. Spadiut, O.; Herwig, C. Production and purification of the multifunctional enzyme horseradish peroxidase. *Pharm. Bioprocess.* **2013**, *1*, 283–295. [CrossRef]
21. Morawski, B.; Lin, Z.; Cirino, P.; Joo, H.; Bandara, G.; Arnold, F.H. Functional expression of horseradish peroxidase in *Saccharomyces cerevisiae* and *Pichia pastoris*. *Protein Eng.* **2000**, *13*, 377–384. [CrossRef] [PubMed]
22. Krainer, F.W.; Dietzsch, C.; Hajek, T.; Herwig, C.; Spadiut, O.; Glieder, A. Recombinant protein expression in *Pichia pastoris* strains with an engineered methanol utilization pathway. *Microb. Cell Fact.* **2012**, *11*, 22. [CrossRef] [PubMed]
23. Smith, A.T.; Santama, N.; Dacey, S.; Edwards, M.; Bray, R.C.; Thorneley, R.N.; Burke, J.F. Expression of a synthetic gene for horseradish peroxidase C in *Escherichia coli* and folding and activation of the recombinant enzyme with Ca^{2+} and heme. *J. Biol. Chem.* **1990**, *265*, 13335–13343. [PubMed]
24. Grigorenko, V.; Chubar, T.; Kapeliuch, Y.; Börchers, T.; Spener, F.; Egorova, A. New approaches for functional expression of recombinant horseradish peroxidase C in *Escherichia coli*. *Biocatal. Biotransform.* **1999**, *17*, 359–379. [CrossRef]
25. Asad, S.; Dabirmanesh, B.; Ghaemi, N.; Etezad, S.M.; Khajeh, K. Studies on the refolding process of recombinant horseradish peroxidase. *Mol. Biotechnol.* **2013**, *54*, 484–492. [CrossRef]
26. Gundinger, T.; Spadiut, O. A comparative approach to recombinantly produce the plant enzyme horseradish peroxidase in *Escherichia coli*. *J. Biotechnol.* **2017**, *248*, 15–24. [CrossRef]
27. Humer, D.; Spadiut, O. Improving the Performance of Horseradish Peroxidase by Site-Directed Mutagenesis. *Int. J. Mol. Sci.* **2019**, *20*, 916. [CrossRef]
28. Kvist, M.; Ryabova, E.S.; Nordlander, E.; Bülow, L. An investigation of the peroxidase activity of *Vitreoscilla* hemoglobin. *J. Biol. Inorg. Chem.* **2007**, *12*, 324–334. [CrossRef]
29. Isarankura-Na-Ayudhya, C.; Yainoy, S.; Tantimongcolwat, T.; Bülow, L.; Prachayasittikul, V. Engineering of a novel chimera of superoxide dismutase and *Vitreoscilla* hemoglobin for rapid detoxification of reactive oxygen species. *J. Biosci. Bioeng.* **2010**, *110*, 633–637. [CrossRef]
30. Isarankura-Na-Ayudhya, C.; Tansila, N.; Worachartcheewan, A.; Bülow, L.; Prachayasittikul, V. Biochemical and cellular investigation of *Vitreoscilla* hemoglobin (VHb) variants possessing efficient peroxidase activity. *J. Microbiol. Biotechnol.* **2010**, *20*, 532–541.
31. Li, W.; Zhang, Y.; Xu, H.; Wu, L.; Cao, Y.; Zhao, H.; Li, Z. pH-induced quaternary assembly of *Vitreoscilla* hemoglobin: The monomer exhibits better peroxidase activity. *Biochim. Biophys. Acta* **2013**, *1834*, 2124–2132. [CrossRef] [PubMed]
32. Giangiacomo, L.; Mattu, M.; Arcovito, A.; Bellenchi, G.; Bolognesi, M.; Ascenzi, P.; Boffi, A. Monomer-dimer equilibrium and oxygen binding properties of ferrous *Vitreoscilla* hemoglobin. *Biochemistry* **2001**, *40*, 9311–9316. [CrossRef] [PubMed]
33. Zhang, Y. I-TASSER server for protein 3D structure prediction. *BMC Bioinform.* **2008**, *9*, 40. [CrossRef] [PubMed]

34. Yainoy, S.; Phuadraksa, T.; Wichit, S.; Sompoppokakul, M.; Songtawee, N.; Prachayasittikul, V.; Isarankura-Na-Ayudhya, C. Production and Characterization of Recombinant Wild Type Uricase from Indonesian Coelacanth (*L. menadoensis*) and Improvement of Its Thermostability by In Silico Rational Design and Disulphide Bridges Engineering. *Int. J. Mol. Sci.* **2019**, *20*, 1269. [CrossRef]
35. Liu, C.Y.; Webster, D.A. Spectral characteristics and interconversions of the reduced oxidized, and oxygenated forms of purified cytochrome o. *J. Biol. Chem.* **1974**, *249*, 4261–4266.
36. Dröse, S.; Galkin, A.; Brandt, U. Chapter 26 Measurement of superoxide formation by mitochondrial complex I of *Yarrowia lipolytica*. *Meth. Enzymol.* **2009**, *456*, 475–490.
37. Ovádi, J.; Tompa, P.; Vértessy, B.; Orosz, F.; Keleti, T.; Welch, G.R. Transient-time analysis of substrate-channelling in interacting enzyme systems. *Biochem. J.* **1989**, *257*, 187–190. [CrossRef]
38. Spivey, H.O.; Ovádi, J. Substrate channeling. *Methods* **1999**, *19*, 306–321. [CrossRef]
39. Liu, M.; He, Y.; Zhou, J.; Ge, Y.; Zhou, J.; Song, G. A "naked-eye" colorimetric and ratiometric fluorescence probe for uric acid based on Ti3C2 MXene quantum dots. *Anal. Chim. Acta* **2020**, *1103*, 134–142. [CrossRef]
40. Wang, X.; Yao, Q.; Tang, X.; Zhong, H.; Qiu, P.; Wang, X. A highly selective and sensitive colorimetric detection of uric acid in human serum based on MoS2-catalyzed oxidation TMB. *Anal. Bioanal. Chem.* **2019**, *411*, 943–952. [CrossRef]
41. Lu, H.-F.; Li, J.-Y.; Zhang, M.-M.; Wu, D.; Zhang, Q.-L. A highly selective and sensitive colorimetric uric acid biosensor based on Cu(II)-catalyzed oxidation of 3,3′,5,5′-tetramethylbenzidine. *Sens. Actuators B Chem.* **2017**, *244*, 77–83. [CrossRef]
42. Zhuang, Q.-Q.; Lin, Z.-H.; Jiang, Y.-C.; Deng, H.-H.; He, S.-B.; Su, L.-T.; Shi, X.-Q.; Chen, W. Peroxidase-like activity of nanocrystalline cobalt selenide and its application for uric acid detection. *Int. J. Nanomed.* **2017**, *12*, 3295–3302. [CrossRef] [PubMed]
43. Lu, J.; Xiong, Y.; Liao, C.; Ye, F. Colorimetric detection of uric acid in human urine and serum based on peroxidase mimetic activity of MIL-53(Fe). *Anal. Methods* **2015**, *7*, 9894–9899. [CrossRef]
44. Suárez, A.S.G.; Stefan, A.; Lemma, S.; Conte, E.; Hochkoeppler, A. Continuous enzyme-coupled assay of phosphate- or pyrophosphate-releasing enzymes. *BioTechniques* **2012**, *53*, 99–103. [CrossRef] [PubMed]
45. George, P. The chemical nature of the second hydrogen peroxide compound formed by cytochrome c peroxidase and horseradish peroxidase. I. Titration with reducing agents. *Biochem. J.* **1953**, *54*, 267–276. [CrossRef]
46. Roskoski, R. Enzyme Assays☆. In *Reference Module in Biomedical Sciences*; Elsevier: Amsterdam, The Netherlands, 2014.
47. Laskowski, R.A.; MacArthur, M.W.; Moss, D.S.; Thornton, J.M. PROCHECK: A program to check the stereochemical quality of protein structures. *J. Appl. Crystallogr.* **1993**, *26*, 283–291. [CrossRef]
48. Bowie, J.U.; Lüthy, R.; Eisenberg, D. A method to identify protein sequences that fold into a known three-dimensional structure. *Science* **1991**, *253*, 164–170. [CrossRef]
49. Lüthy, R.; Bowie, J.U.; Eisenberg, D. Assessment of protein models with three-dimensional profiles. *Nature* **1992**, *356*, 83–85. [CrossRef] [PubMed]
50. Wiederstein, M.; Sippl, M.J. ProSA-web: Interactive web service for the recognition of errors in three-dimensional structures of proteins. *Nucleic Acids Res.* **2007**, *35*, W407–W410. [CrossRef] [PubMed]
51. Wallner, B.; Elofsson, A. Can correct protein models be identified? *Protein Sci.* **2003**, *12*, 1073–1086. [CrossRef] [PubMed]
52. *Schrodinger PyMOL: The PyMOL Molecular Graphics System*; Version 1.8; Schrödinger, LLC: New York, NY, USA, 2015.

Article

Rigorous Model-Based Design and Experimental Verification of Enzyme-Catalyzed Carboligation under Enzyme Inactivation

Dominik Hertweck [1,2,†], Victor N. Emenike [2,3,†], Antje C. Spiess [1,2] and René Schenkendorf [2,3,*]

1 Institute of Biochemical Engineering, TU Braunschweig, 38106 Braunschweig, Germany; d.hertweck@tu-braunschweig.de (D.H.); a.spiess@tu-braunschweig.de (A.C.S.)
2 Center of Pharmaceutical Engineering (PVZ), 38106 Braunschweig, Germany
3 Institute of Energy and Process Systems Engineering, TU Braunschweig, 38106 Braunschweig, Germany; v.emenike@tu-braunschweig.de
* Correspondence: r.schenkendorf@tu-braunschweig.de
† These authors contributed equally to this work.

Received: 6 December 2019; Accepted: 3 January 2020; Published: 9 January 2020

Abstract: Enzyme catalyzed reactions are complex reactions due to the interplay of the enzyme, the reactants, and the operating conditions. To handle this complexity systematically and make use of a design space without technical restrictions, we apply the model based approach of elementary process functions (EPF) for selecting the best process design for enzyme catalysis problems. As a representative case study, we consider the carboligation of propanal and benzaldehyde catalyzed by benzaldehyde lyase from *Pseudomonas fluorescens* (*Pf*BAL) to produce (*R*)-2-hydroxy-1-phenylbutan-1-one, because of the substrate dependent reaction rates and the challenging substrate dependent *Pf*BAL inactivation. The apparatus independent EPF concept optimizes the material fluxes influencing the enzyme catalyzed reaction for the given process intensification scenarios. The final product concentration is improved by 13% with the optimized feeding rates, and the optimization results are verified experimentally. In general, the rigorous model driven approach could lead to selecting the best existing reactor, designing novel reactors for enzyme catalysis, and combining protein engineering and process systems engineering concepts.

Keywords: enzyme catalysis; optimal design; process intensification; elementary process functions; benzaldehyde lyase; 2-hydroxy ketones

1. Introduction

The pharmaceutical industry is considering biocatalytic processes as a possible alternative to chemocatalytic processes [1–3]. This is primarily due to the high stereoselectivity and specificity of biocatalytic processes, which lead to efficient production of high-quality active pharmaceutical ingredients (APIs) in fewer synthesis steps [2,3]. Less complex synthesis is of economic importance to the pharmaceutical industry as it could translate into cost savings and greener pharmaceutical processes [3]. In recent years, the design of biocatalysts for the efficient synthesis of APIs has witnessed significant advancements [1].

To render enzyme processes economically viable, a high product concentration and low enzyme cost must be ensured [2]. To this end, the advancements in protein engineering [1], appropriate reaction engineering concepts [4], and process intensification strategies must be combined [5,6]. A traditional method for performing this optimization is by carrying out numerous experiments in the lab that are cost and time intensive. Therefore, computer aided model based approaches have been proposed to complement laboratory experiments [7,8]. As examples of the application of model based approaches to optimize enzyme catalyzed reactions, Stillger et al. developed an enzyme-membrane continuous stirred tank reactor (CSTR) for the carboligation of benzaldehyde and acetaldehyde catalyzed by benzaldehyde lyase from *Pseudomonas fluorescens* (*Pf*BAL) to produce (*R*)-2-hydroxy-1-phenylpropanone (HPP) [9]. In their work, they performed simulations by using a kinetic model in combination with reactor models to select the best performing reactor. Parallel to this work, Hildebrand et al. [10] investigated the production of HPP by *Pf*BAL in a membrane CSTR using a kinetic model to simulate different reaction engineering strategies to select the best reactor design. In their work, however, *Pf*BA inactivation was observed, but was not included in the kinetic model and, thus, could neither be predicted nor simulated [10]. Begemann et al. used a model based approach to analyze reactors and develop a control strategy for a two phase biocatalytic oxidoreduction system catalyzed by *Candida parapsilosis* carbonyl reductase 2 [11]. Based on simulations of their model, they showed that a fed-batch reactor performed better than a batch reactor for the chosen reaction. Their simulations also showed that controlling the pH could increase conversions and improve productivity. By performing simulations with coupled kinetic models and reactor design equations, Braun et al. minimized the enzyme costs for the biocatalytic production of 12-ketochenodeoxycholic acid in a batch reactor [7]. Marpani et al. [12] utilized a kinetic model and simulation approach to determine the optimal operating conditions for the biocatalytic conversion of formaldehyde to methanol in a batch reactor. Although their work showed excellent agreement between the simulation and experimental results, it involved performing hundreds of simulations [12].

Therefore, even simulation based approaches might be time consuming and could lead to sub-optimal results [13]. Furthermore, there is no guarantee that the design space explored by the simulations would lead to optimal results. Fortunately, model based process optimization methods have been established in process systems engineering, which could lead to (at least locally) guaranteed optimal reactor designs and operating conditions. Furthermore, most existing model based reactor design approaches for enzyme catalysis are based on comparing a limited subset of existing reactor types, which constrains the design space and limits the possibilities of designing novel intensified reactors for enzyme catalysis.

In the last decade, a model based reactor design approach that is based on the concept of elementary process functions (EPF) was proposed by Freund and Sundmacher. The basic premise of the EPF concept is to design reactors based on their essential reaction functions rather than optimizing predefined apparatuses or equipment [14] and, therefore, to formulate the reactor design problem as a reaction function optimization problem. To describe and optimize processes with the EPF approach, a fluid element is considered traveling through the thermodynamic state space of a (bio)chemical reaction that is acted upon by internal and external mass and energy fluxes. These fluxes are then designed to drive the fluid element to the desired state in the thermodynamic state space, thus maximizing a particular objective function or performance metric, such as the final product concentration, space-time yield, or total enzyme turnover number. Material and energy balances of the fluid element are set up from first principles to model the fluid element. They are accompanied by detailed reaction kinetics and thermodynamic relations, which explain the key phenomena taking place within and outside the fluid element.

Given that the fluid element changes with time, a dynamic optimization problem is formulated by using the balances, reaction kinetic and thermodynamic equations, and constraints that are inherent to the particular process. The dynamic optimization problem is then solved for different intensification cases. Based on the corresponding simulations, the best intensification strategy is selected and then technically approximated by using an off-the-shelf reactor or by designing a novel reactor to approximate the optimal fluxes obtained [15]. The EPF approach has been applied extensively to select the best existing reactor or to design novel reactors for the optimal production of bulk chemicals [16–21]. Recently, we worked on extending the EPF approach to biotechnological processes [22,23]. In this paper, we build on our previous work by presenting a clear guideline for applying the EPF approach to enzyme catalysis.

The objective of this paper is to implement and experimentally demonstrate the applicability of the EPF approach for enzyme catalysis problems. As a case study, we chose the *Pf*BAL catalyzed carboligation of propanal and benzaldehyde to form (*R*)-2-hydroxy-1-phenylbutan-1-one, which is representative of biocatalytic carboligations to optically pure 2-hydroxy ketones, key organic intermediates for producing a vast array of APIs [9,10,24]. The *Pf*BAL catalyzed conversion of benzaldehyde in the presence of propanal can, in theory, yield four different enantiopure compounds [25]: the two asymmetric products (*R*)-2-hydroxy-1-phenylbutan-1-one and (*R*)-1-hydroxy-1-phenylbutan-2-one, as well as the two symmetric products (*R*)-benzoin and (*R*)-propioin. However, in the case of *Pf*BAL catalyzed carboligation, the aromatic substrate benzaldehyde is preferred as the donor molecule over the aliphatic aldehyde propanal, thus eliminating two of the possible products [26]. In Figure 1, we show the corresponding reaction system, adapted from Ohs et al. [27].

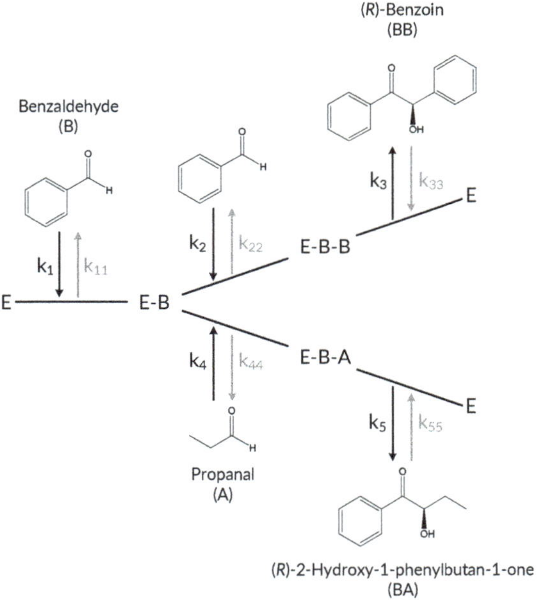

Figure 1. Branched reaction scheme for benzaldehyde lyase from *Pseudomonas fluorescens* (*Pf*BAL) catalyzed carboligations with benzaldehyde and propanal as substrates.

The simplified network can be described as three coupled reaction pathways. In the first step, benzaldehyde (B) forms a covalent bond with the ThDP cofactor in the *Pf*BAL (E) active site. Starting from this first intermediate, the pathway branches into two separate reactions depending on the second substrate: self-carboligation occurs if a second benzaldehyde molecule binds to the enzyme, acting as an acceptor molecule. The resulting (*R*)-benzoin (BB) is regarded as a side product in this study. In the cross-carboligation, propanal (A) acts as the acceptor substrate, leading to the desired product (*R*)-2-hydroxy-1-phenylbutan-1-one (BA). The self-carboligation and the cross-carboligation are modeled as ordered bi-uni-reaction mechanisms. The third reaction pathway describes the transfer reaction between the two carboligation products and is modeled as a ping-pong-bi-bi mechanism.

Because the reaction rate and *Pf*BAL inactivation depend on the substrate concentration [27], the objective is to maximize the final concentration of (*R*)-2-hydroxy-1-phenylbutan-1-one by tuning the fluxes due to substrate feeding. To our knowledge, this work constitutes the first study that designs and experimentally reproduces an optimal enzyme catalyzed carboligation using the model based EPF approach with a kinetic considering enzyme inactivation.

2. Results and Discussion

Following the EPF strategy, three intensification cases were investigated systematically to ascertain the best process intensification scenario for the maximization of the final concentration of (*R*)-2-hydroxy-1-phenylbutan-1-one produced from the *Pf*BAL catalyzed carboligation between propanal and benzaldehyde. A batch reactor was selected as the reference case because it is the most common reactor used for enzyme catalyzed reactions. Due to the rapid enzyme inactivation caused by the substrates (propanal and benzaldehyde), it was hypothesized that dosing concepts could be advantageous over the batch reactor resulting in the following case studies.

Case 1: Dosing of propanal. In this case, the propanal dosing flux was incorporated into the material balance and dynamically optimized. In addition to the dosing flux of propanal, the initial volume and the optimal initial concentrations of propanal and benzaldehyde required for the highest possible concentration of (*R*)-2-hydroxy-1-phenylbutan-1-one were degrees-of-freedom in the optimization.

Case 2: Dosing of benzaldehyde. In a similar fashion to Case 1, the dosing of only benzaldehyde along the reaction coordinate was investigated. The dosing rate of benzaldehyde, the initial volume, and initial concentrations of propanal and benzaldehyde were optimized with the same goal of maximizing the final concentration of the target product.

Case 3: Dosing of propanal and benzaldehyde. In Case 3, all tunable operating conditions were optimized for maximizing the cross-carboligation product concentration, namely the dynamic dosing fluxes of the reactants propanal and benzaldehyde, as well as the initial concentrations of the substrates and the enzyme.

In the subsequent sections, experimental data are presented for the reference case to confirm the validity of the kinetic model derived previously and for the predicted best intensification scenario. However, to carry out the model based optimization, a mathematical model of the underlying reaction phenomena that includes material and energy fluxes is required first. Note that in this study, energy balances are not considered, due to the mild conditions at which the *Pf*BAL catalyzed carboligation is performed and the assumption that these conditions are not energy intensive [2,3]. Only material balances that follow the so-called Lagrangian formulation [15] are considered. The essential terms in these balances are the reaction rate expressions, which are defined by a kinetic model in Section 3.1.2.

2.1. *Reference Case: Batch Reactor*

The concentration profiles of the optimized batch reactor are shown in Figure 2a-i, including the concentration progress of the respective experimental data. For better readability, concentrations of A and E are displayed on the second y-axis and indicated by arrows, and the concentration of active enzyme in the system has been shortened to "enzyme concentration".

The optimized initial concentrations for A and E of 100 $\mathrm{mmol\,L^{-1}}$ and 50 $\mathrm{\mu g\,mL^{-1}}$, respectively, are restricted by the upper bounds, whereas that of B of 5.94 $\mathrm{mmol\,L^{-1}}$ is not (Table 1). The optimization result is because the BA product formation rate is proportional to E and approximately proportional to A and B (Equations (14)–(16)). Note that the enzyme is inactivated much more strongly by benzaldehyde than by propanal, corresponding to the 15 fold higher inactivation rate parameters (Table 2). Due to the increased reaction rate towards BA and the lower inactivation rate parameter of A, a large excess concentration of the aliphatic compound compared to benzaldehyde seemed beneficial for the investigated objective function. The simulated final concentration of the active enzyme reached 1.08 $\mathrm{\mu g\,mL^{-1}}$, indicating nearly full enzyme inactivation of 97.8 % over the 180 min reaction time. The simulated optimal final BA concentration of 5.83 $\mathrm{mmol\,L^{-1}}$ together with the final propanal and benzaldehyde concentrations of 94.17 $\mathrm{mmol\,L^{-1}}$ and 0.11 $\mathrm{mmol\,L^{-1}}$, respectively, showed that propanal and benzaldehyde were converted quantitatively, with a conversion of over 98 % for the limiting substrate benzaldehyde and eliminating benzoin byproduct formation at the reaction endpoint after 180 min.

Table 1. Summary of the optimization results: initial concentrations $C_i(t_0)$, final concentrations $C_i(t_f)$, solution time for the reference case and the three process intensification scenarios, as well as the residual sum of squares (RSS) and mean squared error (MSE) for the experimental validations.

		Ref. Case	Case 1	Case 2	Case 3
$C_A(t_0)$	mmol L^{-1}	100	100	100	100
$C_B(t_0)$	mmol L^{-1}	5.94	6.00	0.71	0.70
$C_E(t_0)$	$\mu\text{g mL}^{-1}$	50	50	50	50
s_A(Dosing of A)	-	0	1	0	1
s_B(Dosing of B)	-	0	0	1	1
$C_{BA}(t_f)$	mmol L^{-1}	5.83	5.88	6.52	6.59
$C_E(t_f)$	$\mu\text{g mL}^{-1}$	1.08	0.95	1.26	1.08
RSS	$\text{mmol}^2\text{ L}^{-2}$	2.51	-	-	3.22
MSE	$\text{mmol}^2\text{ L}^{-2}$	0.05	-	-	0.06
Solution time	s	1.4	2.6	2.5	3.8

Table 2. Kinetic parameters for the *Pf*BAL kinetic model.

Rate Constant	Unit	Value
k_1	$\text{mmol}^{-1}\text{L min}^{-1}$	6184
k_{11}	min^{-1}	93.2
k_2	$\text{mmol}^{-1}\text{L min}^{-1}$	68,621
k_{22}	min^{-1}	7,294,883
k_3	min^{-1}	15,955
k_{33}	$\text{mmol}^{-1}\text{L min}^{-1}$	26,158
k_4	$\text{mmol}^{-1}\text{L min}^{-1}$	1.83
k_{44}	min^{-1}	0.00190
k_5	min^{-1}	41,610
k_{55}	$\text{mmol}^{-1}\text{L min}^{-1}$	373
$k_{\text{inact,A}}$	$\text{mmol}^{-1}\text{L min}^{-1}$	0.000157
$k_{\text{inact,B}}$	$\text{mmol}^{-1}\text{L min}^{-1}$	0.00246
$k_{\text{inact,time}}$	min^{-1}	0.00400

The initial decrease rate of B was higher than that of A for the first 7 min of the reaction, which coincided with the rapid formation of byproduct benzoin (BB) (Figure 2a-ii). These initial rates could be attributed to the higher affinity of the enzyme at the acceptor position to benzaldehyde over propanal [26]. The byproduct benzoin (BB) concentration peak indicated that the net BB formation rate equaled zero and may be explained by the enzyme specificity in the donor position, which was highest for BB (*Pf*BAL has been discovered as lyase [28]) followed by the aromatic benzaldehyde (B). In contrast, the aliphatic propanal was practically not accepted as the substrate in the presence of aromatic compounds. The rate of AB formation exactly mirrored that of A consumption. The experimental data successfully confirmed the predicted model optimum, although minor discrepancies could be observed; see Figure 2a-i. The experimentally determined initial concentrations of B amounted to 6.23 mmol L^{-1} instead of the optimal value of 5.94 mmol L^{-1}. The deviation in the initial concentrations, in turn, led to an initially higher concentration of the byproduct BB during the first 30 min of the reaction time. Up to around 60 min of the reaction time, the measured concentrations of substrate B and product BA were nearly superimposable with the simulation results, but dropped below the predicted values afterward. In addition, the mass balance with respect to B represented in the sum of B, BA, and two times BB was not closed over the reaction period, but instead declined slightly due to evaporation to finally reach 5.47 mmol L^{-1} or a loss of 11 %. The same evaporation loss also applied to Compound A, which, however, was assumed to have a negligible impact on the progress curves due to its high stoichiometric excess. However, the conversion of B to BA after the full reaction time matched the simulation. It is important to note the excellent quality of the optimization prediction to the experimental reproducibility run, in particular given that the kinetic model was obtained using a separate set of experimental data [27] and partially obtained in a different lab using different instrumentation.

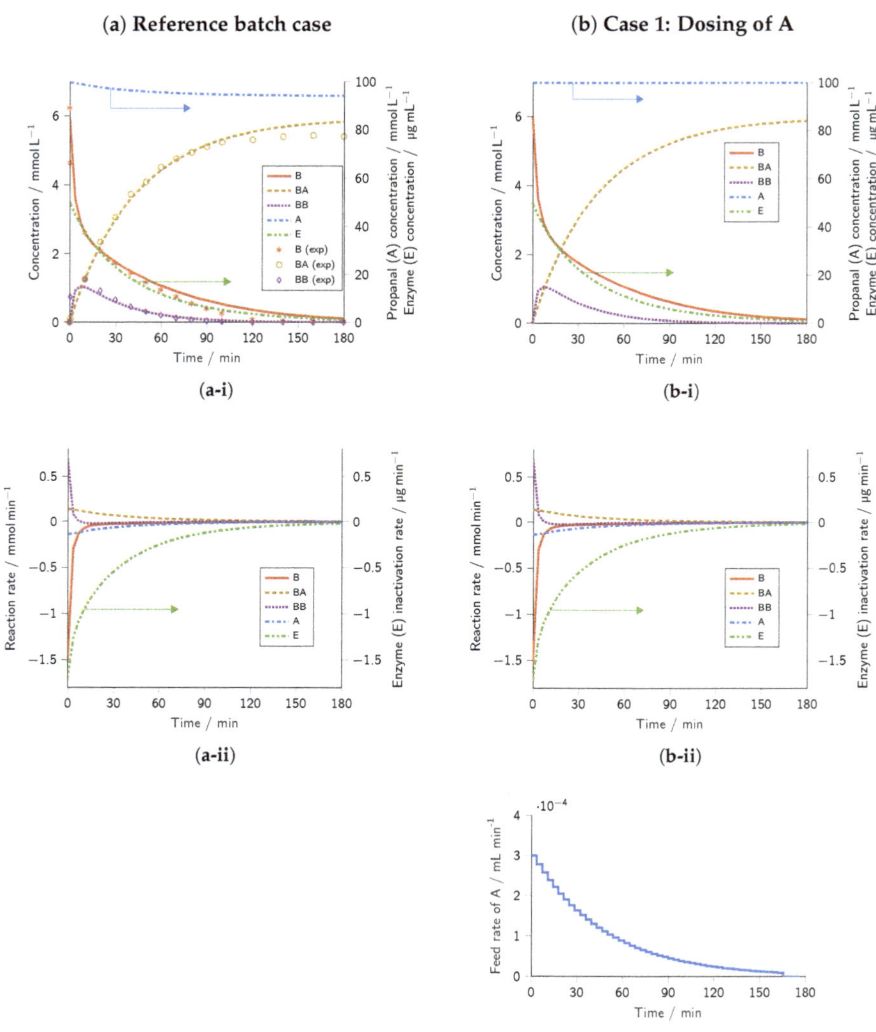

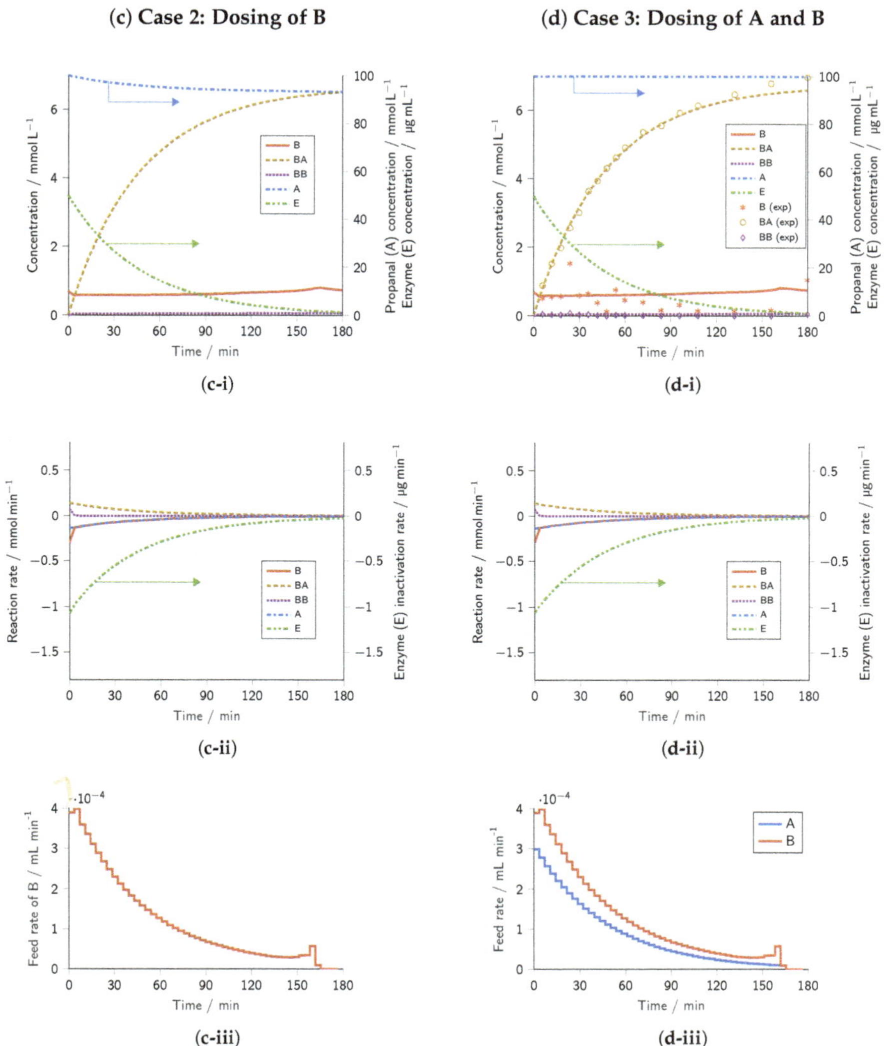

Figure 2. Results of the dynamic optimization and their experimental verification. Columns: (**a**) Batch reference, (**b**) Intensification Case 1 involving the dosing of A, (**c**) Intensification Case 2 involving the dosing of B, and (**d**) Intensification Case 3 involving the dosing of A and B. Rows: (**i**) concentration profiles, (**ii**) reaction rate profiles, and (**iii**) volumetric flow rate profiles. Starting concentrations within the design space were optimized so that the concentration of BA at final time point $t_f = 180\,\text{min}$ was maximized. Lower concentration bounds were defined as $0\,\text{mmol}\,\text{L}^{-1}$ and upper bounds as 100, 149.35, $2.78\,\text{mmol}\,\text{L}^{-1}$, and $50\,\mu\text{g}\,\text{mL}^{-1}$ for A, B, BB, and E, respectively. Experiments were conducted at 30 °C in a reaction volume of 30 mL (70 % v/v TEA buffer with cofactors and 30 % v/v DMSO) at pH = 8.5. Batch reference: $C_{\text{BA,sim}}(t_f) = 5.83\,\text{mmol}\,\text{L}^{-1}$, $C_{\text{BA,exp}}(t_f) = 5.40\,\text{mmol}\,\text{L}^{-1}$, Case 3: $C_{\text{BA,sim}}(t_f) = 6.59\,\text{mmol}\,\text{L}^{-1}$, $C_{\text{BA,exp}}(t_f) = 6.97\,\text{mmol}\,\text{L}^{-1}$.

In summary, optimizing the batch case for a maximum concentration of BA led to nearly full conversion of B with selectivity to the target compound of over 98%, albeit at the cost of almost complete enzyme inactivation.

2.2. Case 1: Dosing of Propanal

The optimization of the dosing of propanal (A) along the reaction coordinate led to a slightly higher allowable initial concentration of B of 6 mmol L^{-1} (those of E and A were again restricted by their upper bounds) and a slightly higher final BA concentration of 5.88 mmol L^{-1} at comparable 98 % conversion (Table 1). The concentration profile of reactant A remained at a constant value of 100 mmol L^{-1} throughout the course of the reaction (Figure 2b-i), which was given by the upper bound defined in Equation (17j). The constant propanal concentration was achieved by a decreasing propanal dosing rate along the reaction coordinate (Figure 2b-iii), which followed its consumption rate (Figure 2b-ii). The maintained optimal concentration of A showed that an increased reaction rate toward BA outweighed the enzyme inactivation caused by A. Note that setting the upper bound to a higher value beyond the design space limit used in this study might further increase the concentration of BA. Except for the concentration of A, all other concentration and reaction rate profiles resembled those of the reference batch case (Figure 2a-i,a-ii). In summary, the only marginal increase in the final concentration of BA for this case was essentially due to the constant concentration of A, but suggested that the added experimental effort for the intensification was probably not warranted.

2.3. Case 2: Dosing of Benzaldehyde

In the case of dosing benzaldehyde (B) (Figure 2c-i), the optimal initial concentration of B was reduced by nearly an order of magnitude to 0.71 mmol L^{-1} with A and E being restricted by their upper bounds. The concentration profile for A resembled that of the reference batch case. A final BA concentration of 6.52 mmol L^{-1} was obtained (Table 1), which corresponded to a 12 % increase over the batch reference.

In Figure 2c-i, we show that the concentration of B was maintained at a relatively constant low concentration. Similar to the propanal dosing case, the almost constant concentration of B was achieved by the dosing profile (Figure 2c-iii) mirroring the benzaldehyde consumption rate (Figure 2c-ii). However, the low B concentration drove the cross-carboligation pathway to maximize the formation of BA while the competing side product BB was formed at a relatively low concentration. In Figure 2c-ii, we highlight this effect, as the reaction rate of BB stays close to zero after a very short initial formation phase. The consumption rates of A and B nearly overlapped, which further supported the assumption that the optimal dosing of A shifted the reaction network significantly toward the conversion of A and B to form the target compound BA. However, the low concentration of B minimized the inactivation of the enzyme due to B, resulting in a significantly reduced enzyme inactivation rate (Figure 2c-ii), which was accompanied by a less steep enzyme activity loss than in the reference batch case (Figure 2a-i) and Case 1 (Figure 2b-i). As a result, a higher concentration of active enzyme was available for the production of BA at any time point. The low concentration of B at each time point, therefore, yielded the optimal balance between reaction kinetics toward BA and the conservation of *Pf*BAL.

2.4. Case 3: Dosing of Propanal and Benzaldehyde

The results for the case of dosing propanal (A) and benzaldehyde (B) simultaneously are shown in Figure 2d-i. The optimal initial concentration of B of 0.70 mmol L^{-1} was only slightly lower than on dosing of benzaldehyde and led to a final BA concentration of 6.59 mmol L^{-1} (Table 1) with a yield of 88 %. This product concentration corresponded to a 13 % increase over the batch reference case and was only slightly higher than the value obtained for Case 2, thus confirming that dosing of B played an essential role in maximizing the final concentration of BA.

The dosing of A and B in Case 3 could be considered as a superposition of Case 1 and Case 2. Thus, the discussions for Cases 1 and 2 also apply here. Dosing of A and B simultaneously led to the highest final BA concentration from the intensification cases. Therefore, dosing of propanal (A) and benzaldehyde (B) was the best intensification strategy among all cases considered in this work and was selected for experimental reproducibility. The experimental setup was technically approximated as a fed-batch reactor.

In Figure 2d-i, we show the overlay of the simulated concentration progress for the optimized initial conditions and feeding profiles with the experimental data, which confirmed the optimization results. The measured concentration of BA nearly perfectly overlayed with the simulation for approximately 120 min and then slightly diverged to higher values by up to 6 %. As the measured initial concentrations were in line with the optimization case, but more benzaldehyde substrate was converted toward the end of the experiment compared to the model prediction, slower enzyme inactivation in the real experiment than in the predicted model seemed a plausible cause for this deviation. Measured concentrations of BB stayed constantly below 0.1 mmol L^{-1}, which was in line with the simulation. The high ratio of BA to BB also confirmed the optimization result that keeping concentrations of B low during the experiment increased the selectivity toward BA and slowed down enzyme inactivation.

In contrast to the progress of BA and BB, the experimental concentration progress of B varied over time, presumably due to the manual sample withdrawal. Although the outlet of the dosing tube was submerged in the reaction solution, non-ideal mixing may have led to increased substrate concentrations near the tube outlet. For the products BA and BB, this issue did not apply as these compounds were homogeneously produced in the liquid body. An additional source of experimental error might exist due to HPLC sample storage in slightly variable vials, where the volatile substrate B may evaporate to the headspace, whereas BA and BB have low vapor pressures at ambient conditions.

In summary, the initial concentrations and feed rates obtained from the model based optimization applied to the laboratory setup could be experimentally reproduced and led to an enhanced final BA concentration, thus demonstrating successfully that the concentration of BA after a total runtime of 180 min could be increased by switching from the optimal batch scenario to an optimized fed-batch setup with time dependent feeding rates of Substrates A and B.

2.5. Data Summary

The optimized initial substrate and enzyme concentrations, together with the predicted final BA concentration and solution times for all cases, as well as the residual sum of squares and mean squared error for the experimental validations are summarized in Table 1. The solution times for all cases computed were between 1 and 4 s and, thus, orders of magnitude faster than the time required to perform laboratory experiments. As laboratory experiments successfully validated the model based predictions, they could significantly reduce the time for the development of enzymatic processes.

The model based prediction quality surpassed the typical results of response-surface-model optimization using statistical experimental design. Thus, this study bolsters the importance and justifies the effort to identify the proper mechanistically based kinetic model.

3. Materials and Methods

3.1. Computational Methods

3.1.1. Material Balances

The dynamic model consisted of the material balances for Species A, B, BA, BB, and E. Instead of using the mole balances, as typically done in the EPF approach, we used a concentration basis because of numerical robustness during optimization. Furthermore, a perfectly mixed fluid element was assumed to avoid additional complexity introduced by the residence time distribution. The corresponding mass balances for the reactants A and B are as follows:

$$\frac{dC_A}{dt} = \frac{j_A}{V} - \frac{C_A}{V}(s_A q_A + s_B q_B) + r_A, \tag{1}$$

$$\frac{dC_{\mathrm{B}}}{dt} = \frac{j_{\mathrm{B}}}{V} - \frac{C_{\mathrm{B}}}{V}(s_{\mathrm{A}}q_{\mathrm{A}} + s_{\mathrm{B}}q_{\mathrm{B}}) + r_{\mathrm{B}}, \tag{2}$$

where:

$$j_A = s_{\mathrm{A}}q_{\mathrm{A}}C_{\mathrm{A}}^{\mathrm{in}}, \tag{3}$$

$$j_B = s_{\mathrm{B}}q_{\mathrm{B}}C_{\mathrm{B}}^{\mathrm{in}}, \tag{4}$$

and where C_{A} and C_{B} denote the concentrations of propanal (A) and benzaldehyde (B) in the reactor, respectively; j, q, and r the corresponding fluxes, feeding rates, and reaction rates, respectively; whereas $C_{\mathrm{A}}^{\mathrm{in}}$ and $C_{\mathrm{B}}^{\mathrm{in}}$ denote the inlet feed concentrations of pure liquid propanal and benzaldehyde of 13.946 M and 9.8 M, respectively. To avoid redundancy and provide a generalized representation of the material balances for all cases considered in this work, the binary terms s_{A} and s_{B} were introduced. That is, the reference batch case and dosing of either A and B were modeled via the binary decision variables, s_{A} and s_{B}. For example, the material balance equations for the reference batch case could be obtained by setting $s_{\mathrm{A}} = 0$ and $s_{\mathrm{B}} = 0$. Moreover, note that the variables C_{A}, C_{B}, j_A, j_B, q_A, q_B, and V were all time varying, but for the sake of readability, the time varying argument was omitted. This also holds for variables C_{BA}, C_{BB}, and C_{E} in the following paragraphs. Similarly, the material balances for the products (*R*)-2-hydroxy-1-phenylbutan-1-one (BA) and benzoin (BB), and the enzyme *Pf*BAL (E), where no dosing occurs, are defined as:

$$\frac{dC_{\mathrm{BA}}}{dt} = -\frac{C_{\mathrm{BA}}}{V}(s_{\mathrm{A}}q_{\mathrm{A}} + s_{\mathrm{B}}q_{\mathrm{B}}) + r_{\mathrm{BA}}, \tag{5}$$

$$\frac{dC_{\mathrm{BB}}}{dt} = -\frac{C_{\mathrm{BB}}}{V}(s_{\mathrm{A}}q_{\mathrm{A}} + s_{\mathrm{B}}q_{\mathrm{B}}) + r_{\mathrm{BB}}, \tag{6}$$

$$\frac{dC_{\mathrm{E}}}{dt} = -\frac{C_{\mathrm{E}}}{V}(s_{\mathrm{A}}q_{\mathrm{A}} + s_{\mathrm{B}}q_{\mathrm{B}}) + r_{\mathrm{E}}. \tag{7}$$

Finally, the change in volume with time due to the amount of A and B added during the reaction is defined as follows:

$$\frac{dV}{dt} = s_{\mathrm{A}}q_{\mathrm{A}} + s_{\mathrm{B}}q_{\mathrm{B}}. \tag{8}$$

3.1.2. Mechanistic Kinetic Model

The mechanistic kinetic model was based on the reaction mechanism shown in Figure 1 and was derived by solving the mass balances for substrates, products, and enzyme species under steady state conditions for the enzyme intermediates. Furthermore, the model considered the rate of inactivation of *Pf*BAL, which was derived previously by combining progress curve analysis and optimal experimental design [27]. In this work, we additionally considered thermal deactivation of the enzyme, described by the reaction rate constant $k_{\mathrm{inact,time}}$. The mechanistic kinetic model is given by:

$$r_{\mathrm{A}} = -\frac{N_{\mathrm{BA}}}{D} \cdot C_{\mathrm{E}} \tag{9}$$

$$r_{\mathrm{B}} = -\frac{2N_{\mathrm{BB}} + N_{\mathrm{BA}}}{D} \cdot C_{\mathrm{E}} \tag{10}$$

$$r_{\mathrm{BA}} = \frac{N_{\mathrm{BA}}}{D} \cdot C_{\mathrm{E}} \tag{11}$$

$$r_{\mathrm{BB}} = \frac{N_{\mathrm{BB}}}{D} \cdot C_{\mathrm{E}} \tag{12}$$

$$r_{\mathrm{E}} = (-k_{\mathrm{inact,A}} \cdot C_{\mathrm{A}} - k_{\mathrm{inact,B}} \cdot C_{\mathrm{B}} - k_{\mathrm{inact,time}})\, C_{\mathrm{E}}. \tag{13}$$

For the ease of representation, the terms N_{BA}, N_{BB}, and D are defined as the following constitutive equations:

$$
\begin{aligned}
N_{BA} = {} & k_{22} \cdot k_{33} \cdot k_4 \cdot k_5 \cdot C_A \cdot C_{BB} \\
& - k_2 \cdot k_3 \cdot k_{44} \cdot k_{55} \cdot C_B \cdot C_{BA} \\
& + k_1 \cdot (k_{22} + k_3) \cdot k_4 \cdot k_5 \cdot C_A \cdot C_B \\
& - k_{11} \cdot (k_{22} + k_3) \cdot k_{44} \cdot k_{55} \cdot C_{BA},
\end{aligned} \tag{14}
$$

$$
\begin{aligned}
N_{BB} = {} & k_2 \cdot k_3 \cdot k_{44} \cdot k_{55} \cdot C_B \cdot C_{BA} \\
& - k_{22} \cdot k_{33} \cdot k_4 \cdot k_5 \cdot C_A \cdot C_{BB} \\
& + k_1 \cdot k_2 \cdot k_3 \cdot (k_{44} + k_5) \cdot C_B^2 \\
& - k_{11} \cdot k_{22} \cdot k_{33} \cdot (k_{44} + k_5) \cdot C_{BB},
\end{aligned} \tag{15}
$$

$$
\begin{aligned}
D = {} & k_{11} \cdot (k_{22} + k_3) \cdot (k_{44} + k_5) \\
& + k_1 \cdot (k_{22} + k_3) \cdot k_4 \cdot C_A \cdot C_B \\
& + (k_{22} + k_3) \cdot k_4 \cdot k_{55} \cdot C_A \cdot C_{BA} \\
& + k_{33} \cdot k_4 \cdot (k_{22} + k_5) \cdot C_A \cdot C_{BB} \\
& + k_1 \cdot k_2 \cdot (k_{44} + k_5) \cdot C_B^2 \\
& + k_2 \cdot (k_3 + k_{44}) \cdot k_{55} \cdot C_B \cdot C_{BA} \\
& + k_2 \cdot k_{33} \cdot (k_{44} + k_5) \cdot C_B \cdot C_{BB} \\
& + (k_{22} + k_3) \cdot k_4 \cdot k_5 \cdot C_A \\
& + k_1 \cdot (k_{22} + k_3) \cdot (k_{44} + k_5) \cdot C_B \\
& + k_2 \cdot k_3 \cdot (k_{44} + k_5) \cdot C_B \\
& + (k_{11} + k_{44}) \cdot (k_{22} + k_3) \cdot k_{55} \cdot C_{BA} \\
& + (k_{11} + k_{22}) \cdot k_{33} \cdot (k_{44} + k_5) \cdot C_{BB}.
\end{aligned} \tag{16}
$$

Note that the original parameter values in [27] were not directly applicable as these parameters were derived for a kinetic model without thermal *Pf*BAL inactivation. However, in this study, all kinetic parameters in Equations (9)–(16) were estimated using the methods and data presented in [27] and are summarized in Table 2.

3.1.3. Dynamic Optimization Problem and Solution Strategy

The dynamic optimization problem aimed to maximize the final concentration of BA with respect to the time varying material balances, the volume change, and the reaction kinetic equations and is defined for the various intensification cases as:

$$\begin{aligned}
\underset{\substack{j_A(t),\, j_B(t),\, C_{A0},\\ C_{B0},\, C_{E0},\, V_0}}{\text{maximize}} \quad & C_{BA}(t_f) && \text{(17a)}\\
\text{subject to} \quad & \text{Material balances: Equations (1) – (7),} && \text{(17b)}\\
& \text{Volume change: Equation (8),} && \text{(17c)}\\
& \text{Reaction kinetics: Equations (9) – (16),} && \text{(17d)}\\
& 0 \leq q_i(t) \leq 0.1\,\mathrm{L\,min^{-1}}, \quad \text{for all } i \text{ in } \{A, B\}, && \text{(17e)}\\
& 0 \leq V_0 \leq 3 \times 10^{-2}\,\mathrm{L}, && \text{(17f)}\\
& 0 \leq V(t) \leq 3 \times 10^{-2}\,\mathrm{L}, && \text{(17g)}\\
& 0 \leq C_{i,0} \leq C_{i,0}^{UB}, \quad \text{for all } i \text{ in } \{A, B, BB, E\}, && \text{(17h)}\\
& 0 \leq C_i(t) \leq C_i^{UB}, \quad \text{for all } i \text{ in } \{A, B, BB, E\}, && \text{(17i)}\\
& C_{A,0}^{UB} = C_A^{UB} = 100\,\mathrm{mmol\,L^{-1}}, && \text{(17j)}\\
& C_{B,0}^{UB} = C_B^{UB} = 149.35\,\mathrm{mmol\,L^{-1}}, && \text{(17k)}\\
& C_{BB,0}^{UB} = C_{BB}^{UB} = 2.78\,\mathrm{mmol\,L^{-1}}, && \text{(17l)}\\
& C_{E,0}^{UB} = C_E^{UB} = \frac{50}{\mathrm{Mwt_E}} = 8.4862 \times 10^{-4}\,\mathrm{mmol\,L^{-1}}, && \text{(17m)}\\
& t_f = 180\,\mathrm{min}, && \text{(17n)}
\end{aligned}$$

The upper bounds on $q(t)$ were set to 100 $\mathrm{mL\,min^{-1}}$, which was the maximum pumping rate of the dosing pumps (Equation (17e)); those for V_0 and $V(t)$ were set to 3×10^{-2} L, which was the maximum volume of the reactor used in the experimental setup (Equations (17f) and (17g)).

The inequality constraints (Equations (17h) and (17i)) represent the bounds for the initial and time varying concentrations $C_{i,0}$ and $C_i(t)$, respectively. The corresponding upper bounds were set to their respective maximum values used when estimating the kinetic parameters with C_A^{UB} of 100 $\mathrm{mmol\,L^{-1}}$ and C_E^{UB} of 50 $\mathrm{\mu g\,mL^{-1}}$, with $\mathrm{Mwt_E}$ = 58919 g/mol (Equations (17j) and (17m)). The upper bounds for B and BB were set to their solubility limits (Equations (17k) and (17l)). Finally, the maximum reaction time t_f was set to 180 min for the experimental duration time.

The dynamic optimization problem (Equation (17)) was solved by using the direct simultaneous approach, because of its ability to efficiently handle numerical instabilities and path constraints [29,30]. Specifically, the dynamic optimization problem for all cases was transcribed into algebraic equations by using the method of orthogonal collocation on finite elements [31]. The resulting system of algebraic equations falls into a class of optimization problems called nonlinear programming (NLP) [30]. The NLP was implemented in the MATLAB API for CasADi 3.4.0, a framework for automatic differentiation and numerical optimization [32]. Furthermore, 50 finite elements and three collocation points were used to discretize the dynamic optimization problems for all intensification cases considered. The number of 50 finite elements selected was found to provide a good trade-off between numerical accuracy and computational cost; while the three collocation points were chosen as recommended in [30]. The resulting NLP was solved by using IPOPT, an interior point solver designed for large scale NLPs [33], in combination with the sparse symmetric linear solver MA57 [34,35]. All computations were performed on a Linux computer running a CentOS 7 operating system with an Intel(R) Core(TM) i7-4789 processor at 3.60 GHz and 16 GB RAM.

3.2. Experimental Setup

3.2.1. Materials

All chemicals used in this study were of analytical grade and used as purchased. Benzoin (racemic), propanal, thiamine diphosphate (ThDP), and triethanolamine (TEA) were purchased from Sigma Aldrich; acetonitrile (ACN), dimethyl sulfoxide (DMSO), and Bradford reagent from Carl Roth GmbH + Co. KG; 3,5-dimethoxy-benzaldehyde (DMBA), benzaldehyde, and bovine serum albumin (BSA) from Thermo Fisher Scientific; and trichloroacetic acid (TCA) from PanReac AppliChem. No further purification steps were performed before the experiment. Pure (*R*)-2-hydroxy-1-phenylbutan-1-one was provided by the Institute of Bio- and Geosciences at Forschungszentrum Jülich, Germany. *Pf*BAL was kindly provided by the Institute of Bio- and Geosciences at Forschungszentrum Jülich as well; the His tagged enzyme was expressed in *Escherichia coli* SG13009, subsequently purified via immobilized metal affinity chromatography on a Ni-NTA-column, and finally desalted with size exclusion chromatography [36]. The pooled fractions were diluted to a concentration of 1 mg mL^{-1} with a desalting buffer (10 mmol L^{-1} TEA, 2.5 mmol L^{-1} $MgSO_4$, 0.5 mmol L^{-1} ThDP, adjusted to pH 7.5 with 1 mol L^{-1} HCl), lyophilized for several days, and then kept at −20 °C in an airtight container.

3.2.2. Preparation of *Pf*BAL Solution

Before each experiment, the enzyme solution and the reactant solution were freshly prepared in separate flasks. For the enzyme solution, the *Pf*BAL lyophilizate was added to 10–20 mL of a prepared TEA buffer solution (50 mmol L^{-1} TEA, adjusted to pH 8.5 with 1 mol L^{-1} HCl) containing cofactors (0.72 mmol L^{-1} ThDP and 3.56 mmol L^{-1} Mg^{2+}). The *Pf*BAL concentration in the stock solution was determined with a standard Bradford assay [37] using a Tecan Spark photometer. The enzyme stock solution was appropriately diluted with 50 mmol L^{-1} TEA buffer solution containing cofactors to the concentration desired for the specific reaction.

*Pf*BAL activity was measured via the *Pf*BAL catalyzed ligation of fluorescent DMBA to (*R*)-3,3′,5,5′-tetramethoxy-benzoin, causing a decrease in fluorescence over time [38]. Three different solutions with 70 % v/v TEA buffer with cofactors and 30 % v/v DMSO were prepared for the activity test. Solution I was obtained by diluting the enzyme stock solution with TEA buffer with cofactors and DMSO to a final *Pf*BAL concentration of 15 µg mL^{-1}. For Solution II, 2 mmol L^{-1} DMBA was dissolved, and Solution III remained without further additions. Four samples and four blanks were prepared in a 96 well microtiter plate by mixing 20 µL of Solution I with 180 µL of either Solution II or III. The plate was placed inside the preheated plate reader (Tecan Spark) at a temperature of 30 °C, and the emission intensity at 460 nm with an excitation at 362 nm was measured in intervals of 5 s for a total time of 90 s. The activity of *Pf*BAL was calculated with an existing calibration curve for the fluorescence of DMBA at varying concentrations.

For the reactant solution, benzaldehyde and propanal were dissolved in DMSO. This step was done immediately before starting the experiment to avoid losses of any volatile substances.

3.2.3. Progress Curve Experiments

The results of dynamic optimization for the intensification cases were experimentally reproduced by performing progress curve experiments under the respective optimal conditions. The experiments were conducted in a magnetically stirred glass cylinder, which was placed in a water bath for temperature control at 30 °C. To minimize the evaporation of volatile Substances A and B during the experimental procedure, a PTFE lid was placed inside the glass cylinder right above the fluid level. Small holes in the PTFE lid gave access for the feeding tubes. The position of the lid was manually adjusted to correct for volume changes caused by sampling and, to a lesser extent, dosing.

The reaction was initiated by pouring both solutions (9 mL of reactant solution and 21 mL of enzyme solution) into the glass cylinder. The liquid was homogenized with the magnetic stirrer, and the lid was placed into position. After 10 s, a 100 µL zero sample was taken from the mixture and transferred to a 1.1 mL high performance liquid chromatography (HPLC) vial containing 50 µL of 1.25 % TCA solution to quench the enzymatic reaction immediately. About 950 µL of ACN were added to dilute the sample for the following analysis and to minimize the headspace inside the HPLC vial. Samples were taken at 6 to 10 min intervals for the first hour. Then, the sampling frequency was reduced to minimize the overall extraction of the reaction volume. All samples were prepared for analysis in HPLC on the same day or stored in a freezer at −20 °C until further processing.

In the case of the fed-batch experiments, two glass syringes were filled with A and B, respectively, before the start of the experiment and positioned in high precision neMESYS 290N syringe pumps (Cetoni GmbH, Korbussen). PTFE tubes connected to the syringes were threaded through the holes of the PTFE lid into the reaction vessel and submerged in the reaction mixture. The feeding profile obtained from the dynamic optimizations for both substrates was individually programmed via the neMESYS UserInterface software. Dosings of A and B were initiated simultaneously with the zero sample.

3.2.4. HPLC Analysis

All samples from the progress curve experiments were filtered through 0.2 µm PTFE syringe filters and were subsequently analyzed in reversed phase chromatography (RP-HPLC). About 30 µL of the sample volume were injected onto an RP8-column (LiChrospher 100 RP-8 (5 µm), Merck KGaA) at 40 °C. The eluent was a mixture of ultrapure water and ACN with a flow rate of $1.2\,\mathrm{mL\,min^{-1}}$. Initially, the eluent composition was 82 % v/v ultrapure water and 18 % v/v ACN. After 5.5 min, the composition was gradually changed within 2 min to 100 % ACN and maintained for 1 min. Then, the composition was gradually changed back to 82 % v/v ultrapure water and 18 % v/v ACN within 2 min. The column was then flushed for 3 min to conclude the method, with a total runtime of 13.5 min. Peaks of the individual compounds appeared after retention times of 6.1 min for B, 8.5 min for BA, and 9.1 min for BB. HPLC calibration was performed using standard samples of B, BB, and BA dissolved in 70 % v/v TEA buffer solution and 30 % v/v DMSO.

4. Conclusions

This contribution presents the elementary process functions (EPF) approach as a viable model based tool for designing optimal operating conditions for enzyme catalyzed reactions. For *Pf*BAL catalyzed cross-carboligation of benzaldehyde and propanal, the final concentration of the product (*R*)-2-hydroxy-1-phenylbutan-1-one was chosen as the objective function to be optimized, and three process intensification scenarios were evaluated making use of the EPF concept.

The best final product concentration was obtained when dosing both aldehyde substrates, leading to a 13 % increase compared to the optimized batch case that was used as the reference scenario. It was found that dosing propanal only served in replenishing the amount converted during the reaction so that the concentration constantly stayed at the upper limit specified in the design space. In contrast, dosing of benzaldehyde was found to be optimal when the concentration was set to a low initial value and never reached higher than $0.81\,\mathrm{mmol\,L^{-1}}$. This strategy slowed down the *Pf*BAL inactivation considerably, as well as the competing reaction to the byproduct benzoin. Although fed-batch processes are widely used in cases where substrates inhibit or even inactivate enzymes, determining time dependent dosing strategies is feasible only using model based approaches as presented. The value of the model based design becomes even more significant for optimization problems where the objective function is less intuitive than the product concentration. In the case of dosing propanal and benzaldehyde simultaneously and the derived optimal feeding profiles, experimental data confirmed the predicted results in the lab.

The apparatus independent EPF approach presented in this work could serve as a tool that enables process engineers to design novel processes and reactors systematically for enzyme catalyzed reaction systems in general. Additional reaction conditions, such as temperature, pH value, and composition of the reaction medium, might be incorporated into the kinetic model and open up the possibility for further improvements. Once the effects of protein engineering become tractable by appropriate structure–function relations, EPF based optimization of biocatalytic reactions may contribute to combining protein engineering and process systems engineering concepts [6,39].

Author Contributions: Conceptualization, V.N.E., D.H., R.S., and A.C.S.; investigation, D.H.; software, V.N.E.; writing, original draft preparation, V.N.E. and D.H.; writing, review and editing, R.S. and A.C.S.; visualization, V.N.E. and D.H.; supervision, R.S. and A.C.S. All authors have read and agreed to the published version of the manuscript.

Funding: This research was funded by the German Research Foundation (DFG), Research Group for Diversity of Asymmetric Thiamine Catalysis, Grant/Award Number: 128900243.

Acknowledgments: We thank the Institute of Bio- and Geosciences at Forschungszentrum Jülich, especially Martina Pohl, for providing (*R*)-2-hydroxy-1-phenylbutan-1-one, as well as benzaldehyde lyase lyophilizate. The author V.N.E. is also affiliated with the International Max Planck Research School for Advanced Methods in Process and Systems Engineering (IMPRS-ProEng), Magdeburg, Germany. Moreover, we thank Ulrike Krewer for helpful comments on a previous version of the manuscript and Rüdiger Ohs for his support regarding experimental and modeling procedures. We also acknowledge support from the German Research Foundation and the Open Access Publication Funds of the Technische Universität Braunschweig.

Conflicts of Interest: The authors declare no conflict of interest. The funders had no role in the design of the study; in the collection, analyses, or interpretation of data; in the writing of the manuscript; nor in the decision to publish the results.

Abbreviations

Mwt	molecular weight ($g\,mol^{-1}$)
E. coli	*Escherichia coli*
*Pf*BAL	benzaldehyde lyase from *Pseudomonas fluorescens*
A	propanal
ACN	acetonitrile
B	benzaldehyde
BA	(*R*)-2-hydroxy-1-phenylbutan-1-one
BB	(*R*)-benzoin
CSTR	continuous stirred tank reactor
DMBA	3,5-dimethoxy-benzaldehyde
DMSO	dimethyl sulfoxide
E	enzyme
EPF	elementary process functions
HPLC	high performance liquid chromatography
HPP	(*R*)-2-hydroxy-1-phenylpropanone
NLP	nonlinear programming
PTFE	polytetrafluoroethylene
RP-HPLC	reversed phase chromatography
TEA	triethanolamine
ThDP	thiamine diphosphate
Indices, subscripts and superscripts	
f	final
i	species or component index
0	initial
in	inlet feed
UB	upper bound
Latin symbols	
C	concentration ($mmol\,L^{-1}$)
D	denominator term for the reaction rate
j	dosing flux ($mmol\,min^{-1}$)
k	reaction rate constant

N numerator term for the reaction rate
q feeding rate (L min^{-1})
r reaction rate (mmol L^{-1} min^{-1})
s binary variable (−)
t reaction time (min^{-1})
V volume (L)

References

1. Lalonde, J. Highly engineered biocatalysts for efficient small molecule pharmaceutical synthesis. *Curr. Opin. Biotechnol.* **2016**, *42*, 152–158. [CrossRef] [PubMed]
2. Pollard, D.J.; Woodley, J.M. Biocatalysis for pharmaceutical intermediates: The future is now. *Trends Biotechnol.* **2007**, *25*, 66–73. [CrossRef] [PubMed]
3. Woodley, J.M. New opportunities for biocatalysis: Making pharmaceutical processes greener. *Trends Biotechnol.* **2008**, *26*, 321–327. [CrossRef] [PubMed]
4. Ringborg, R.H.; Woodley, J.M. The application of reaction engineering to biocatalysis. *React. Chem. Eng.* **2016**, *1*, 10–22. [CrossRef]
5. Kiss, A.A.; Grievink, J.; Rito-Palomares, M. A systems engineering perspective on process integration in industrial biotechnology. *J. Chem. Technol. Biotechnol.* **2015**, *90*, 349–355. [CrossRef]
6. Woodley, J.M. Bioprocess intensification for the effective production of chemical products. *Comput. Chem. Eng.* **2017**, *105*, 297–307. [CrossRef]
7. Braun, M.; Link, H.; Liu, L.; Schmid, R.D.; Weuster-Botz, D. Biocatalytic process optimization based on mechanistic modeling of cholic acid oxidation with cofactor regeneration. *Biotechnol. Bioeng.* **2011**, *108*, 1307–1317. [CrossRef]
8. Schenkendorf, R. Supporting the shift towards continuous pharmaceutical manufacturing by condition monitoring. In Proceedings of the Conference on Control and Fault-Tolerant Systems, SysTol, Barcelona, Spain, 7–9 September 2016; pp. 593–598.
9. Stillger, T.; Pohl, M.; Wandrey, C.; Liese, A. Reaction engineering of benzaldehyde lyase from *Pseudomonas fluorescens* catalyzing enantioselective C-C bond formation. *Org. Process Res. Dev.* **2006**, *10*, 1172–1177. [CrossRef]
10. Hildebrand, F.; Kühl, S.; Pohl, M.; Vasic-Racki, D.; Müller, M.; Wandrey, C.; Lütz, S. The production of (R)-2-hydroxy-1-phenyl-propan-1-one derivatives by benzaldehyde lyase from Pseudomonas fluorescens in a continuously operated membrane reactor. *Biotechnol. Bioeng.* **2007**, *96*, 835–843. [CrossRef]
11. Begemann, J.; Ohs, R.B.; Ogolong, A.B.; Eberhard, W.; Ansorge-Schumacher, M.B.; Spiess, A.C. Model based analysis of a reactor and control concept for oxidoreductions based on exhaust CO2-measurement. *Process Biochem.* **2016**, *51*, 1397–1405. [CrossRef]
12. Marpani, F.; Sárossy, Z.; Pinelo, M.; Meyer, A.S. Kinetics based reaction optimization of enzyme catalyzed reduction of formaldehyde to methanol with synchronous cofactor regeneration. *Biotechnol. Bioeng.* **2017**, *114*, 2762–2770. [CrossRef] [PubMed]
13. Emenike, V.N.; Schenkendorf, R.; Krewer, U. A systematic reactor design approach for the synthesis of active pharmaceutical ingredients. *Eur. J. Pharm. Biopharm.* **2018**, *126*, 75–88. [CrossRef] [PubMed]
14. Freund, H.; Sundmacher, K. Towards a methodology for the systematic analysis and design of efficient chemical processes. *Chem. Eng. Process. Process Intensif.* **2008**, *47*, 2051–2060. [CrossRef]
15. Peschel, A.; Freund, H.; Sundmacher, K. Methodology for the design of optimal chemical reactors based on the concept of elementary process functions. *Ind. Eng. Chem. Res.* **2010**, *49*, 10535–10548. [CrossRef]
16. Hentschel, B.; Peschel, A.; Freund, H.; Sundmacher, K. Simultaneous design of the optimal reaction and process concept for multiphase systems. *Chem. Eng. Sci.* **2014**, *115*, 69–87. [CrossRef]
17. Jokiel, M.; Kaiser, N.M.; Kováts, P.; Mansour, M.; Zähringer, K.; Nigam, K.D.P.; Sundmacher, K. Helically coiled segmented flow tubular reactor for the hydroformylation of long-chain olefins in a thermomorphic multiphase system. *Chem. Eng. J.* **2019**. [CrossRef]
18. Kaiser, N.M.; Jokiel, M.; McBride, K.; Flassig, R.J.; Sundmacher, K. Optimal reactor design via flux profile analysis for an integrated hydroformylation process. *Ind. Eng. Chem. Res.* **2017**, *56*, 11507–11518. [CrossRef]

19. Maußner, J.; Dreiser, C.; Wachsen, O.; Freund, H. Systematic model based design of tolerant chemical reactors. *J. Adv. Manuf. Process.* **2019**. [CrossRef]
20. Peschel, A.; Karst, F.; Freund, H.; Sundmacher, K. Analysis and optimal design of an ethylene oxide reactor. *Chem. Eng. Sci.* **2011**, *66*, 6453–6469. [CrossRef]
21. Freund, H.; Maußner, J.; Kaiser, M.; Xie, M. Process intensification by model based design of tailor-made reactors. *Curr. Opin. Chem. Eng.* **2019**, *26*, 46–57. [CrossRef]
22. Emenike, V.N.; Schenkendorf, R.; Krewer, U. Model based optimization of biopharmaceutical manufacturing in *Pichia pastoris* based on dynamic flux balance analysis. *Comput. Chem. Eng.* **2018**, *118*, 1–13. [CrossRef]
23. Emenike, V.N.; Xie, X.; Schenkendorf, R.; Spiess, A.C.; Krewer, U. Robust dynamic optimization of enzyme catalyzed carboligation: A point estimate based back-off approach. *Comput. Chem. Eng.* **2019**, *121*, 232–247. [CrossRef]
24. Müller, M.; Sprenger, G.A.; Pohl, M. C–C bond formation using ThDP dependent lyases. *Curr. Opin. Chem. Biol.* **2013**, *17*, 261–270. [CrossRef]
25. Demir, A.S.; Pohl, M.; Janzen, E.; Müller, M. Enantioselective synthesis of hydroxy ketones through cleavage and formation of acyloin linkage. enzymatic kinetic resolution via C–C bond cleavage. *J. Chem. Soc. Perkin Trans.* **2001**, *1*, 633–635. [CrossRef]
26. Pohl, M.; Gocke, D.; Müller, M. Thiamine based enzymes for biotransformations. In *Handbook of Green Chemistry*; Wiley-VCH: Weinheim, Germany, 2009.
27. Ohs, R.; Fischer, K.; Schoepping, M.; Spiess, A.C. Derivation and identification of a mechanistic model for a branched enzyme catalyzed carboligation. *Biotechnol. Prog.* **2019**, *e2868*. [CrossRef] [PubMed]
28. González, B.; Vicuña, R. Benzaldehyde lyase, a novel thiamine PP-requiring enzyme, from *Pseudomonas fluorescens* biovar 1. *J. Bacteriol.* **1989**, *171*, 2401–2405. [CrossRef] [PubMed]
29. Biegler, L.T. An overview of simultaneous strategies for dynamic optimization. *Chem. Eng. Process. Process Intensif.* **2007**, *46*, 1043–1053. [CrossRef]
30. Biegler, L.T. *Nonlinear Programming*; Society for Industrial and Applied Mathematics: Philadelphia, PA, USA, 2010.
31. Cuthrell, J.E.; Biegler, L.T. On the optimization of differential-algebraic process systems. *AIChE J.* **1987**, *33*, 1257–1270. [CrossRef]
32. Andersson, J.A.E.; Gillis, J.; Horn, G.; Rawlings, J.B.; Diehl, M. CasADi: A software framework for nonlinear optimization and optimal control. *Math. Program. Comput.* **2019**, *11*, 1–36. [CrossRef]
33. Wächter, A.; Biegler, L.T. On the implementation of an interior-point filter line-search algorithm for large-scale nonlinear programming. *Math. Program.* **2006**, *106*, 25–57. [CrossRef]
34. Duff, I.S. MA57—A code for the solution of sparse symmetric definite and indefinite systems. *ACM Trans. Math. Softw.* **2004**, *30*, 118–144. [CrossRef]
35. HSL. A Collection of Fortran Codes for Large Scale Scientific Computation. Available online: http://www.hsl.rl.ac.uk (accessed on 29 September 2017).
36. Ohs, R.; Leipnitz, M.; Schöpping, M.; Spiess, A.C. Simultaneous identification of reaction and inactivation kinetics of an enzyme catalyzed carboligation. *Biotechnol. Prog.* **2018**, *34*, 1081–1092. [CrossRef] [PubMed]
37. Bradford, M.M. A rapid and sensitive method for the quantitation of microgram quantities of protein utilizing the principle of protein-dye binding. *Anal. Biochem.* **1976**, *72*, 248–254. [CrossRef]
38. Zavrel, M.; Schmidt, T.; Michalik, C.; Ansorge-Schumacher, M.; Marquardt, W., Büchs, J.; Spiess, A.C. Mechanistic kinetic model for symmetric carboligations using benzaldehyde lyase. *Biotechnol. Bioeng.* **2008**, *101*, 27–38. [CrossRef] [PubMed]
39. Woodley, J.M. Integrating protein engineering with process design for biocatalysis. *Philos. Trans. R. Soc. A Math. Phys. Eng. Sci.* **2018**, *376*. [CrossRef] [PubMed]

Article

Conversion of Shrimp Head Waste for Production of a Thermotolerant, Detergent-Stable, Alkaline Protease by *Paenibacillus* sp.

Chien Thang Doan [1,2], Thi Ngoc Tran [1,2,3,4], I-Hong Wen [1], Van Bon Nguyen [2], Anh Dzung Nguyen [5] and San-Lang Wang [1,6,*]

1 Department of Chemistry, Tamkang University, New Taipei City 25137, Taiwan; doanthng@gmail.com (C.T.D.); tranngoctnu@gmail.com (T.N.T.); winie_mayday@hotmail.com (I.H.W.)

2 Department of Science and Technology, Tay Nguyen University, Buon Ma Thuot 630000, Vietnam; bondhtn@gmail.com

3 Doctoral Program in Applied Sciences, College of Science, Tamkang University, New Taipei City 25137, Taiwan

4 Department of Chemical and Materials Engineering, Tamkang University, New Taipei City 25137, Taiwan

5 Institute of Biotechnology and Environment, Tay Nguyen University, Buon Ma Thuot 630000, Vietnam; nadzungtaynguyenuni@yahoo.com.vn

6 Life Science Development Center, Tamkang University, New Taipei City 25137, Taiwan

* Correspondence: sabulo@mail.tku.edu.tw; Tel.: +886-2-2621-5656; Fax: +886-2-2620-9924

Received: 14 August 2019; Accepted: 23 September 2019; Published: 24 September 2019

Abstract: Fishery processing by-products have been of great interest to researchers due to their beneficial applications in many fields. In this study, five types of marine by-products, including demineralized crab shell, demineralized shrimp shell, shrimp head, shrimp shell, and squid pen, provided sources of carbon and nitrogen nutrition by producing a protease from *Paenibacillus* sp. TKU047. Strain TKU047 demonstrated the highest protease productivity (2.98 U/mL) when cultured for two days on a medium containing 0.5% of shrimp head powder (SHP). The mass of TKU047 protease was determined to be 32 kDa (approximately). TKU047 protease displayed optimal activity at 70–80 °C and pH 9, with a pH range of stability from 6 to 11. TKU047 protease also showed stability in solutions containing surfactants and detergents. Based on its excellent properties, *Paenibacillus* sp. TKU047 protease may be a feasible candidate for inclusion in laundry detergents.

Keywords: marine chitinous by-products; *Paenibacillus*; protease; shrimp heads

1. Introduction

Every year, the seafood processing industry discards a large amount of by-products, including viscera, shells, heads, squid pens, fins, and bones, even though they could be recycled to produce bioactive compounds like gelatin [1–4], enzymes [4–17], chitin [8,18–26], chitin oligomers [7,11,12], α-glucosidase inhibitors (aGI) [27–31], carotenoids [32,33], and bioactive peptides [34–40]. Consequently, much research has gone into converting these by-products into bioactive products that have potential applications in biotechnological, agricultural, nutritional, pharmaceutical, and biomedical industries [1,4,5,17]. Marine chitinous byproducts, such as shells of crabs and shrimps, and pens (gladius) of squids, are a great source of chitin. However, these materials also contain a significant amount of mineral salts and proteins [4,8,17]. As a result, strong alkalis and acids are typically used for deproteinization and demineralization during the production of chitin and chitosan. However, there are several drawbacks to the use of chemical procedures. With a more environmentally friendly production process, these marine byproducts could be used as carbon/nitrogen sources for microorganism bioconversion to various high-value products, including chitinase/chitosanase [6,11–13,15,16,20], proteases [13,14], exopolysaccharides [41–43], and tyrosinase inhibitors [44].

Proteases break down proteins to release peptides and amino acid bases upon catalyzing the hydrolytic reaction [45]. Since there is a huge demand for proteases in many fields, including detergent, brewing, meat, photography, leather, and dairy industries, research on the production and application of proteases is ongoing [46,47]. Compared to plant and animal sources, proteases from microorganisms have more advantages as they can be cultured at an industrial scale, have a short fermentation period, and are easy to obtain [45]. In addition, microorganisms could utilize a wide range of C/N sources, including various types of fishery by-products, to produce proteases [45], which would help reduce the cost of protease production and make their application more attractive.

Paenibacillus is a useful bacterial family employed in the medical, industrial, and agricultural fields [48]. Recently, when fishery processing by-products were used as the nutrition source (carbon and nitrogen), several strains of *Paenibacillus* exhibited excellent abilities in the production of bioactive compounds, including α-glucosidase inhibitors [49], exopolysaccharides [43], anti-inflammatory medication [50,51], antioxidants [30], and chitosanases [11–13]. However, few reports have examined proteases derived from *Paenibacillus*, especially using shrimp heads. While several *Paenibacillus* strains demonstrated protease productivity on media containing squid pen or demineralized crab shell, the properties of their proteases were not explored [13].

In the current study, a proteolytic strain screened from soil samples at Tamkang University, *Paenibacillus* sp. TKU047, used squid pen powder (SPP) as the main source for providing carbon, as well as nitrogen. The optimal conditions for protease production on five types of fishery processing by-products including SPP, shrimp shell powder (SSP), demineralized crab shell powder (deCSP), shrimp head powder (SHP), and demineralized shrimp shell powder (deSSP), as well as the enzyme characteristics were investigated. In order to determine its industrial potential, the detergent compatibility of *Paenibacillus* sp. TKU047 was also explored.

2. Results and Discussions

2.1. Screening of a Proteolytic Strain

More than 60 bacterial strains were obtained from soil samples using a medium containing 1% SPP. Among them, TKU047 showed the highest protease activity (1.97 U/mL). In studies aimed to find a bacterial strain which efficiently converted fishery processing by-products into worthwhile bioactive compounds, TKU047 showed good potential for protease production. Thus, TKU047 was selected for further experiments. TKU047 was proven to be a member of *Paenibacillus* by morphological, biochemical, and 16S rDNA sequencing methods. However, its identified profile using the API 50 CHB/E kit did not match with any specific *Paenibacillus* species. As such, TKU047 was simply identified as *Paenibacillus* sp. The biosynthesis of proteases by *Paenibacillus* have rarely been reported [52–54], especially on fishery processing by-products [13], therefore the current study shows promise in seeking to discover a novel protease.

2.2. Screening the C/N Source for Protease Production

Finding the best carbon/nitrogen (C/N) source was a necessary step for protease production in numerous reports [10–16]. In the current study, five kinds of fishery processing byproducts were used to grow *Paenbacillus* sp. TKU047, including SHP, SPP, SSP, deSSP, and deCSP. Each material was added to a medium containing only basal mineral salts (0.05% $MgSO_4$ and 0.1% K_2HPO_4) at 1% concentration. As shown in Figure 1, protease productivity achieved the highest value at days 4 and 5, with the exception of SHP (only 3–4 days). The greatest production was found with SHP (2.93 U/mL on day 3 and 2.94 on day 4) and SPP (2.90 U/mL on day 5), followed by deCSP (2.63 U/mL on day 4), deSSP (2.16 U/mL on day 5), and SSP (0.95 U/mL on day 4). Although there was no significant difference between SHP and SPP in maximum protease productivity, TKU047 required a shorter incubation period with SHP than SPP (3 versus 5 days). As a result, SHP was chosen as the best source for producing protease by TKU047. Compared to other bacterial strains producing protease, TKU047 had higher protease productivity than *Bacillus*, *Paenibacillus*, and *Serratia* [13]. This suggested that *Paenibacillus* sp. TKU047 had potential for the conversion of fishery processing by-products for protease production.

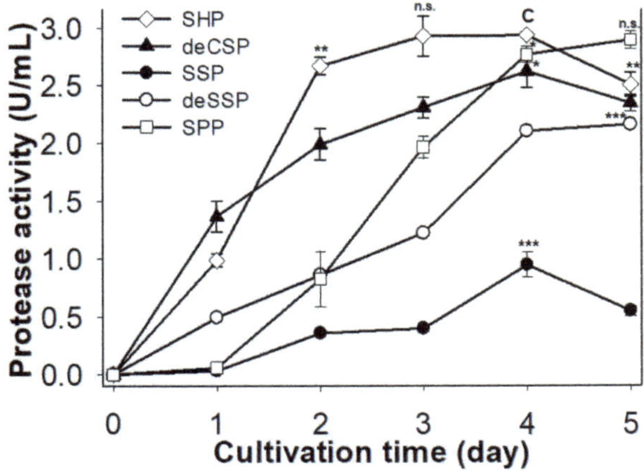

Figure 1. Screening of fishery processing byproducts as the C/N source for protease production in *Paenibacillus* sp. TKU047. All points on the graph indicate the mean of each experiment (with $n = 3$) and standard deviation (error bar). The letters n.s., *, **, and *** indicate not significantly different or significantly different at $p < 0.05$, $p < 0.01$, or $p < 0.001$, respectively, in comparison with the control group (C) based on one-way ANOVA analysis. The data point of SHP at day 4 was used to denote the control group.

In many reports, protease production required extra sources, such as glucose [55], lactose [56], beef extract [57], peptone [58], or yeast extract [44]. Due to their higher costs, these chemicals would increase the price of enzyme production. By-products from agriculture and fishery processing could overcome this drawback by providing the sole C/N source for the microorganisms [8,13,14,45]. Among these, shrimp heads contain a high ratio of protein and chitin [13], which makes them a good C/N source for producing various bioactive compounds via microbial fermentation [11,27,29].

2.3. *Effect of Shrimp Head Powder Concentration*

The effect of SHP concentration on producing protease was investigated herein. Different SHP concentrations (0.1%, 0.5%, 1%, 1.5%, and 2%) were added to the basal medium (0.05% $MgSO_4$, and 0.1% K_2HPO_4) to determine optimal concentration. As shown in Figure 2, maximum protease activity was found on days 2–4, but lower SHP concentrations achieved maximum protease activity in a shorter period of time than higher concentrations of SHP (2 days with 0.1% SHP and 0.5% SHP, 3 days with 1% SHP and 1.5% SHP, and 4 days with 2% SHP). The highest protease activity was generated with 0.5% SHP (2.98 U/mL, 2 days) and 1% SHP (2.97 U/mL on day 3 and 2.96 U/mL on day 4). This result suggests that lower SHP concentrations may achieve a better result than higher concentrations. Similarly, it was found that *Paenibacillus* sp. TKU042 expressed its highest protease activity at lower SPP concentrations of 0.5–2% [13]. Due to its shorter cultivation time and the lack of significant difference in protease activity compared to 1% SHP, 0.5% (*w/v*) was chosen as the best SHP concentration for producing protease.

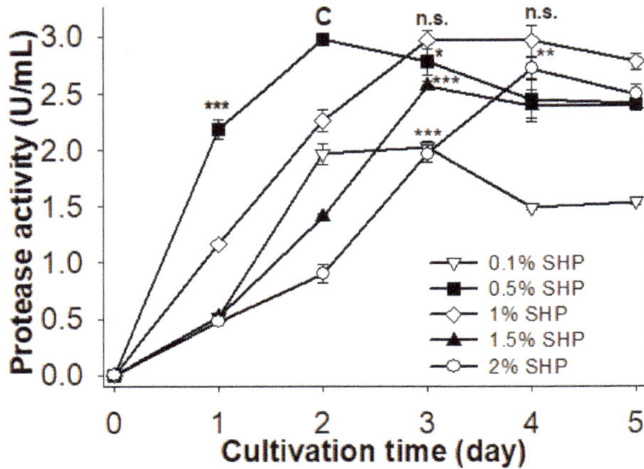

Figure 2. Effect of SHP concentration on protease production in *Paenibacillus* sp. TKU047. All points on the graph indicate the mean of each experiment (with $n = 3$) and standard deviation (error bar). The letters n.s., *, **, and *** indicate not significantly different or significantly different at $p < 0.05$, $p < 0.01$, or $p < 0.001$ (respectively) in comparison with the control group (C), based on one-way ANOVA analysis. Data point of 0.5% SHP on day 2 was used to signify the control group.

2.4. Isolation of TKU047 Protease

TKU047 protease was purified from culture broth by the following steps: EtOH precipitation, ion chromatography, and size-exclusion chromatography. Table 1 shows the purification profile. The enzyme was efficiently concentrated by EtOH with 96.44% of activity retained and eluted on a Macro-Prep High S column using a NaCl gradient from 0 M to 1 M. As shown in Figure 3, only one activity peak appeared at the elution stage. The activity fractions were concentrated by lyophilization for later use. A high performance liquid chromatography (HPLC) system with KW802.5 column was used to purify TKU047 protease, and 7.83 g of TKU047 protease was collected with a recovery yield of 34.70% and a specific activity of 16.18 U/mg. It had been reported that *Paenibacillus* sp. TKU042 and *P. tezpurensis* sp. nov. AS-S24-II had higher values of specific activity than the TKU047 protease. However, the enzyme from TKU047 had the best recovery yield among the three [13,52].

Table 1. Purification of *Paenibacillus* sp. TKU047 protease.

Step	Total Protein (mg)	Total Activity (U)	Specific Activity (U/mg)	Recovery (%)	Purification (Fold)
Cultural supernatant	1972.83	365.00	0.19	100.00	1.00
EtOH precipitation	125.56	352.00	2.80	96.44	15.15
Macro-Prep High S	88.88	272.30	3.06	74.60	16.56
KW-802.5	7.83	126.67	16.18	34.70	87.44

The purification of *Paenibacillus* sp. TKU047 was also analyzed by sodium dodecyl sulfate-polyacrylamide gel electrophoresis (SDS-PAGE). As shown in Figure 4, the band of TKU047 protease clearly appears at the line of the culture supernatant, indicating that TKU047 may release protease into the medium as one of its main products in the presence of SHP. According to protein markers, the mass of TKU047 protease was 32 kDa, which was very similar to *Paenibacillus* sp. TKU042 protease (35 kDa) [13], but different from other *Paenibacillus* proteases, such as *P. tezpurensis* sp. nov. AS-S24-II (43 kDa) [52], *P. larvae* strain 44 (59 kDa) [53], and *P. lautus* CHN26 (52 kDa) [54].

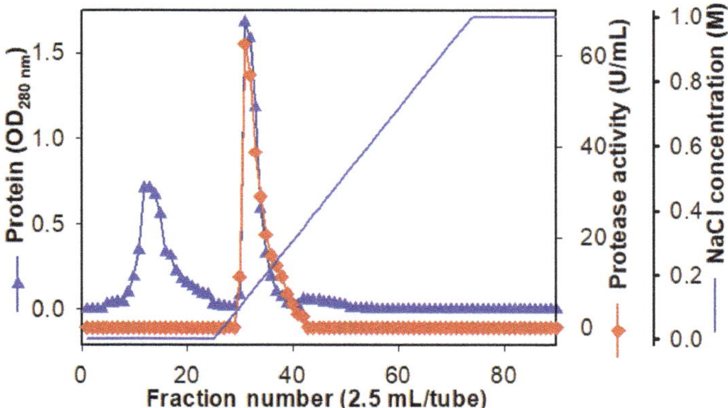

Figure 3. A typical chromatogram of Macro-prep High S column of *Paenibacillus* sp. TKU047 protease.

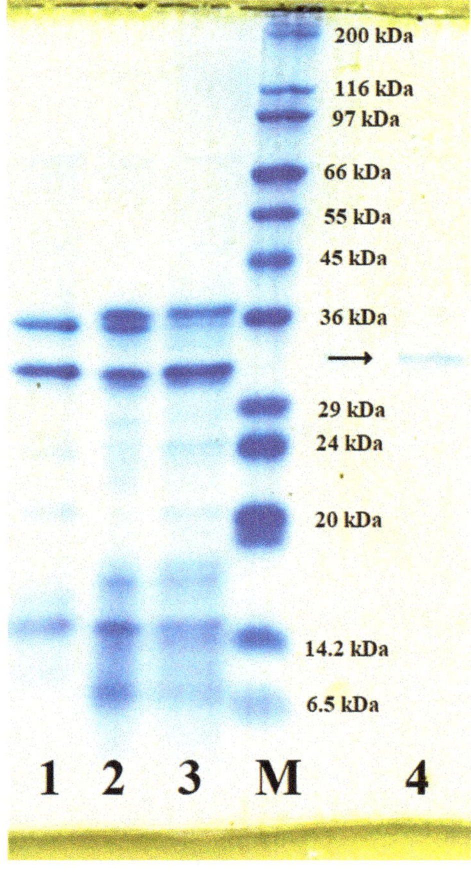

Figure 4. Sodium dodecyl sulfate-polyacrylamide gel electrophoresis (SDS-PAGE) profile of the TKU047 protease. M: Protein markers. 1 and 2: Crude enzyme after EtOH precipitation. 3: Adsorbed on Macro-prep High S column. 4: Purified enzyme after HPLC analysis on KW802.5 column. →: Purified enzyme.

2.5. Effects of Temperature and pH on TKU047 Protease

As shown in Figure 5, TKU047 protease displayed optimal activity between 70–80 °C. TKU047 protease also expressed thermal stability, while retaining over 80% of its activity at 60 °C. These results suggest that TKU047 protease is a thermostable enzyme. Compared to other reports, TKU047 showed a higher optimal temperature than *P. tezpurensis* sp. nov. AS-S24-II protease (45–50 °C) [52] or *P. lautus* protease (20–30 °C) [54], and could be comparable with thermostable proteases from *Bacillus* strains [45]. A higher optimal temperature and thermostability are important factors for industrial applications; therefore, *Paenibacillus* sp. TKU047 protease has a great advantage over other proteases.

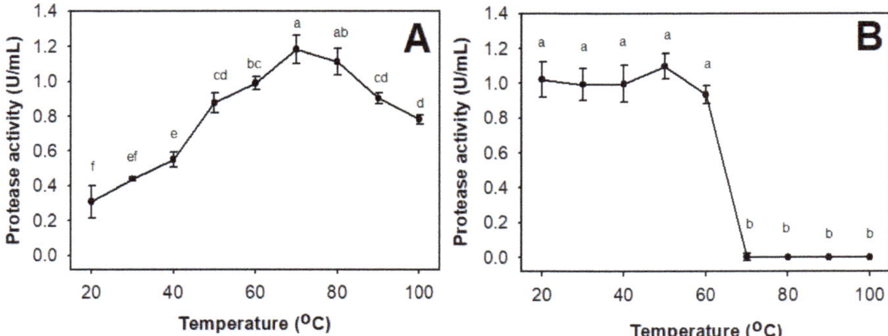

Figure 5. Effect of temperature on the activity (**A**) and stability (**B**) of *Paenibacillus* sp. TKU047 protease. All points on the graph indicate the mean of each experiment (with $n = 3$) and standard deviation (error bar). Different letters (a, b, c, d, e, and f) indicate significant difference based on Tukey's test with $p < 0.05$.

The effect of pH on the activity and stability of TKU047 protease was also investigated. The optimal pH of TKU047 was 9 and the enzyme was stable in a pH range of 6–11 (Figure 6). With over 80% of enzyme activity retained at alkaline pH range (7–11), TKU047 protease was consequently defined as an alkaline protease. The vast majority of *Paenibacillus* proteases express higher activity under alkaline conditions; the optimal pH is approximately 9 [52,54]. Alkaline proteases have received much attention due to their many applications in the detergent, food, pharmaceutical, and leather industries [45]. It is therefore worthwhile to investigate *Paenibacillus* sp. TKU047 protease for other valuable properties.

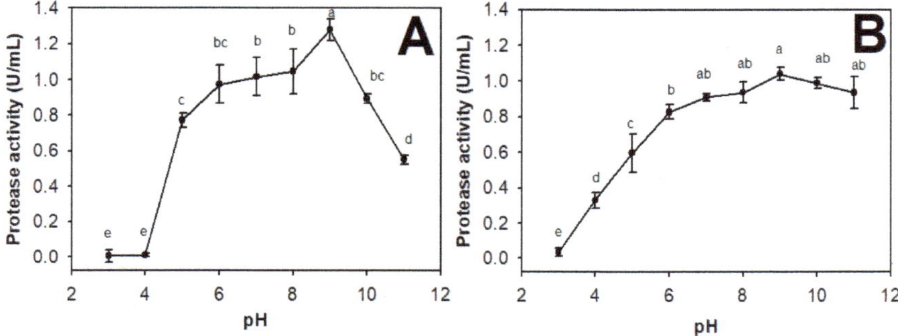

Figure 6. Effect of pH on activity (**A**) and stability (**B**) of *Paenibacillus* sp. TKU047 protease. All points on the graph indicate the mean of each experiment (with $n = 3$) and standard deviation (error bar). Different letters (a, b, c, d, and e) indicate significant difference based on Tukey's test with $p < 0.05$.

2.6. Substrate Specificity

As shown in Figure 7, TKU047 protease expressed the greatest activity in order of casein > fibrinogen > keratin > albumin > gelatin > elastin > collagen. It suggested that TKU047 could exhibit good caseinolytic and fibrinolytic ability compared to other activities. This result is different from the protease produced by *P. tezpurensis* sp. nov. AS-S24-II strain, which demonstrated its best activity on keratin [52].

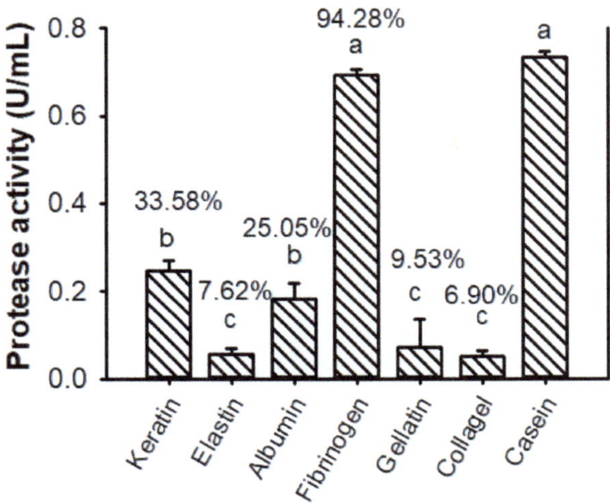

Figure 7. Substrate specificity of *Paenibacillus* sp. TKU047 protease. All points on the graph indicate the mean of each experiment (with $n = 3$) and standard deviation (error bar). Different letters (a, b, c, and d) indicate significant difference based on Tukey's test with $p < 0.05$. The values on top of the bars (%) are the ratios of the activities of TKU047 protease on different substrates to those on casein.

2.7. Effect of Metal Ions, Protease Inhibitors, and Surfactants

Among the tested ion metals, Fe^{2+}, Na^{+}, and Ba^{2+} significantly increased the activity of TKU047 protease by 25%, 44%, and 56%, respectively; other metal ions showed only a slight effect on enzyme activity (Figure 8). Na^{+} has been reported as a factor that enhances protease activity [8]; however, the results for Fe^{2+} and Ba^{2+} were different in other reports, which suggests that those metal ions inhibited protease activity [45,54]. In the presence of surfactants, TKU047 protease showed good resistance, with the retained activity higher than the control group (112% on tween 40, 116% on tween 20, and 161% on SDS) with only Triton X-100 being the exception (retaining 54% of initial activity). Since no activity was lost after incubating with 2-mercaptoethanol (2-ME), the thiol group may not contribute to the activity of the TKU047 protease. In addition, the results of adding ethylenediaminetetraacetic acid (EDTA) and phenylmethylsulfonyl fluoride (PMSF) revealed that TKU047 was only strongly inhibited by EDTA. This indicates that TKU047 protease is part of the metallo-protease family. This is consistent with reports that proteases from *Paenibacillus* strains are metallo- or serine-proteases [52,53].

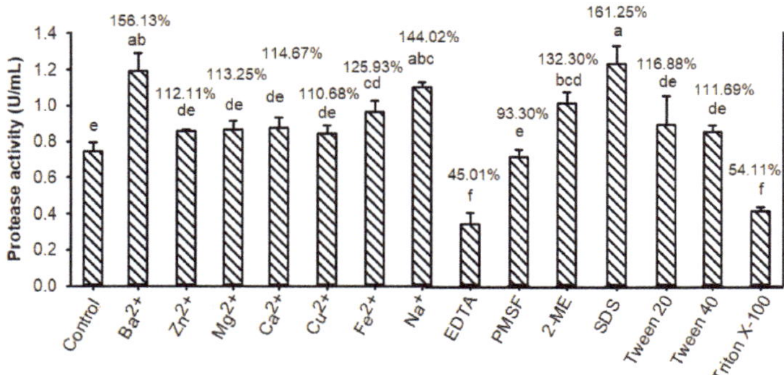

Figure 8. Effect of metal ions, protease inhibitors, and surfactants on *Paenibacillus* sp. TKU047 protease. All points on the graph indicated the mean of each experiment (with $n = 3$) and standard deviation (error bar). Different letters (a, b, c, d, e, and f) indicate significant difference based on Tukey's test with $p < 0.05$. The values on top of the bars (%) are the ratios of the residual activities to the protease activity of the enzyme in the phosphate buffer (control).

2.8. Compatibility with Commercial Detergents

According to the above results, *Paenibacillus* sp. TKU047 protease expressed several valuable properties, such as high productivity when converting shrimp heads, thermal stability, alkaline tolerance, and resistance to surfactants. As such, TKU047 protease could be considered a good candidate for a detergent additive. However, the detergent stability of this protease needed to be investigated before any further steps were taken. To explore the compatibility of TKU047 protease with detergent, four kinds of commercially available Taiwanese detergents: EKOS, Yigan Sanjing (YGSJ), AMAH concentrated lavender laundry detergent (AMAH), and YEUHYANG eco-friendly laundry powder (YEUHYANG), were used at 1% concentration. Before testing, all the detergents were examined for protease activity, but none exhibited proteolytic ability on casein. As shown in Figure 9, TKU047 protease had excellent compatibility with YEUHYANG detergent, with no protease activity lost after treatment. The compatibility of TKU047 protease with EKOS, YGSJ, and AMAH detergents was also acceptable, with over 70% of residual activity observed. It has been reported that some alkaline proteases from bacteria express good detergent compatibility. For instance, *Sardinella longiceps* protease retained over 90% of activity with Arial, Bahar, Tide, and Bonux detergents [59], while *B. licheniformis* RP1 retained over 93% of activity with Axion and Dixan detergents [60] and *Bacillus* sp. SSR1 exhibited over 70% of activity with tested detergents [61]. Due to its high detergent compatibility, TKU047 protease has potential as a detergent additive.

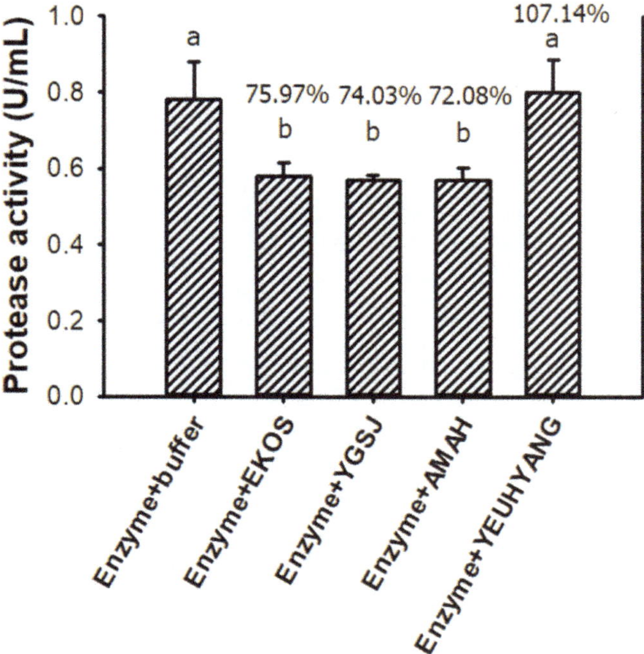

Figure 9. Compatibility of *Paenibacillus* sp. TKU047 protease with some commercial detergents (0.5%, *w/v*). All points on the graph indicate the mean of each experiment (with $n = 3$) and standard deviation (error bar). Different letters (a and b) indicate significant difference based on Turkey's test with $p < 0.05$. The values on top of the bars (%) are the ratios of the residual activities to the protease activity of the enzyme in the phosphate buffer.

3. Materials and Methods

3.1. Materials

Fishery processing byproducts, excepting shrimp heads (from Fwu-Sow Industry, Taichun, Taiwan), were from Shin-Ma Frozen Food Co. (I-Lan, Taiwan) [7]. Crab shells and shrimp shells were treated by acid to remove the mineral contents [11]. The resin of ion chromatography (Macro-prep High S) was bought from BioRad (Hercules, CA, USA). HPLC columns (KW-802.5) were bought from Showa Denko K. K (Tokyo, Japan). Other reagents were all laboratory chemicals with high quality.

3.2. Protease Activity Assay

The assay for determining protease activity followed the previous report with some modification [12]. Briefly, a 50 μL sample was dropped into an Eppendorf containing 50 μL substrate (2% casein in 50 mM phosphate buffer, pH 7). Later on, the Eppendorf was gently vortexed and kept on an incubator at 37 °C for 30 min. The enzyme activity was eliminated by the addition of 300 μL trichloroacetic acid (TCA) solution (5%). The solution was collected by centrifuging the Eppendorf at 11,000× *g* for 15 min, and then transferred to a mixture of Folin–Ciocalteu reagent and 20% Na_2CO_3 at a ratio of 8:2:15, respectively. After 20 min, the color of the solution was quantified based on 660 nm wavelength. Protease activity was described by the amount of required enzyme for liberating 1 μmol of tyrosine during 1 min of proteolytic reaction.

3.3. Screening of a Proteolytic Bacteria

The method of bacterial isolation was described in a previous report [8]. After isolation, the bacterial strains were cultivated in a liquid medium containing SPP (1%, *w/v*), $MgSO_4 \cdot 7H_2O$ (0.05%, *w/v*), and K_2HPO_4 (0.1%, *w/v*) at 37 °C (150 rpm) for 3 days. Later on, the culture supernatants were separated from bacterial cells and other sediments by centrifuging the medium (9000× *g* for 30 min). This supernatant was collected to measure protease activity, as described below. The screening step was done by selecting the strain which exhibited the highest protease productivity. The selected strain's morphological and biochemical properties were characterized by light microscopy and API 50 CHB system (bioMérieux, Inc., Durham, NC, USA). The scientific name of the bacterial strain was identified by DNA sequencing method.

3.4. Screening C/N Source for Protease Production

Five fishery byproducts: deSSP, deCSP, SPP, SHP, and SSP, were supplemented to the basal medium containing $MgSO_4 \cdot 7H_2O$ (0.05%, *w/v*), and K_2HPO_4 (0.1%, *w/v*) at the same concentration (1%, *w/v*). Cultivation was started by injecting 1% of seed to the medium and maintained at 37 °C, 150 rpm for 3 days. The protease productivity of bacterial strain on different C/N sources were tested every 24 h.

3.5. Isolation of TKU047 Protease

Paenibacillus sp. TKU047 was cultivated as per the conditions described above. Two hundred milliliters of culture supernatant was concentrated with EtOH at a ratio of 1:4. The protein precipitate was obtained by centrifuging the miscellaneous solution of culture supernatant and EtOH for 30 min at 12,000× *g*, and then dissolved in sodium phosphate buffer (50 mM, pH 7). The first enzyme isolation step was carried out by a Macro-Prep High S column. The fractions, which exhibited protease activity, were pooled for the next isolation step via HPLC method (column, KW802.5; injection volume, 20 µL; mobile phase, 50 mM sodium phosphate buffer; detector, ultraviolet 280 nm; flowrate, 1 mL/min; temperature, 20 °C). The molecular mass determination of the isolated protease was done by SDS-PAGE method.

3.6. Effects of pH and Temperature

The optimal temperature of TKU047 protease was investigated by incubating the mixture of enzyme (50 µL) and substrate (50 µL) at different temperature points (20–100 °C) for 30 min. Thermal stability of TKU047 protease was determined by pre-incubating the enzyme solution in a range of temperature (up to 100 °C) for 30 min. After heat treatment, the enzyme solution was measured for activity under the protease activity assay as mentioned above. In order to investigate the optimum pH of TKU047 protease, several buffers (50 mM) were used to prepare enzyme solutions as well as substrate solutions, including sodium decarbonate–carbonate buffer (pH 9–11), sodium phosphate buffer (pH 6–8), sodium acetate buffer (pH 3.6–5), and glycine HCl buffer (pH 2–3.6). The pH stability of TKU047 protease was carried out by pre-incubating the enzyme at different pH points (pH 2–7) for 30 min. Later on, the pH of the enzyme solutions was adjusted to pH 7 by using sodium phosphate buffer (0.2 M). The treated enzyme solutions were measured for activity under protease activity assay as mentioned above.

3.7. Effect of Metal ions, Protease Inhibitors, and Surfactants

TKU047 protease was prepared in various chemical solutions, including: Ba^{2+}, Zn^{2+}, Fe^{2+}, Cu^{2+}, Ca^{2+}, Mg^{2+}, Na^{+}, EDTA, PMSF, Triton X-100, tween 40, 2-ME, and SDS. The final concentration of metal ions, and protease inhibitors in the solutions were 5 mM, whereas those of surfactants was 10%. The mixtures were kept in an incubator at 37 °C for 30 min. Residual protease activity was then measured, using the methods described above.

3.8. Substrate Specificity

TKU047 protease was prepared in sodium phosphate buffer (50 mM) with numerous substrates. These included casein, fibrinogen, keratin, albumin, gelatin, elastin, and collagen. Later on, the mixtures were kept in an incubator under the following conditions of protease assay as mentioned above.

3.9. Detergent Compatibility

TKU047 protease was prepared in a sodium phosphate buffer (50 mM) with four kinds of commercial detergents at 30 °C for 60 min. These included: EKOS from Maobao Co. (Hsinchu, Taiwan), Yigan Sanjing from Maobao Co. (Hsinchu, Taiwan), AMAH concentrated lavender laundry detergent from Magic Amah household Co. (New Taipei, Taiwan), and YEUHYANG eco-friendly laundry powder from YEUHYANG Manufacture Technology (Hsinchu, Taiwan).

4. Conclusions

Paenibacillus has several strains with the potential to convert fishery processing by-products into various bioactive compounds [11–13,30,43,48–51]. This study continued to reveal more abilities of *Paenibacillus*, which could contribute to the biotechnological field. In the current study, protease was produced by *Paenibacillus* sp. TKU047 using shrimp heads, a marine byproduct, as the sole source of carbon and nitrogen. Purified *Paenibacillus* sp. TKU047 protease had a mass of 32 kDa. Characterization revealed that the TKU047 enzyme was a thermotolerant, detergent-stable, alkaline protease. The excellent characteristics of *Paenibacillus* sp. TKU047 suggest that it may be applied as a detergent additive.

Author Contributions: S.-L.W. and C.T.D. conceptualized and designed the study; C.T.D., T.N.T., and I.H.W. performed the experiment; C.T.D., T.N.T., I.H.W., V.B.N., and A.D.N. compiled and analyzed the data; S.-L.W. and C.T.D. wrote the paper; S.-L.W. acquired funding and supervised the project.

Funding: This work was supported in part by a grant from the Ministry of Science and Technology, Taiwan (MOST 106-2320-B-032-001-MY3).

Conflicts of Interest: The authors declare no conflicts of interest.

References

1. Halim, N.R.A.; Yusof, H.M.; Sarbon, N.M. Functional and bioactive properties of fish protein hydolysates and peptides: A comprehensive review. *Trends Food Sci. Technol.* **2016**, *51*, 24–33. [CrossRef]
2. Kuo, J.M.; Lee, G.C.; Liang, W.S.; Yang, J.I. Process optimization for production of antioxidant gelatin hydrolysates from tilapia skin. *J. Fish Soc. Taiwan* **2009**, *36*, 15–28.
3. Lin, L.; Lv, S.; Li, B. Angiotensin-I-converting enzyme (ACE)-inhibitory and antihypertensive properties of squid skin gelatin hydrolysates. *Food Chem.* **2013**, *131*, 225–230. [CrossRef]
4. Wang, C.H.; Doan, C.T.; Nguyen, A.D.; Nguyen, V.B.; Wang, S.L. Reclamation of Fishery Processing Waste: A Mini-Review. *Molecules* **2019**, *24*, 2234. [CrossRef] [PubMed]
5. Wang, S.L.; Liang, T. Microbial reclamation of squid pens and shrimp shells. *Res. Chem. Intermed.* **2017**, *43*, 3445–3462. [CrossRef]
6. Wang, S.L.; Yu, H.T.; Tsai, M.H.; Doan, C.T.; Nguyen, V.B.; Nguyen, A.D. Conversion of squid pens to chitosanases and dye adsorbents via *Bacillus cereus*. *Res. Chem. Intermed.* **2018**, *44*, 4903–4911. [CrossRef]
7. Tran, T.N.; Doan, C.T.; Nguyen, V.B.; Nguyen, A.D.; Wang, S.L. The isolation of chitinase from *Streptomyces thermocarboxydus* and its application in the preparation of chitin oligomers. *Res. Chem. Intermed.* **2019**, *45*, 727–742. [CrossRef]
8. Doan, C.T.; Tran, T.N.; Nguyen, V.B.; Vo, T.P.K.; Nguyen, A.D.; Wang, S.L. Chitin extraction from shrimp waste by liquid fermentation using an alkaline protease-producing strain, *Brevibacillus parabrevis*. *Int. J. Biol. Macromol.* **2019**, *131*, 706–715. [CrossRef]
9. Wang, S.L.; Chio, S.H. Deproteination of shrimp and crab shell with the protease of *Pseudomonas aeruginosa* K-187. *Enzym. Microb. Technol.* **1998**, *22*, 629–633. [CrossRef]
10. Yang, J.K.; Shih, I.L.; Tzeng, Y.M.; Wang, S.L. Production and purification of protease from a *Bacillus subtilis* that can deproteinize crustacean wastes. *Enzym. Microb. Technol.* **2000**, *26*, 406–413. [CrossRef]

11. Doan, C.T.; Tran, T.N.; Nguyen, V.B.; Nguyen, A.D.; Wang, S.L. Production of a thermostable chitosanase from shrimp heads via *Paenibacillus mucilaginosus* TKU032 conversion and its application in the preparation of bioactive chitosan oligosaccharides. *Mar. Drugs* **2019**, *17*, 217. [CrossRef] [PubMed]
12. Doan, C.T.; Tran, T.N.; Nguyen, V.B.; Nguyen, A.D.; Wang, S.L. Reclamation of marine chitinous materials for chitosanase production via microbial conversion by *Paenibacillus macerans*. *Mar. Drugs* **2018**, *16*, 429. [CrossRef] [PubMed]
13. Doan, C.T.; Tran, T.N.; Nguyen, V.B.; Nguyen, A.D.; Wang, S.L. Conversion of squid pens to chitosanases and proteases via *Paenibacillus* sp. TKU042. *Mar. Drugs* **2018**, *16*, 83. [CrossRef] [PubMed]
14. Doan, C.T.; Tran, T.N.; Nguyen, M.T.; Nguyen, V.B.; Nguyen, A.D.; Wang, S.L. Anti-α-glucosidase activity by a protease from *Bacillus licheniformis*. *Molecules* **2019**, *24*, 691. [CrossRef] [PubMed]
15. Wang, S.L.; Tseng, W.N.; Liang, T.W. Biodegradation of shellfish wastes and production of chitosanases by a squid pen-assimilating bacterium, *Acinetobacter calcoaceticus* TKU024. *Biodegradation* **2011**, *22*, 939–948. [CrossRef] [PubMed]
16. Nguyen, A.D.; Huang, C.C.; Liang, T.W.; Nguyen, V.B.; Pan, P.S.; Wang, S.L. Production and purification of a fungal chitosanase and chitooligomers from *Penicillium janthinellum* D4 and discovery of the enzyme activators. *Carbohydr. Polym.* **2014**, *108*, 331–337. [CrossRef] [PubMed]
17. Wang, S.L. Microbial reclamation of squid pen. *Biocatal. Agric. Biotechnol.* **2012**, *1*, 177–180. [CrossRef]
18. Srinivasan, H.; Velayutham, K.; Ravichandran, R. Chitin and chitosan preparation from shrimp shells *Penaeus monodon* and its human ovarian cancer cell line, PA-1. *Int. J. Biol. Macromol.* **2018**, *107*, 662–667. [CrossRef] [PubMed]
19. Hamed, I.; Özogul, F.; Regenstein, J.M. Industrial applications of crustacean by-products (chitin, chitosan, and chitooligosaccharides): A review. *Trends Food Sci. Technol.* **2016**, *48*, 40–50. [CrossRef]
20. Kumar, A.; Kumar, D.; George, N.; Sharma, P.; Gupta, N. A process for complete biodegradation of shrimp waste by a novel marine isolate *Paenibacillus* sp. AD with simultaneous production of chitinase and chitin oligosaccharides. *Int. J. Biol. Macromol.* **2018**, *109*, 263–272. [CrossRef]
21. Jaworska, M.M.; Górak, A. New ionic liquids for modification of chitin particles. *Res. Chem. Intermed.* **2018**, *44*, 4841–4854. [CrossRef]
22. Hiranpattanakul, P.; Jongjitpissamai, T.; Aungwerojanawit, S.; Tachaboonyakiat, W. Fabrication of a chitin/chitosan hydrocolloid wound dressing and evaluation of its bioactive properties. *Res. Chem. Intermed.* **2018**, *44*, 4913–4928. [CrossRef]
23. Chang, F.S.; Chin, H.Y.; Tsai, M.L. Preparation of chitin with puffing pretreatment. *Res. Chem. Intermed.* **2018**, *44*, 4939–4955. [CrossRef]
24. Deepthi, S.; Venkatesan, J.; Kim, S.K.; Bumgardner, J.D.; Jayakumar, R. An overview of chitin or chitosan/nano ceramic composite scaffolds for bone tissue engineering. *Int. J. Biol. Macromol.* **2016**, *93*, 1338–1353. [CrossRef] [PubMed]
25. Vázquez, J.A.; Ramos, P.; Mirón, J.; Valcarcel, J.; Sotelo, C.G.; Pérez-Martín, R.I. Production of chitin from *Penaeus vannamei* by-products to pilot plant scale using a combination of enzymatic and chemical processes and subsequent optimization of the chemical production of chitosan by response surface methodology. *Mar. Drugs* **2017**, *15*, 180. [CrossRef] [PubMed]
26. Lopes, C.; Antelo, L.T.; Franco-Uría, A.; Alonso, A.A.; Pérez-Martín, R. Chitin production from crustacean biomass: Sustainability assessment of chemical and enzymatic processes. *J. Clean Prod.* **2018**, *172*, 4140–4152. [CrossRef]
27. Wang, S.L.; Su, Y.C.; Nguyen, V.B.; Nguyen, A.D. Reclamation of shrimp heads for the production of α-glucosidase inhibitors by *Staphylococcus* sp. TKU043. *Res. Chem. Intermed.* **2018**, *44*, 4929–4937. [CrossRef]
28. Nguyen, V.B.; Wang, S.L. Reclamation of marine chitinous materials for the production of α-glucosidase inhibitors *via* microbial conversion. *Mar. Drugs* **2017**, *15*, 350. [CrossRef] [PubMed]
29. Hsu, C.H.; Nguyen, V.B.; Nguyen, A.D.; Wang, S.L. Conversion of shrimp heads to α-glucosidase inhibitors via co-culture of *Bacillus mycoides* TKU040 and *Rhizobium* sp. TKU041. *Res. Chem. Intermed.* **2018**, *44*, 4597–4607. [CrossRef]
30. Nguyen, V.B.; Nguyen, T.H.; Doan, C.T.; Tran, T.N.; Nguyen, A.D.; Kuo, Y.H.; Wang, S.L. Production and bioactivity-guided isolation of antioxidants with α-glucosidase inhibitory and anti-NO properties from marine chitinous materials. *Molecules* **2018**, *23*, 1124. [CrossRef]

31. Nguyen, V.B.; Nguyen, A.D.; Wang, S.L. Utilization of fishery processing byproduct squid pens for *Paenibacillus* sp. fermentation on producing potent α-glucosidase inhibitors. *Mar. Drugs* **2017**, *15*, 274. [CrossRef] [PubMed]
32. Parjikolaei, B.R.; El-Houri, R.B.; Fretté, X.C.; Christensen, K.V. Influence of green solvent extraction on carotenoid yield from shrimp (*Pandalus borealis*) processing waste. *J. Food Eng.* **2015**, *155*, 22–28. [CrossRef]
33. Cahúa, T.B.; Santos, S.D.; Mendes, A.; Córdula, C.R.; Chavante, S.F.; Carvalho, L.B., Jr.; Nader, H.B.; Bezerra, R.S. Recovery of protein, chitin, carotenoids and glycosaminoglycans from Pacific white shrimp (*Litopenaeus vannamei*) processing waste. *Process Biochem.* **2012**, *47*, 570–577. [CrossRef]
34. He, S.; Franco, C.; Zhang, W. Fish protein hydrolysates: Application in deep-fried food and food safety analysis. *J. Food Sci.* **2015**, *80*, E108–E115. [CrossRef] [PubMed]
35. Morales-Medina, R.; Pérez-Gálvez, R.; Guadix, A.; Guadix, E.M. Multiobjective optimization of the antioxidant activities of horse mackerel hydrolysates produced with protease mixtures. *Process Biochem.* **2017**, *52*, 149–158. [CrossRef]
36. Slizyte, R.; Rommi, K.; Mozuraityte, R.; Eck, P.; Five, K.; Rustad, T. Bioactivities of fish protein hydrolysates from defatted salmon backbones. *Biotechnol. Rep.* **2016**, *11*, 99–109. [CrossRef] [PubMed]
37. Vázquez, J.A.; Blanco, M.; Massa, A.E.; Amado, I.R.; Pérez-Martín, R.I. Production of fish protein hydrolysates from *Scyliorhinus canicula* discards with antihypertensive and antioxidant activities by enzymatic hydrolysis and mathematical optimization using response surface methodology. *Mar. Drugs* **2017**, *15*, 306. [CrossRef]
38. Blanco, M.; Fraguas, J.; Sotelo, C.G.; Pérez-Martín, R.I.; Vázquez, J.A. Production of chondroitin sulphate from head, skeleton and fins of *Scyliorhinus canicula* by-products by combination of enzymatic, chemical precipitation and ultrafiltration methodologies. *Mar. Drugs* **2015**, *13*, 3287–3308. [CrossRef]
39. Šližytė, R.; Rustad, T.; Storrø, I. Enzymatic hydrolysis of cod (*Gadus morhua*) by-products: Optimization of yield and properties of lipid and protein fractions. *Process Biochem.* **2005**, *40*, 3680–3692. [CrossRef]
40. Silva, J.F.X.; Ribeiro, K.; Silva, J.F.; Cahú, T.B.; Bezerra, R.S. Utilization of tilapia processing waste for the production of fish protein hydrolysate. *Anim. Feed Sci. Technol.* **2012**, *196*, 96–106. [CrossRef]
41. Liang, T.W.; Wang, S.L. Recent advances in exopolysaccharides from *Paenibacillus* spp.: Production, isolation, structure, and bioactivities. *Mar. Drugs* **2015**, *13*, 1847–1863. [CrossRef]
42. Liang, T.W.; Wu, C.C.; Cheng, W.T.; Chen, Y.C.; Wang, C.L.; Wang, I.L.; Wang, S.L. Exopolysaccharides and antimicrobial biosurfactants produced by *Paenibacillus macerans* TKU029. *Appl. Biochem. Biotechnol.* **2014**, *172*, 933–950. [CrossRef]
43. Liang, T.W.; Tseng, S.C.; Wang, S.L. Production and characterization of antioxidant properties of exopolysaccharides from *Paenibacillus mucilaginosus* TKU032. *Mar. Drugs* **2016**, *14*, 40. [CrossRef]
44. Hsu, C.H.; Nguyen, A.D.; Chen, Y.W.; Wang, S.L. Tyrosinase inhibitors and insecticidal materials produced by *Burkholderia cepacian* using squid pen as the sole carbon and nitrogen source. *Res. Chem. Intermed.* **2014**, *40*, 2249–2258. [CrossRef]
45. Sharma, K.M.; Kumar, R.; Panwar, S.; Kumar, A. Microbial alkaline proteases: Optimization of production parameters and their properties. *J. Genet. Eng. Biotechnol.* **2017**, *15*, 115–126. [CrossRef]
46. Contesini, F.J.; Melo, R.R.; Sato, H.H. An overview of *Bacillus* proteases: From production to application. *Crit. Rev. Biotechnol.* **2018**, *38*, 321–334. [CrossRef]
47. Kalisz, M.H. Microbial proteases. *Adv. Biochem. Eng. Biotechnol.* **1988**, *38*, 1–65.
48. Grady, E.N.; MacDonald, J.; Liu, L.; Richman, A.; Yuan, Z.C. Current knowledge and perspectives of *Paenibacillus*: A review. *Microb. Cell Fact.* **2016**, *15*, 203. [CrossRef]
49. Nguyen, V.B.; Wang, S.L. New novel α-glucosidase inhibitors produced by microbial conversion. *Process Biochem.* **2018**, *65*, 228–232. [CrossRef]
50. Wang, S.L.; Li, H.T.; Zhang, L.J.; Lin, Z.H.; Kuo, Y.H. Conversion of squid pen to homogentisic acid via *Paenibacillus* sp. TKU036 and the antioxidant and anti-inflammatory activities of homogentisic acid. *Mar. Drugs* **2016**, *14*, 183. [CrossRef]
51. Liang, T.W.; Chen, W.T.; Lin, Z.H.; Kuo, Y.H.; Nguyen, A.D.; Pan, P.S.; Wang, S.L. An amphiprotic novel chitosanase from *Bacillus mycoides* and its application in the production of chitooligomers with their antioxidant and anti-inflammatory evaluation. *Int. J. Mol. Sci.* **2016**, *17*, 1302. [CrossRef]
52. Rai, S.K.; Roy, J.K.; Mukherjee, A.K. Characterisation of a detergent-stable alkaline protease from a novel thermophilic strain *Paenibacillus tezpurensis* sp. nov. AS-S24-II. *Appl. Microbiol. Biotechnol.* **2010**, *85*, 1437–1450. [CrossRef]

53. Antúnez, K.; Arredondo, D.; Anido, M.; Zunino, P. Metalloprotease production by *Paenibacillus larvae* during the infection of honeybee larvae. *Microbiology* **2011**, *157*, 1474–1480. [CrossRef]
54. Li, Y.; Pan, Y.; She, Q.; Chen, L. A novel carboxyl-terminal protease derived from *Paenibacillus lautus* CHN26 exhibiting high activities at multiple sites of substrates. *BMC Biotechnol.* **2013**, *13*, 89. [CrossRef]
55. Sharma, A.K.; Sharma, V.; Saxena, J.; Yadav, B.; Alam, A.; Prakash, A. Optimization of protease production from bacteria isolated from soil. *Appl. Res. J.* **2015**, *1*, 388–394.
56. Dodia, M.S.; Joshi, R.H.; Patel, R.K.; Singh, S.P. Characterization and stability of extracellular alkaline proteases from halophilic and alkaliphilic bacteria isolated from saline habitat of coastal Gujarat, India. *Braz. J. Microbiol.* **2006**, *37*, 276–282. [CrossRef]
57. Shafee, N.; Aris, S.N.; Rahman, R.N.Z.A.; Basri, M.; Salleh, A.B. Optimization of environmental and nutritional conditions for the production of alkaline protease by a newly isolated bacterium Bacillus cereus strain 146. *J. Appl. Sci. Res.* **2005**, *1*, 1–8.
58. Naidu, K.S.B.; Devi, K.L. Optimization of thermostable alkaline protease production from species of *Bacillus* using rice bran. *Afr. J. Biotechnol.* **2005**, *4*, 724–726.
59. Ramkumar, A.; Sivakumar, N.; Gujarathi, A.M.; Victor, R. Production of thermotolerant, detergent stable alkaline protease using the gut waste of *Sardinella longiceps* as the substrate: Optimization and characterization. *Sci. Rep.* **2018**, *8*, 12442. [CrossRef]
60. Sellami-Kamoun, A.; Haddar, A.; Ali, N.E.H.; Ghorbel-Frikha, B.; Kanoun, S.; Nasri, M. Stability of thermostable alkaline protease from *Bacillus licheniformis* RP1 in commercial solid laundry detergent formulations. *Microbiol. Res.* **2008**, *163*, 299–306. [CrossRef]
61. Singh, J.; Batra, N.; Sobti, R.C. Serine alkaline protease from a newly isolated *Bacillus* sp. SSR1. *Process Biochem.* **2011**, *36*, 781–785. [CrossRef]

Article

Catalytic Activities of Multimeric G-Quadruplex DNAzymes

Raphael I. Adeoye [1], Dunsin S. Osalaye [1], Theresia K. Ralebitso-Senior [2], Amanda Boddis [2], Amanda J. Reid [2], Amos A. Fatokun [2], Andrew K. Powell [2], Sylvia O. Malomo [1] and Femi J. Olorunniji [2,*]

1 Department of Biochemistry, University of Ilorin, Ilorin 240003, Kwara State, Nigeria
2 School of Pharmacy and Biomolecular Sciences, Faculty of Science, Liverpool John Moores University, Liverpool L3 3AF, UK
* Correspondence: F.J.Olorunniji@ljmu.ac.uk; Tel.: +44-0151-231-2116

Received: 9 July 2019; Accepted: 15 July 2019; Published: 19 July 2019

Abstract: G-quadruplex DNAzymes are short DNA aptamers with repeating G4 quartets bound in a non-covalent complex with hemin. These G4/Hemin structures exhibit versatile peroxidase-like catalytic activity with a wide range of potential applications in biosensing and biotechnology. Current efforts are aimed at gaining a better understanding of the molecular mechanism of DNAzyme catalysis as well as devising strategies for improving their catalytic efficiency. Multimerisation of discrete units of G-quadruplexes to form multivalent DNAzyes is an emerging design strategy aimed at enhancing the peroxidase activities of DNAzymes. While this approach holds promise of generating more active multivalent G-quadruplex DNAzymes, few examples have been studied and it is not clear what factors determine the enhancement of catalytic activities of multimeric DNAzymes. In this study, we report the design and characterisation of multimers of five G-quadruplex sequences (AS1411, Bcl-2, c-MYC, PS5.M and PS2.M). Our results show that multimerisation of G-quadruplexes that form parallel structure (AS1411, Bcl-2, c-MYC) leads to significant rate enhancements characteristic of cooperative and/or synergistic interactions between the monomeric units. In contrast, multimerisation of DNA sequences that form non-parallel structures (PS5.M and PS2.M) did not exhibit similar levels of synergistic increase in activities. These results show that design of multivalent G4/Hemin structures could lead to a new set of versatile and efficient DNAzymes with enhanced capacity to catalyse peroxidase-mimic reactions.

Keywords: G-quadruplex; DNAzymes; peroxidase; multimers; synergism

---

1. Introduction

Haem peroxidases use protein scaffolds that activate haem to react with H_2O_2 [1]. The reaction mechanism and properties of peroxidases have been extensively studied and they have several practical uses in research, bioanalysis and industrial applications. More recently, it was discovered that certain DNA aptamers have the ability to catalyse reactions similar to those carried out by haem peroxidases [2,3]. These DNA aptamers are Guanine-rich DNA oligonucleotides that form complexes with hemin to form G-quadruplexes (often referred to as G4/Hemin) with catalytic activities that mimic peroxidases [4–6]. The oligonucleotide sequences that form G-quartets are organised into G-quadruplexes in the presence of cations to form high-affinity binding sites for hemin turning these nanostructures into powerful catalysts that activate H_2O_2 for peroxidase-like oxidation reactions (Figure 1). These complexes (often referred to as DNAzymes) are more stable to extreme reaction conditions than protein enzymes, and are easier to synthesise and modify via chemical processes.

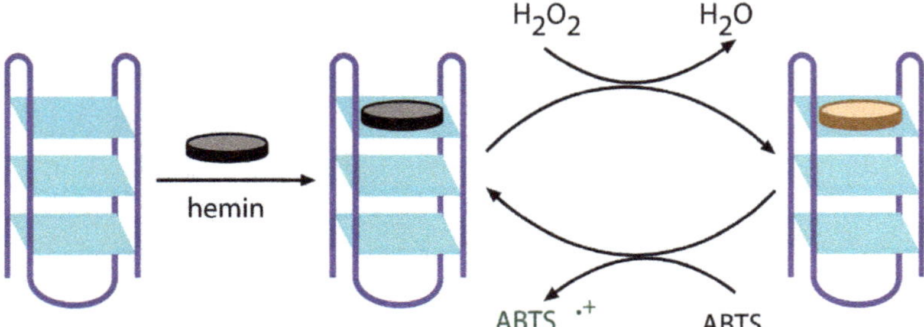

Figure 1. Assembly of G-quadruplex-hemin DNAzymes and H_2O_2-activated oxidation of ABTS. Guanine-rich DNA oligonucleotides fold to form G-quartet interactions (blue planes) which bind to hemin (grey discs). The resulting DNAzyme complex activates H_2O_2 to oxidise a donor substrate (in this instance ABTS) using a mechanism similar to haem peroxidase-catalysed reactions. The reduced haem (brown disc) can react with H_2O_2 and initiate another round of ABTS oxidation. The reaction product ($ABTS^{\bullet}$) has a bright green colour with absorption maximum around 412–422 nm.

G-quadruplex DNAzymes have the potential for development as nanodevices for applications in biosensors, diagnostics, and bioremediation and other aspects of biotechnology [7,8]. However, their relatively low catalytic activity in comparison to protein peroxidases currently restricts further development and applications of these systems. In order to develop DNAzymes into useful applications, it is necessary to characterize the structural and functional sequence parameters that influence the efficiency of DNAzyme reactions and their functionalities in applied devices.

Several strategies are being explored to improve the catalytic efficiency and reaction properties of peroxidase G-quadruplex DNAzymes [8]. One such approach is the finding that addition of polycationic amines such as spermine, spermidine, and putrescine significantly increases the activities of DNAzymes by favouring catalytically active conformations [9–11]. Similar rate-enhancements were observed by the addition of the nucleotide ATP to DNAzyme reactions [12,13]. It has also been reported that the conjugation of hemin with the G4-quadruplex moiety via covalent linkage [14] or with cationic peptides [15] can enhance the activities of DNAzymes. Other studies have focused on the rate-enhancing effects of flanking adenine or cytosine nucleotides on G-quadruplex activities [16–19].

Multimerisation of discrete units of G-quadruplexes to form multivalent DNAzymes is an emerging design strategy aimed at enhancing the peroxidase activities of DNAzymes [20,21]. In these multiple active site structures, individual DNAzyme molecular units are joined together by polynucleotide linkers that do not participate directly in the reactions. Some studies show that the linker length between individual DNAzyme molecules can be a determining factor in the activities of multimeric DNAzymes [17,19]. It is thought that this rate enhancement results from synergistic interactions between the individual subunits. In these arrangements, a multivalent DNAzyme sequence would be more active than the sum activities of an equal number of monomeric sequence units [20,21]. One hypothesis that explains the rate-enhancing effect of mulitimerisation is that multimeric DNAzyme structures potentially create additional high-affinity binding sites for haem incorporation, thereby resulting in more than one active site per DNAzyme monomer. However, in certain cases, there is no direct correlation between the number of units and the activity of the multimer [20,21]. This suggests that certain structural features do not necessarily support activity enhancement in multimers. Gaining a good understanding of the properties of such multivalent G-quadruplex DNAzymes would provide more insight into strategies for rational design of highly active DNAzymes.

While multimerisation has the potential to generate more active DNAzymes, only a handful of DNAzyme multimers have been reported [20–22]. It is not entirely clear what factors determine the catalytic properties of multimers when compared to their monomeric G-quadruplexes. For example, several key questions remained unanswered. Does multimerisation lead to cooperative and synergistic rate enhancement of activity in all G-quadruplexes that exhibit peroxidase activities? Is there a direct correlation between the number of monomer subunits and catalytic efficiency? Do certain multimeric combinations result in limited rate enhancement or potential inhibition of catalytic activity? Does the number of G-quartets in a G-quadruplex aptamer determine catalytic efficiency? Can contiguous multimerisation result in a redistribution of catalytic units or generation of high-activity pockets?

In an effort to understand these questions, we started by designing dimers and trimers of selected DNAzymes in order to study their catalytic properties in detail. In this report, we describe the activities of multimers of five different DNAzymes with their corresponding monomers. Using the dinucleotide TT as the linker sequence, we made dimers and trimers of AS1411, also known as AGRO100 [23], Bcl-2 [11,24], c-MYC [11], PS5.M and PS2.M [23]. Using the standard H_2O_2/ABTS oxidation assay, we determined the effects of multimerisation on five G-quadruplex DNAzymes. Our findings indicate that multivalency results in synergistic rate enhancements for some DNAzyme sequences, while multimers of certain aptamers resulted in reduced activity per haem binding site.

2. Results and Discussion

2.1. Design of Multimeric DNAzymes

To study the properties of multimeric G-quadruplex DNAzymes, we selected five different DNAzyme sequences (Table 1) that have been described in the literature. We selected AS1411, a highly-active DNAzyme [11,23] that is also reported as an anti-cancer agent owing to its ability to target the protein nucleolin which is highly expressed on the surface of cancer cells [25]. PS5.M and its sequence-derivative PS2.M [3] were also chosen for their high activities. In addition, this group of DNAzymes has been extensively studied and reported in the literature [26].

Table 1. Sequences of DNAzymes used in this study. The DNA sequences are written in 5′ to 3′ direction.

DNAzyme	Sequence
AS1411	GGTGGTGGTGGTTGTGGTGGTGGTGG
Bcl-2	GGGCGCGGGAGGAAGGGGGCGGG
c-MYC	GAGGGTGGGGAGGGTGGGGAAG
PS5.M	GTGGGTCATTGTGGGTGGGTGTGG
PS2.M	GTGGGTAGGGCGGGTTGG

c-MYC and Bcl-2 are less-characterised DNAzymes in comparison to the three described above. The assumption is that these low-activity DNAzymes would be an ideal model system to test any potential rate enhancement due to G-quadruplex multimerisation. c-MYC is derived from a G-rich region in the promoter sequence of the human *c-myc* gene [27], while Bcl-2 was identified as a G-rich segment of the Bcl-2 P1 promoter and shown to form mixed intramolecular G-quadruplex structures [28,29].

In each case, the multimers (dimers and trimers) were made by joining the nucleotide at the 3′ end of the DNAzyme sequence to the one on the 5′ end of the same sequence using a TT dinucleotide linker (Figure 2F). It is known that the identity of the base composition and length of the linker sequence can influence the activities of monovalent and multivalent DNAzymes [17]. To remove the potential for linker-dependent effects on our analyses, we chose to use the relatively simple TT linker sequence in order to obtain a relatively unconfounded effect of multimerisation on DNAzyme activity [20,21].

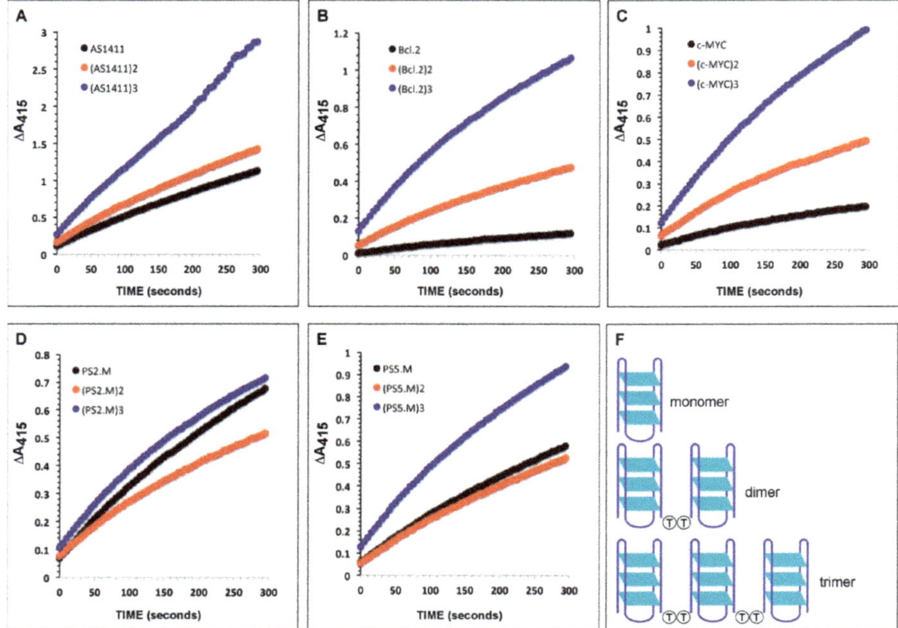

Figure 2. Time course of DNAzyme-catalysed reaction between H_2O_2 and ABTS (**A**) AS1411 and multimers; (**B**) Bcl-2 and multimers; (**C**) c-MYC and multimers; (**D**) PS5.M and multimers, (**E**) PS2.M and multimers. The numerals 2 and 3 behind each DNAzyme name refer to dimer and trimer respectively e.g., (AS1411)2 and (AS1411)3. Reactions were carried out in 25 mM HEPES-NaOH pH 7.4, 20 mM KCl, 200 mM NaCl, 0.05% Triton X-100, 1% DMSO. The activities of 0.25 μM DNAzyme were assayed using ABTS (2.5 mM) and H_2O_2 (0.425 mM). Each reaction was started by the addition of H_2O_2, and change in A_{415} was monitored immediately after starting the reaction for 5 min. The absorbance values shown were zeroed with those obtained in reactions that did not include the DNAzyme. Each trace is representative of three different reactions. (**F**) Schematic illustration of the design of G-quadruplex DNAzyme dimers and trimers. In the multimers, each monomeric unit is joined to the next by a TT dinucleotide.

2.2. Catalytic Activities of DNAzymes

Several buffer systems have been used to assay the activities of G4/Hemin DNAzymes, and it is clear that the reaction buffer pH as well as its composition play important role in the reaction rate and the stability of the product radical [3]. A common issue in peroxidation reactions is disproportionation of the reaction product [13], with certain buffers supporting the stability of the reaction product more than others. We used 25 mM HEPES-NaOH, pH 7.4, as the reaction buffer in these experiments since this buffer system supports both the actual peroxidase reaction as well as stability of the product [3,13,23,30].

All the 15 G-quadruplexes tested in this study displayed ABTS oxidation activities (Figure 2) and preliminary assays showed that a linear relationship exists between the concentration of G-quadruplex DNAzymes and peroxidase activities. In all cases, initial rates of ABTS oxidation increased linearly with DNAzyme concentration providing the concentrations of the substrates ABTS and H_2O_2 are not limiting (Figure 2).

AS1411: Consistently, AS1411 performed better in the reaction conditions used here than the other 4 DNAzymes tested. Similar high activities for AS1411 have been reported [23]. Their initial rate measurements showed that AS1411 is about 1.5-fold more active than PS5.M, and 2-fold more active than PS2.M. Our results generally agree with these measurements, both in initial rate and turnover numbers (Figure 3). Initial rate measurements showed that the multimeric forms of AS1411 were more active than the monomer. While there was a slight increase in the activity of the dimer (819 nM/s) over the monomer (639 nM/s), a more remarkable effect was seen in the activity of the trimer (1472 nM/s). Similar trends were seen in the amount of ABTS• radical formed after 5 min of reaction (Figure 3B).

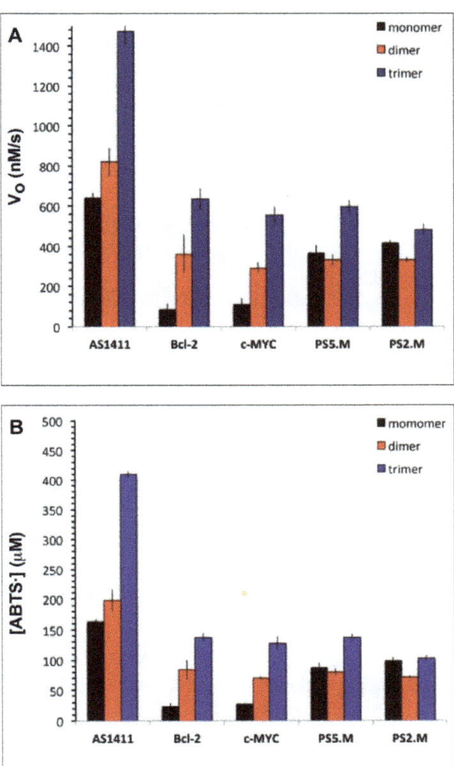

Figure 3. DNAzyme activities of DNAzymes. (**A**) Initial Rates (expressed as V_O, nM/s) were calculated from the slope of the initial linear portion (typically, the first 30 s) of the reaction curves shown in Figure 2. (**B**) The amount of ABTS• radical produced after 300 s from the activity of 1 µM DNAzyme were determined from the endpoint of the reaction time course shown in Figure 2. See *Materials and Methods*.

Bcl-2: In comparison to AS1411, Bcl-2 is a relatively low activity DNAzyme [11,24]. As shown in Figure 2B, multimerisation had a remarkable rate-enhancing effect on Bcl-2 activity. The initial rate measurements (monomer, 83 nM/s; dimer, 360 nM/s; trimer, 634 nM/s) showed a 4-fold enhancement effect from dimerisation and nearly 8-fold from trimerisation (Figure 3A). Similar trends were observed in the total amount of ABTS• product formed (Figure 3B).

c-MYC: In a pattern similar to that seen with Bcl-2, multimerisation significantly enhanced the ABTS oxidation activity of c-MYC. (Figure 2C). The initial rate measurements obtained were monomer (111 nM/s), dimer (291 nM/s), and trimer (556 nM/s). The amount of product formed showed a similar trend (Figure 3). These values represent approximately 3- and 5-fold increases as a result of dimerisation and trimerisation, respectively.

PS5.M and PS2.M: PS5.M and PS2.M are two of the best-studied peroxidase DNAzymes and have been used as model systems for characterising aspects of G-quadruplex DNAzymes catalytic mechanism. Indeed, PS5.M is one of the first DNA oligonucleotide aptamers shown to have the ability to bind hemin and acquire peroxidatic catalytic activities [2,3]. The 18-nucleotide PS2.M was derived from rational redesign of the 24-nucleotide PS5.M to achieve better catalytic activity [3]. In our experiments, the activity of PS2.M (416 nM/s) (Figure 2D,E) was higher than that of PS5.M (364 nM/s) (Figure 3A).

In contrast to the DNAzymes described above (AS1411, Bcl-2, and c-MYC), PS5.M and PS2.M dimers were slightly less active than their corresponding monomers (Figure 2D,E and Figure 3A). The initial rate values determined for PS5.M monomer (365 nM/s) and dimer (333 nM/s) were similar, an indication that dimerisation of this aptamer did not have a synergistic effect on activity. Furthermore, the initial rate value determined for the trimer (597 nM/s) is about 1.5-fold higher than that of the monomer. Rather than having a rate-enhancing effect, these results suggest that multimerisation resulted in reduced activity per haem binding site in PS5.M. Similar effects were seen in the activities of PS2.M constructs. The dimer was 20% less active than the monomer, while the activity of the trimer (485 nM/s) was about 16% higher than that of the monomer (416 nM/s) (Figure 3A).

2.3. AS1411 Multimers Display High Activities at Low DNAzyme Concentrations

The results presented above (Figures 2 and 3) show that AS1411 DNAzymes were generally more active than the other aptamers tested in this study. The high activity seen with AS1411 trimer indicates that this construct could be the starting point for further optimisation experiments aimed at designing more active aptamers. From a catalytic perspective, it is desirable to have DNAzymes that display high activities at low concentrations, i.e., those with high turnover per unit catalyst concentration.

We compared the activities of AS1411 monomer, dimer and trimer in reactions that contained 0.1, 0.25, 0.5 and 1.0 μM DNAzyme concentrations. We included PS5.M for comparison with the AS1411 constructs since it is one of the most studied high activity DNAzymes reported to date. The results (Figure 4) show that all four DNAzymes are active at lower concentrations, and that there is a linear relationship between DNAzyme concentration and ABTS oxidation activity. At 0.1 μM, the lowest concentration studied, the initial rate for AS1411 trimer was 139 nM/s, which was approximately 4.5 times the value for PS5.M (31 nM/s) at the same DNAzyme concentration. The high activity of AS1411 trimer at nanomolar concentrations could have practical use in applications where low concentrations of the G4/Hemin DNAzyme are desirable.

2.4. Effects of Reactants' Concentrations on Activities of AS1411 Multimers

To further understand the properties of the highly active AS1411 multimers, we examined the effects of ABTS and H_2O_2 concentrations on the activities of the DNAzymes. We chose a range of ABTS and H_2O_2 concentrations that will allow us to determine the nature of the relationship between reactant concentration and DNAzyme activities. Assays were carried out in reactions that contained 0.25, 0.5, 0.75, 1.0, 1.5, 2.0, and 2.5 mM ABTS, representing a 10-fold increase from the lowest to the highest concentration. The effect of H_2O_2 concentration was investigated in reactions that contain 0.106, 0.213, 0.319, 0.425, 1.063, 2.125, 3.188, and 4.250 mM, representing a 40-fold increase from the lowest to the highest concentration.

The results (Figure 5) show that the rates of DNAzyme reactions are effectively controlled by the concentration of H_2O_2 in the reaction mixture. The initial rates remained largely unchanged over a 10-fold increase in ABTS concentration (Figure 5G), an indication that the activities of the DNAzymes do not depend on the concentration of the donor substrate. However, as shown in Figure 5A–C, the amount of reaction product formed, as indicated by the endpoint after 5 min of reaction, is dependent on the concentration of ABTS. It is observed that the time course curves show complete depletion of the substrate at lower concentrations. The rate of this depletion is faster in the more active multimers.

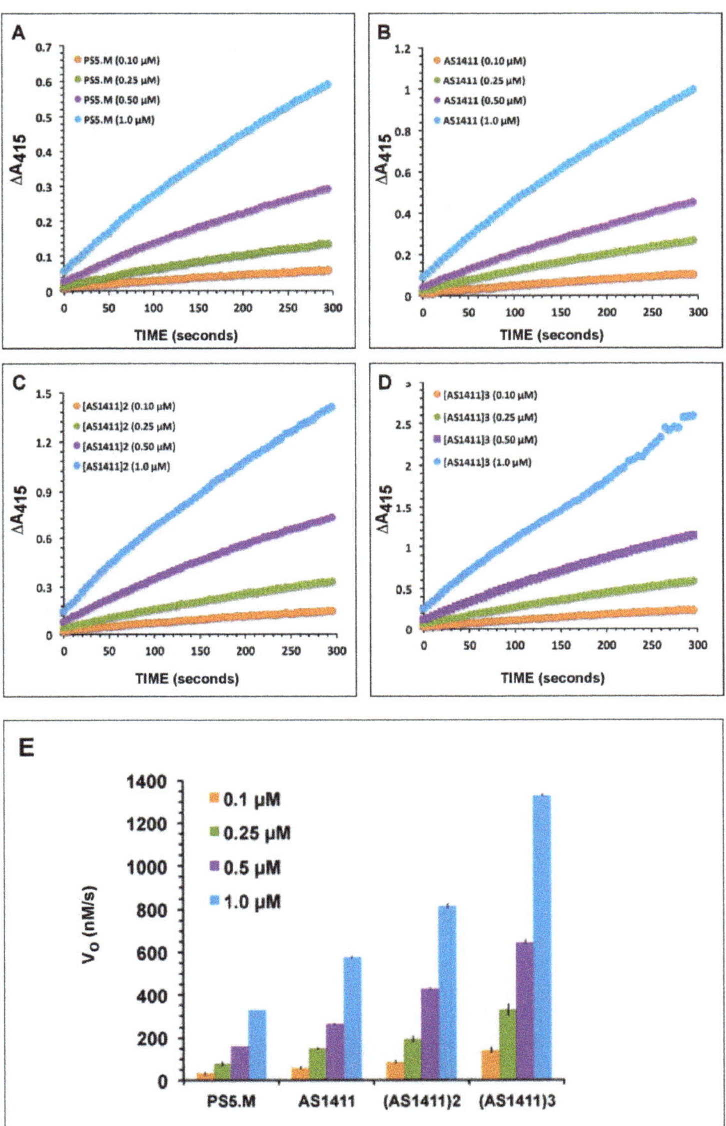

Figure 4. Effect of DNAzyme concentrations on activities of AS1411 multimers and PS5.M. (**A**) PS5.M; (**B**) AS1411 monomer; (**C**) AS1411 dimer; (**D**) AS1411 trimer. In (**A**–**D**), the curves show time courses of ABTS oxidation by 0.1, 0.25, 0.5, and 1.0 μM DNAzyme. (**E**) Initial Rates (expressed as V_O, nM/s) were calculated from the slope of the initial linear portion of the curves shown in (**A**–**D**).

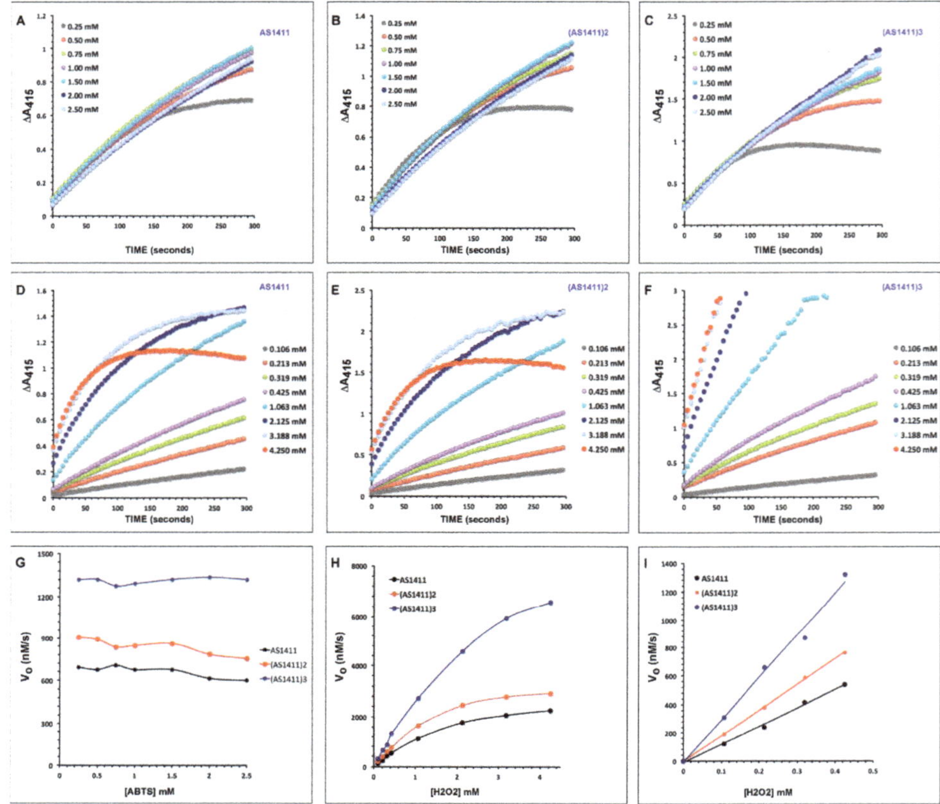

Figure 5. Effects of ABTS and H_2O_2 concentrations on activities of AS1411 multimers. The effects of ABTS concentrations on the time course of DNAzyme activities are shown in Panels (**A–C**). The final ABTS concentrations in the reaction mixtures are 0.25, 0.5, 0.75, 1.0, 1.5, 2.0, and 2.5 mM. Reactions were carried out at a fixed H_2O_2 concentration (0.425 mM), and 1.0 μM DNAzyme. Similarly, Panels (**D–F**) show the effects of H_2O_2 concentrations on the time course of reactions at a fixed ABTS concentration (2.5 mM), and 1.0 μM DNAzyme. The final H_2O_2 concentrations are 0.106, 0.213, 0.319, 0.425, 1.063, 2.125, 3.188, and 4.250 mM. The relationship between substrate concentrations and Initial Rates are shown in Panel (**G**) (ABTS) and Panel (**H**) (H_2O_2). Panel (**I**) shows the linear relationship between Initial Rates and H_2O_2 concentration at lower concentrations of the reactant (0.106, 0.213, 0.319, and 0.425 mM). Initial Rates (expressed as V_o, nM/s) were calculated from the slope of the initial linear portion of the curves in (**A–F**).

In contrast to the effect of ABTS concentration, results shown in Figure 5D,E indicate that the initial rates of the three DNAzymes are dependent on the concentration of H_2O_2. Figure 5H shows the dependence of initial rates on H_2O_2 concentration. The pattern indicates a linear correlation between H_2O_2 concentration and initial rates at the lower H_2O_2 concentration (Figure 5I). The trend changed to what appears at first to be a saturation kinetics at higher H_2O_2 concentration, especially with AS1411 and (AS1411)2. However, the time course of the reactions at high H_2O_2 concentrations (over 2 mM) shows an early plateau of activity, an indication of inactivation of the DNAzymes at high oxidant concentration. Suicide inactivation of natural protein peroxidases and G-quadruplex peroxidase DNAzymes at high H_2O_2 concentration is a well-known phenomenon that limits the use of peroxidases in certain applications [1,6,8]. In addition, high concentration of H_2O_2 is known to accelerate the disproportionation of ABTS• radical cation [23]. The response of the activities of the DNAzymes to changes in ABTS and H_2O_2 concentrations observed here has also been reported in a previous study where it was shown that DNAzymes behave differently from HRP, a natural protein peroxidase [23].

Interestingly, the plot of Initial Rates versus H_2O_2 concentration (Figure 5H) shows that the trimer, (AS141)3, appears to be more resistant to the reduction in activity at high H_2O_2 concentration. This could be due to presence of multiple active sites or as a consequence of a more stable structural arrangement that protects the hemin at the reaction centre. This property could make (AS1411) particularly useful in reaction systems that require the use of H_2O_2 at high concentrations. Further structural studies will provide more insights into the basis for this stability.

2.5. Determinants of Rate Enhancement by Multivalent DNAzymes

The goal in designing multivalent G4/Hemin DNAzymes is to achieve increased catalytic performance. While it is expected that molecular coupling of DNAzyme units as dimers, trimers or tetramers could have synergistic effects by resulting in a single molecule with activity higher than that from the sum of the parts, this may not necessarily be the case. One of the key objectives of this study is to gain an understanding of the factors that determine the outcome of multimerising G-quadruplexes in terms of rate-enhancement or lack of it. The activities of G4/Hemin structures are strongly related to the topological structures formed by the DNA sequence [31].

Generally, DNAzymes with parallel G-quadruplex structures tend to have higher activities than those with antiparallel arrangements. It is suggested that the lower activities of antiparallel structures are due to reduced hemin binding as a result of the steric hindrance caused by protruding loops at their outer G-quartets, while parallel G-quadruplexes have no such constraints [32]. Based on several studies, the activities of the different topological forms of G-quadruplexes follow the pattern: parallel > mixed parallel/antiparallel > antiparallel [8,17,26]. The topology of a G-quadruplex sequence depends on the length of the loops that connect two adjacent G-tracts. G-tracts with short loops tend to form parallel structures, while long loops give antiparallel topologies [33,34]. In addition, the nature of cations (Na^+, K^+, Mg^{2+}, NH^{4+}) present in the G-quadruplex structure can determine its topological arrangement [8].

Our results show that the G-quadruplexes that are known to form predominantly parallel structures (AS1411, Bcl-2, and c-MYC) showed increased activity upon multimerisation. AS1411 forms a parallel bimolecular quadruplex structure that consists of four pairs of stacked G-tetrads [25]. Similarly, c-MYC forms a parallel structure [11,35]. These parallel structural arrangements favour external stacking of the hemin ligand [36]. Earlier studies using NMR showed that Bcl-2 forms mixed parallel/antiparallel structure [28,29]. More recently, it was shown that Bcl-2 formed intramolecular parallel G4 structures in experiments where CD spectra were measured in a buffer containing 10 mM NaH_2PO_4/Na_2HPO_4, 100 mM KCl, and 2 mM $MgCl_2$ [11]. It follows that certain buffer conditions might favour a predominance of parallel G4 arrangements, which could explain the high rate enhancement caused by multimerisation of Bcl-2 aptamers into multivalent unimolecular structures.

In contrast, PS2.M forms mixed parallel/antiparallel structure and its structural state seems to be dependent on the nature of the cation used in inducing the folding of the sequence into the G-quadruplex structure [3,32,37]. Hence, it is likely that a tendency toward antiparallel topology may not support favourable binding of hemin thereby explaining the relatively modest rate enhancement in multimeric PS2.M and PS5.M. Further mechanistic studies are required to understand why the multimers of PS2. M and PS5.M behaved differently from those of the parallel topology-forming AS1411, Bcl-2 and c-MYC.

It is anticipated that highly active DNAzymes, such as the ones reported in this study, will have applications in designing highly sensitive sensors based on G-quadruplex peroxidase system. The notably high activities of AS1411 multimers would require detailed mechanistic investigation considering that AS1411 forms intermolecular G4 structures. Detailed spectroscopic and structural insights into the basis for the high activities of the multimers could provide valuable information on how to design better DNAzymes. Such studies, which are beyond the scope of the work reported here, will be a focus of our future investigation into the properties and applications of multimeric G4/Hemin DNAzymes.

The effect of multimerisation on Bcl-2 and c-MYC shows that the strategy can be used to enhance the activities of G4/Hemin sequences with low activity but which possess other desirable properties. Certain DNAzymes have been designed for optimal activities in conditions of extreme pH, high temperature or other non-standard conditions [8,17,19].

It is worth noting that a different buffer condition, especially one with a different composition of cations, could yield a result pattern that is different from what we observed in this study. Careful variations in ABTS and H_2O_2 concentration as well as the concentration of the DNAzyme could reveal further aspects of the reaction properties of multivalent G4/Hemin structures. Further efforts will focus on how flanking residues, linker length and sequence, and rate-enhancing boosting agents (ATP and polyamines) affect the catalytic properties of multivalent G-quadruplex DNAzymes.

3. Materials and Methods

3.1. Reagents and DNA Oligonucleotides

HPLC-purified DNA oligonucleotides were purchased from Integrated DNA Technologies, IDT (Leuven, Belgium) without further purification. These were dissolved in TE (1×) buffer (10 mM Tris-HCl pH 7.5, 0.1 mM EDTA) as 100 µM solutions and stored at −20 °C when not in use. DNA concentrations were determined using A_{260} and the molar extinction coefficient calculated for each oligonucleotide according to its base sequence. All reagents and substrates were purchased from Sigma-Aldrich, Dorset, UK, and were of analytical grade. The buffer reagent HEPES (2-[4-(2-hydroxyethyl)-1-piperazinyl]ethanesulfonic acid) was prepared as a 200 mM stock solution and stored at 4 °C. Buffers were prepared freshly each week and stored at 22 °C. The standard peroxidase substrate, ABTS (2,2′-azino-bis(3-ethylbenzothiazoline-6-sulphonic acid) was freshly prepared in water as 50 mM solution, and H_2O_2 was prepared freshly daily from a 30% aqueous stock solution (Sigma-Aldrich, UK).

3.2. Preparation of G4/hemin DNAzyme

G4/hemin DNAzymes were prepared as 10 µM solutions and stored at −20 °C. Briefly, 50 µL of 100 µM oligonucleotide solution in TE (1×) was added to 450 µL of a buffer that contains 25 mM HEPES-NaOH pH 7.4, 20 mM KCl, 200 mM NaCl, 0.05% Triton X-100, 1% DMSO. The sample was heated at 95 °C for 10 min, and cooled rapidly by dipping in ice. The sample was left at 22 °C for 30 min after which equivalent concentration of hemin was added and mixed thoroughly. The sample was kept at 22 °C for a further one hour after which it was stored at −20 °C.

3.3. Determination of G4/hemin DNAzyme Activity

The activities of the G4/Hemin DNAzymes were determined in reactions that contained 25 mM HEPES-NaOH pH 7.4, 20 mM KCl, 200 mM NaCl, 0.05% Triton X-100, 1% DMSO, 0.25 µM DNAzyme, ABTS (2.5 mM) and H_2O_2 (0.425 mM). Each reaction was started by the addition of H_2O_2, and change in A_{415} (characteristic for the reaction product $ABTS^{\bullet}$) was monitored immediately after starting the reaction over 5 min on a CLARIOstar Plate Reader (BMG LABTECH). Absorbance changes were converted to the amount of $ABTS^{\bullet}$ formed using a molar extinction coefficient of 36000 M^{-1} cm^{-1} for $ABTS^{\bullet}$ [38]. The initial reaction rate, V_o, expressed as nM/s, and the amount of $ABTS^{\bullet}$ formed after 5 min reaction, expressed as µM, were determined from the time course of the reaction. The absorbance intensity at $\lambda = 415$ nm (corresponding to the amount of $ABTS^{\bullet}$ formed) was measured as a function of time. The initial rates (V_o) were calculated from the slope of the initial (30 s) linear portion of the increase in absorbance [15,39].

4. Conclusions

In this study, we have shown that multimeric G4/Hemin peroxidase DNAzymes derived from parallel-forming G-quadruplexes show synergistic and cooperative catalytic activities when compared to their monomeric units. The striking activity of AS1411 trimer makes it a candidate for further designs with potential applications in sensor technologies and other aspects of biocatalysis. The simple multimeric constructs reported in this study provides the basis for further investigation into how this approach could be used to build more efficient catalytic DNA nanodevices. The evidence from this work suggests that parallel topology-forming G-quadruplexes are more likely to form highly active multimers. However, different linker lengths and additional rate-enhancing features could improve the activities of multimers derived from antiparallel G-quadruplexes. Future studies on kinetic characterisation and structural analyses will provide more insights into how multimeric G4/Hemin systems fold and activate peroxidase catalysis.

Author Contributions: Conceptualisation, F.J.O.; Methodology, F.J.O, R.I.A., A.K.P.; Validation, F.J.O.; Formal Analysis, F.J.O.; Investigation, R.I.A., D.S.O., T.K.R.-S., A.B., A.J.R.; Data Curation, F.J.O.; Writing—Original Draft Preparation, F.J.O.; Writing—Review and Editing, F.J.O., R.I.A., A.A.F., T.K.R.-S.; Visualisation, F.J.O; Supervision, F.J.O., S.O.M.; Project Administration, F.J.O.

Funding: This research was supported by institutional funding from Liverpool John Moores University, Liverpool U.K., and MRC Proximity to Development grant MC_PC_16077.

Conflicts of Interest: The authors declare no conflict of interest.

References

1. Singh, P.K.; Iqbal, N.; Sirohi, H.V.; Bairagya, H.R.; Kaur, P.; Sharma, S.; Singh, T.P. Structural basis of activation of mammalian heme peroxidases. *Prog. Biophys. Mol. Biol.* **2018**, *133*, 49–55. [CrossRef] [PubMed]
2. Li, Y.; Sen, D. A catalytic DNA for porphyrin metallation. *Nat. Struct. Biol.* **1996**, *3*, 743–747. [CrossRef] [PubMed]
3. Travascio, P.; Li, Y.; Sen, D. DNA-enhanced peroxidase activity of a DNA aptamer-hemin complex. *Chem. Biol.* **1998**, *5*, 505–517. [CrossRef]
4. Kosman, J.; Juskowwiak, B. Peroxidase-mimicking DNAzymes for biosensing applications: A review. *Anal. Chim. Acta* **2011**, *707*, 7–17. [CrossRef] [PubMed]
5. Tang, Y.; Ge, B.; Sen, D.; Yu, H.Z. Functional DNA switches: Rational design and electrochemical Signaling. *Chem. Soc. Rev.* **2014**, *43*, 518–529. [CrossRef] [PubMed]
6. Silverman, S.K. Catalytic DNA: Scope, applications, and biochemistry of deoxyribozymes. *Trends Biochem. Sci.* **2010**, *41*, 595–609. [CrossRef] [PubMed]
7. Pelossof, G.; Tel-Vered, R.; Elbaz, J.; Willner, I. Amplified biosensing using the horseradish peroxidase-mimicking DNAzyme as an electrocatalyst. *Anal. Chem.* **2010**, *82*, 4396–4402. [CrossRef]
8. Kong, D.M. Factors influencing the performance of G-quadruplex DNAzyme-based sensors. *Methods* **2013**, *64*, 199–204. [CrossRef]
9. Keniry, M.A. A comparison of the association of spermine with duplex and quadruplex DNA by NMR. *FEBS Lett.* **2003**, *542*, 153–158. [CrossRef]
10. Fu, Y.; Wang, X.; Zhang, J.; Xiao, Y.; Li, W.; Wang, J. Orderly microaggregates of G-/C-Rich oligonucleotides associated with spermine. *Biomacromolecules* **2011**, *12*, 747–756. [CrossRef]
11. Qi, C.; Zhang, N.; Yan, J.; Liu, X.; Bing, T.; Mei, H.; Shangguan, D. Activity enhancement of G-quadruplex/hemin DNAzyme by spermine. *RSC Adv.* **2014**, *4*, 1441–1448. [CrossRef]
12. Kong, D.M.; Xu, J.; Shen, H.X. Positive Effects of ATP on G-quadruplex-Hemin DNAzyme-mediated reactions. *Anal. Chem.* **2010**, *82*, 6148–6153. [CrossRef]
13. Stefan, L.; Denat, F.; Monchaud, D. Insights into how nucleotide supplements enhance the peroxidase-mimicking DNAzyme activity of the G-quadruplex/hemin system. *Nucleic Acids Res.* **2012**, *40*, 8759–8772. [CrossRef]
14. Kosman, J.; Zukowski, K.; Juskowiak, B. Comparison of characteristics and DNAzyme activity of G4/Hemin conjugates obtained via two hemin attachment methods. *Molecules* **2018**, *23*, 1400. [CrossRef]

15. Xiao, L.; Zhou, Z.; Feng, M.; Tong, A.; Xiang, Y. Cationic peptide conjugation enhances the activity of peroxidase-mimicking DNAzymes. *Bioconj. Chem.* **2016**, *27*, 621–627. [CrossRef]
16. Chang, T.; Gong, H.; Ding, P.; Liu, X.; Li, W.; Bing, T.; Cao, Z.; Shangguan, D. Activity enhancement of G-quadruplex/Hemin DNAzyme by flanking d(CCC). *Chem. Eur. J.* **2016**, *22*, 4015–4021. [CrossRef]
17. Chen, J.; Guo, Y.; Zhou, J.; Ju, H. The effect of Adenine repeats on G-quadruplex/hemin peroxidase mimicking DNAzyme activity. *Chem. Eur. J.* **2017**, *23*, 4210–4215. [CrossRef]
18. Li, W.; Li, Y.; Liu, Z.; Lin, B.; Yi, H.; Xu, F.; Nie, Z.; Yao, S. Insight into G-quadruplex-hemin DNAzyme/RNAzyme: Adjacent adenine as the intramolecular species for remarkable enhancement of enzymatic activity. *Nucleic Acids Res.* **2016**, *44*, 7373–7384. [CrossRef]
19. Guo, Y.; Chen, J.; Cheng, M.; Monchaud, D.; Zhou, J.; Ju, H. A thermophilic tetramolecular G-quadruplex/Hemin DNAzyme. *Angew. Chem. Int. Ed.* **2017**, *56*, 16636–16640. [CrossRef] [PubMed]
20. Stefan, L.; Denat, F.; Monchaud, D. Deciphering the DNAzyme activity of multimeric quadruplexes: Insights into their actual role in the telomerase activity evaluation assay. *J. Am. Chem. Soc.* **2011**, *133*, 20405–20415. [CrossRef]
21. Yang, D.K.; Kuo, C.J.; Chen, L.C. Synthetic multivalent DNAzymes for enhanced hydrogen peroxide catalysis and sensitive colorimetric glucose detection. *Anal. Chim. Acta* **2015**, *856*, 96–102. [CrossRef]
22. Liu, S.; Xu, N.; Tan, C.; Fang, W.; Tan, Y.; Jiang, Y. A sensitive colorimetric aptasensor based on trivalent peroxidase mimic DNAzyme and magnetic nanoparticles. *Anal. Chim. Acta* **2018**, *1018*, 86–93. [CrossRef] [PubMed]
23. Li, T.; Dong, S.; Wang, E. G-Quadruplex aptamers with peroxidase-like DNAzyme functions: Which is the best and how does it work? *Chem. Asian J.* **2009**, *4*, 918–922. [CrossRef] [PubMed]
24. Liu, B.; Li, D.; Shang, H. General peroxidase activity of a parallel G-quadruplex-hemin DNAzyme formed by Pu39WT—A mixed G-quadruplex forming sequence in the Bcl-2 P1 promoter. *Chem. Cent. J.* **2014**, *8*, 43. [CrossRef] [PubMed]
25. Girvan, A.C.; Teng, Y.; Casson, L.K.; Thomas, S.D.; Juliger, S.; Ball, M.W.; Klein, J.B.; Pierce, W.M., Jr.; Barve, S.S.; Bates, J.P. AGRO100 inhibits activation of nuclear factor-κB (NF-κB) by forming a complex with NF-κB essential modulator (NEMO) and nucleolin. *Mol. Cancer Ther.* **2006**, *5*, 1790–1799. [CrossRef]
26. Kong, D.M.; Yang, W.; Wu, J.; Li, C.X.; Shen, H.X. Structure–function study of peroxidase-like G-quadruplex-hemin complexes. *Analyst* **2010**, *135*, 321–326. [CrossRef] [PubMed]
27. Simonsson, T.; Pecinka, P.; Kubista, M. DNA tetraplex formation in the control region of c-myc. *Nucleic Acids Res.* **1998**, *26*, 1167–1172. [CrossRef]
28. Dai, J.; Chen, D.; Jones, R.A.; Hurley, L.H.; Yang, D. NMR solution structure of the major G-quadruplex structure formed in the human BCL2 promoter region. *Nucleic Acids Res.* **2006**, *34*, 5133–5144. [CrossRef]
29. Dai, J.; Dexheimer, T.S.; Chen, D.; Carver, M.; Ambrus, A.; Jones, R.A.; Yang, D. An intramolecular G-quadruplex structure with mixed parallel/antiparallel G-strands formed in the human BCL-2 promoter region in solution. *J. Am. Chem. Soc.* **2006**, *128*, 1096–1098. [CrossRef]
30. Liu, Z.; He, K.; Li, W.; Liu, X.; Xu, X.; Nie, Z.; Yao, S. DNA G-quadruplex-based assay of enzyme activity. *Methods Mol. Biol.* **2017**, *1500*, 133–151.
31. Kwok, C.K.; Merrick, C.J. G-quadruplexes: Prediction, characterization, and biological application. *Trends Biotechnol.* **2017**, *35*, 997–1013. [CrossRef]
32. Kosman, J.; Juskowwiak, B. Hemin/G-quadruplex structure and activity alteration induced by magnesium cations. *Int. J. Biol. Macromol.* **2016**, *85*, 555–564. [CrossRef]
33. Rachwal, P.A.; Brown, T.; Fox, K.R. Effect of G-Tract length on the topology and stability of intramolecular DNA quadruplexes. *Biochemistry* **2007**, *46*, 3036–3044. [CrossRef]
34. Bugaut, A.; Balasubramanian, S. A Sequence-independent study of the influence of short loop lengths on the stability and topology of intramolecular DNA G-Quadruplexes. *Biochemistry* **2008**, *47*, 689–697. [CrossRef]
35. Phan, A.T.; Modi, Y.S.; Patel, D.J. Propeller-type parallel-stranded G-quadruplexes in the human c-myc promoter. *J. Am. Chem. Soc.* **2004**, *126*, 8710–8716. [CrossRef]
36. Li, T.; Shi, L.; Wang, E.; Dong, S. Multifunctional G-quadruplex aptamers and their application to protein detection. *Chem. Eur. J.* **2009**, *15*, 1036–1042. [CrossRef]
37. Liu, W.; Zhu, H.; Zheng, B.; Cheng, S.; Fu, Y.; Li, W.; Lau, T.C.; Liang, H. Kinetics and mechanism of G-quadruplex formation and conformational switch in a G-quadruplex of PS2.M induced by Pb^{2+}. *Nucleic Acids Res.* **2012**, *40*, 4229–4236. [CrossRef]

38. Cheng, X.H.; Liu, X.J.; Bing, T.; Cao, Z.H.; Shangguan, D.H. General peroxidase activity of G-quadruplex–Hemin complexes and its application in ligand screening. *Biochemistry* **2009**, *48*, 7817–7823. [CrossRef]
39. Nakayama, S.; Wang, J.; Sintim, H.O. DNA-based peroxidation catalyst–what is the exact role of topology on catalysis and is there a special binding site for catalysis? *Chemistry* **2011**, *17*, 5691–5698. [CrossRef]

Article

Biotransformation with a New *Acinetobacter* sp. Isolate for Highly Enantioselective Synthesis of a Chiral Intermediate of Miconazole

Yanfei Miao, Yuewang Liu, Yushu He and Pu Wang *

College of Pharmaceutical Science, Zhejiang University of Technology, Hangzhou 310014, China; miaoyanfeizjut@163.com (Y.M.); liuyuewangzjut@163.com (Y.L.); heyushuzjut@163.com (Y.H.)
* Correspondence: wangpu@zjut.edu.cn; Tel.: +86-571-88320389

Received: 30 April 2019; Accepted: 15 May 2019; Published: 20 May 2019

Abstract: (*R*)-2-Chloro-1-(2,4-dichlorophenyl) ethanol is a chiral intermediate of the antifungal agent Miconazole. A bacterial strain, ZJPH1806, capable of the biocatalysis of 2-chloro-1-(2,4-dichlorophenyl) ethanone, to (*R*)-2-chloro-1-(2,4-dichlorophenyl) ethanol with highly stereoselectivity was isolated from a soil sample. It was identified as the *Acinetobacter* sp., according to its morphological observation, physiological-biochemical identification, and 16*S* rDNA sequence analysis. After optimizing the key reaction conditions, it was demonstrated that the bioreduction of 2-chloro-1-(2,4-dichlorophenyl) ethanone was effectively transformed at relatively high conversion temperatures, along with glycerol as cosubstrate in coenzyme regeneration. The asymmetric reduction of the substrate had reached 83.2% yield with an enantiomeric excess (ee) of greater than 99.9% at 2 g/L of 2-chloro-1-(2,4-dichlorophenyl) ethanone; the reaction was conducted at 40 °C for 26 h using resting cells of the *Acinetobacter* sp. ZJPH1806 as the biocatalyst. The yield had increased by nearly 2.9-fold (from 28.6% to 83.2%). In the present study, a simple and novel whole-cell-mediated biocatalytic route was applied for the highly enantioselective synthesis of (*R*)-2-chloro-1-(2,4-dichlorophenyl) ethanol, which allowed the production of a valuable chiral intermediate method to be transformed into a versatile tool for drug synthesis.

Keywords: (*R*)-2-chloro-1-(2,4-dichlorophenyl) ethanol; whole-cell catalysis; *Acinetobacter* sp.; isolation; (*R*)-miconazole

---

1. Introduction

Azoles have been widely performed in the treatment of immunocompromised patients suffering from undergoing invasive surgery or AIDS, graft reception, or anti-cancer therapy [1]. Pharmacological studies have found that some of the antifungal activity of *R*-enantiomer is superior to that of the corresponding *S*-isomer and its racemate, such as sertaconazole, which can significantly improve drug efficacy [2]. Miconazole is a broad-spectrum antifungal agent and its mechanism of action is to inhibit ergosterol biosynthesis and cause toxic methylated sterol levels, thereby inhibiting the growth of fungi [3]. (*R*)-miconazole (Figure 1) appeared to account for a greater degree of the biological activity than did racemic miconazole [2]. (*R*)-2-Chloro-1-(2,4-dichlorophenyl) ethanol is applied in the synthesis of many antifungal intermediates, such as miconazole, econazole, and sertaconazol. Thus, it is necessary to focus on the synthesis of (*R*)-2-chloro-1-(2,4-dichlorophenyl) ethanol in the preparation of (*R*)-miconazole.

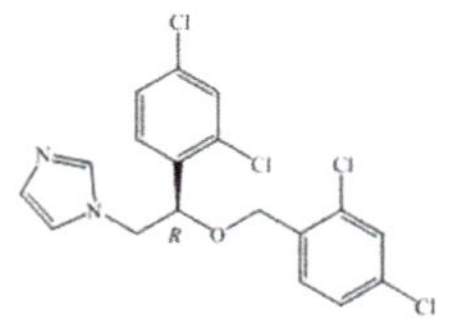

Figure 1. Structure of *R*-Miconazole.

To replace the above racemates, single enantiomers can be produced by chemical methods. de Mattos reported that rhodium, ruthenium, and iridium were the most popular catalysts in the hydrogenation of ketones. Sharpless dihydroxylation and oxazaborolidine catalysts were also used to synthesize enantiomerically enriched single enantiomers [4]. Zhang sought to provide an industrialized production method for (*R*)-2-chloro-1-(2,4-dichlorophenyl) ethanol using borane complexes and diphenyl proline in organic solvents, with 2-chloro-1-(2,4-dichlorophenyl) ethanone as the substrate; the reaction was able to achieve a yield of 92.4% with 99.3% enantiomeric excess (ee) [5]. Luo et al. presented their chemical method using ruthenium and (1*S*, 2*S*)-(+)-N-(4-toluenesulfonyl)-1, 2-diphenylethylenediamine to reduce 2-chloro-1-(2,4-dichlorophenyl) ethanone, achieving 96.4% yield and 99.2% ee [6]. The enantioselectivity of the product is related to the complexity of the ligands. Some transition metals need additional enantiomeric ligands to achieve high enantioselectivity, and the cost of transition metals limit the use of industrialization processes [7].

The asymmetric reduction of prochiral ketones is considered as an essential technique in preparing optically active compounds [8]. Numerous works have been studied concerning the preparation of single enantiomers employing biocatalysis in the culturing of plant, yeast, and bacterial products, which mostly focus on the preparation of *R*-configuration [9–11]. The biological process for the production of (*R*)-2-chloro-1-(2,4-dichlorophenyl)-ethanol was also reported. JuanMangas-Sanchez related that commercially available purified alcohol dehydrogenase ADH A, using NADH as cofactor, can convert 1.0 mM 2-chloro-1-(2,4-dichlorophenyl) ethanone in 24 h with a yield of 74%. Under the same conditions, ADH T used expensive NADPH as cofactor, and the yield was more than 99% [2]; additionally, Tang et al. employed an enzymatic reduction process to produce (*R*)-2-chloro-1-(2,4-dichlorophenyl) ethanol by the recombinant expression of *Candida macedoniensis* AKU 4588, with the ee value they attained being 99% [12]; Yue-Peng Shang explored some technique without using additional cofactors, demonstrating that a ketoreductase from *Scheffersomyces stipitis* CBS 6045, SsCR, in lyophilized cells was able to reduce 67 g/L 2-chloro-1-(2,4-dichlorophenyl) ethanone, at 88.2% yield and 99.9% ee [7]. Biocatalysts took advantage of high chemo-, regio-, unsurpassed selectivity and observed the green principles [13].

However, current reports on the synthesis of (*R*)-2-chlorophenyl-(2,4-dichlorophenyl) ethanol are mainly focused on chemical and pure enzyme synthesis [12]. Whole-cell catalysis, rather than purified enzymes, may be more suitable for the large-scale production, because microbial whole-cells used in production as catalysts contain multiple active enzymes and better protects desired enzymes against inactivation. Therefore, whole-cell biocatalysis provides an attractive alternative to selectively producing corresponding single enantiomers. In addition, the advantage of the in situ recycle of cofactors reduces the process cost, making it an increasingly attractive method for biocatalysis [14–16]. In the present study, isolated ZJPH1806 was employed as a whole-cell biocatalyst in the asymmetric bioreduction of 2-chloro-1-(2,4-dichlorophenyl) ethanone to (*R*)-2-chloro-1-(2,4-dichlorophenyl) ethanol in an aqueous medium using glycerol 20% (*w*/*v*) as cosubstrate; *Acinetobacter* sp. ZJPH1806 was maintained at an enantiomeric excess (ee) of greater than 99.9% (Figure 2). The effects of some key reaction parameters were studied to increase the yield for the production of single enantiomer alcohol. The strain identified as *Acinetobacter* sp. ZJPH1806 had been newly isolated from a soil sample for the enantioselective preparation of (*R*)-2-chloro-1-(2,4-dichlorophenyl) ethanol. The strain exhibited excellent enantioselectivity and thermotolerance at a wide catalytic range during enantiopure alcohol synthesis. Thermotolerant carbonyl reductase is a promising biocatalyst for single enantiomers production and is consequently beneficial for industrial scale manufacturing.

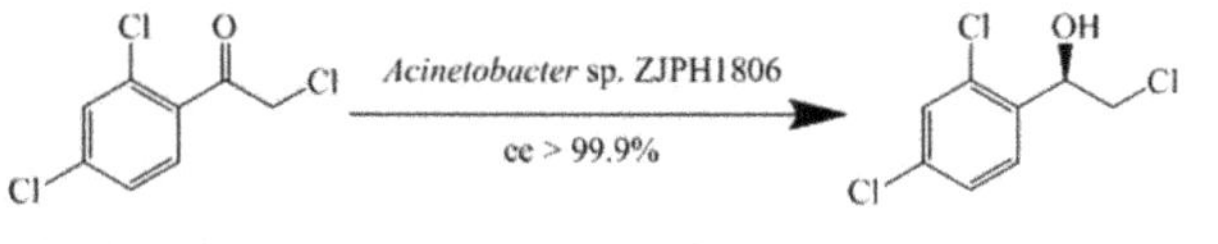

Figure 2. Asymmetric synthesis of (*R*)-2-chloro-1-(2,4-dichlorophenyl) ethanol.

2. Results

2.1. The Isolation of Strains for the Production of (R)-2-chloro-1-(2,4-dichlorophenyl) Ethanol

A variety of microorganisms including bacteria, yeasts, and molds were isolated from the soil in order to obtain strains with excellent catalytic performance. Nine strains were thus discovered, which displayed the required product yield. Among these, strain ZJPH1806 demonstrated a powerful enantioselectivity and yield (Table 1). The results, determined through the use of high-performance liquid chromatography (HPLC), are displayed in Figure S1.

Table 1. The results of the screening strain.

Strain	Yield (%)	ee (%)	Stereoselectivity
NB1-2	17.9	57.1	*S*
CP2-3	23.4	79.6	*S*
SX4-7	15.8	99.9	*S*
BJ-1	9.0	72.1	*S*
AF-4	20.1	54.3	*S*
JX1-3	23.9	59.0	*R*
HZ1-6	12.1	99.9	*R*
XM1-1	15.0	99.9	*R*
ZJPH1806	28.6	99.9	*R*

2.2. Identification and Characterization of Strain ZJPH1806

Following cultivation on nutrient agar plate for three days at 30 °C, the isolate, ZJPH1806, grew white colonies with diameters of around 0.2 cm. Physiological-biochemical identification results revealed that the species can use the tyrosine aromatic amine enzyme, citrate (Na), to alkalize *L*-lactate and succinate. The colony morphology of ZJPH1806 was illustrated in Figure S2.

The partial 16*S* rDNA sequence of ZJPH1806 (1467 bp) was determined and deposited in the GenBank database (Accession no. MK784893), and related species of similar DNA sequence were selected for the conduction of a homologous analysis with strain ZJPH1806. The phylogenetic tree of *Acinetobacter* sp. ZJPH1806, based on 16S rDNA sequencing, is shown in Figure S3. Strain ZJPH1806 had the highest homology with *Actinetobacter* sp. PAMU-1.11 (GenBank Accession no. 44885679), sharing a high sequence identity of 95%. Based on the results of phylogenetic analysis and morphological characterization, this wild strain was named *Acinetobacter* sp. ZJPH1806 and preserved in the China Centre for Type Culture Collection (CCTCC), with the Accession code of CCTCC M 2019214.

2.3. The Growth Curve of Acinetobacter sp. ZJPH1806

The enzyme production conditions of *Acinetobacter* sp. ZJPH1806 were optimized. The composition of the optimized medium was 27.63 g/L glucose, 57.35 g/L corn steep liquor, and 0.9 g/L KH_2PO_4, pH 8.0. The inoculation amount was 6% at 250 mL, with the Erlenmeyer flasks containing 90 mL of fermentation medium and cultured for 24 h on a rotary shaker at 30 °C and 200 rpm. The curves of growth and enzyme production of the strain were displayed in Figure 3. The ^{1}H NMR nuclear magnetic spectrum of product is shown in Figure 4, and the substrate's spectrum is shown in Figure S4.

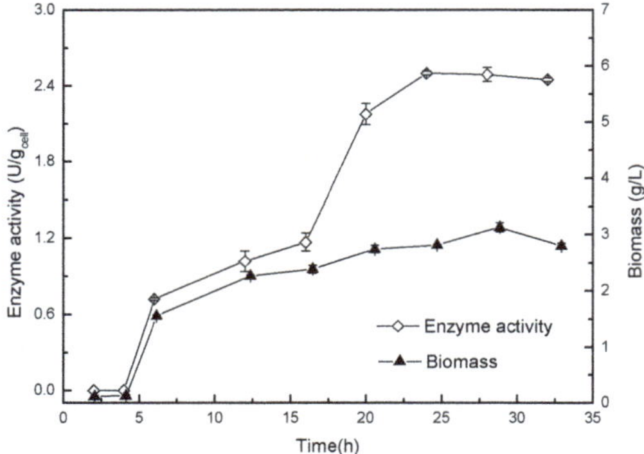

Figure 3. Curves of growth and enzyme production for *Acinetobacter* sp. ZJPH1806.

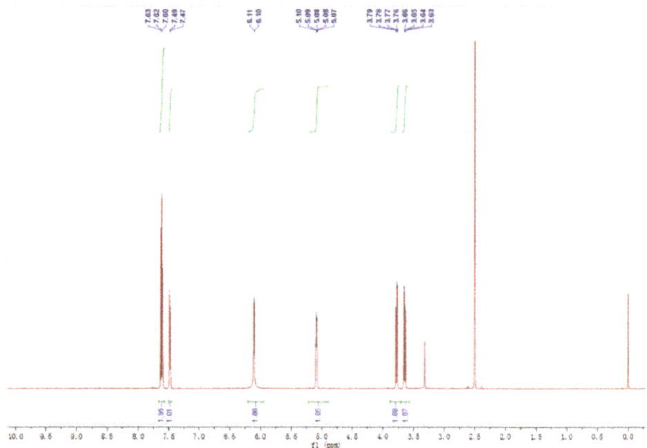

Figure 4. The ^{1}H NMR spectrum of the product.

It can be concluded from Figure 3 that the strain was in a lag phase within 4 h, with a 5 to 15 h logarithmic growth period, then entered a stable phase after 24 h, at which time the enzymatic activity and biomass of the bacterial strain had stabilized.

2.4. The Cosubstrate-Coupled System for Cofactors Regeneration

The cosubstrate-coupled, enzyme-coupled, photochemical, and electrochemical approaches are the most effective means by which the cofactors (especially NAD(H) and NADP(H)) can be regenerated in asymmetric biological reduction reactions [17]. Cosubstrate coupling was considered the method of choice. Its advantages were its low cost and ease of operation. The study selected several sugars and alcohols, including glucose, maltose, sucrose, glycerol, alcohol, methanol, and isopropanol, as cosubstrates for coenzyme regeneration (Table 2). It was indicated that glycerol reached the highest yield at 50.1%, significantly higher than the control (at a 32.6% yield). Enantioselectivity remained above 99.9%.

Table 2. Effect of different cosubstrates on the asymmetric reduction.

Cosubstrate	Yield (%)	ee (%)
Glucose	49.4	99.9
Sucrose	40.6	99.9
Maltose	48.6	99.9
Glycerol	50.1	99.9
Alcohol	33.3	99.9
Isopropanol	40.4	99.9
Methanol	9.6	99.9
Control	32.6	99.9

Various glycerol concentrations, ranging from 5% to 25% (*w*/*v*), were used to improve the asymmetric reduction of 2-chloro-1-(2,4-dichlorophenyl) ethanone to (*R*)-2-chloro-1- (2,4-dichlorophenyl) ethanol with the *Acinetobacter* sp. ZJPH1806 as biocatalyst. As shown in Figure 5, the yield rose remarkably with an increase in the glycerol concentration of up to 20% (*w*/*v*). Following that, the concentration of glycerol was further increased, decreasing the efficiency of the reaction. This reduction may be due to glycerol's high viscosity causing problems in transfer efficiency [18]. Asami's results coincide with the above research, in that glycerol can be used to regenerate cofactors in the bioreduction of 1-[3,5-bis(dimethylcarbamoyloxy)phenyl]-2-chloroethanone by *Williopsis californica* JCM 3600 [19].

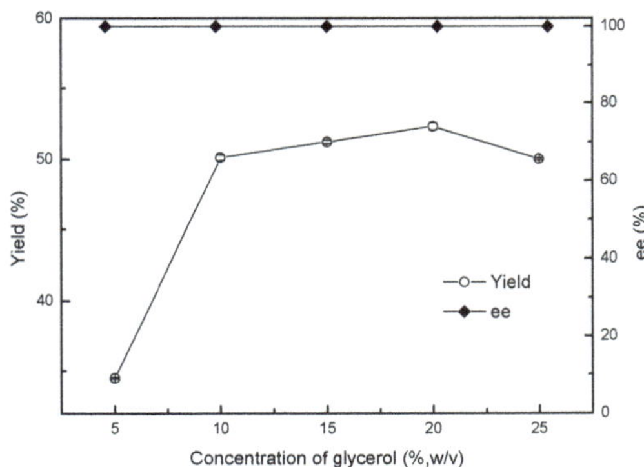

Figure 5. Effect of different glycerol addition on the asymmetric reduction.

2.5. The Effects of Buffer pH and Ionic Strength on the Asymmetric Reduction of 2-chloro-1-(2,4-dichlorophenyl) Ethanone

Varying the pH of the reaction system affects the permeability of the cell membrane, and the dissociation state of the groups necessary for the enzyme activity center and substrate, thus affecting the catalytic activity, stereoselectivity, and catalytic efficiency of the enzyme [20]. In this experimental design, the *Acinetobacter* sp. ZJPH1806 mediated reduction of 2-chloro-1-(2,4-dichlorophenyl) ethanone was investigated at different buffer pH levels within the range of 6.0 to 8.0. The results are shown in Table 3. Stereoselectivity remained above 99.9% within this range, but the yield was greatly affected. The yield was 56.2% when the pH was 7.6, but it did not significantly improve when the ionic strength of the phosphate buffer was changed from 0.05 to 0.2 M at the same pH of 7.6, indicating that *Acinetobacter* sp. ZJPH1806 can adapt to a wide range of ionic phosphate strengths. Reduction could maintain a high yield and ee value under relatively wide pH conditions. This differed strikingly from Guo's account in which the optimized pH with *Acinetonacter* sp. SC13874 was 5.5 with 0.1 M potassium phosphate [21].

Table 3. Effect of buffer pH on the asymmetric reduction.

	Buffer Solution Condition	Yield (%)	ee (%)
pH [a]	6.0, 0.1 M	10.4	99.9
	6.4, 0.1 M	37.8	99.9
	6.8, 0.1 M	53.3	99.9
	7.2, 0.1 M	55.3	99.9
	7.6, 0.1 M	56.2	99.9
	8.0, 0.1 M	50.1	99.9
pH 7.6	7.6, 0.05 M	55.8	99.9
	7.6, 0.2 M	55.0	99.9

[a] A series of pH from 6.0 to 8.0.

2.6. The Effect of Temperature on the Asymmetric Reduction of 2-chloro-1-(2,4-dichlorophenyl) Ethanone

Temperature, likewise, plays an important role in catalytic reduction mediated by microbial cells, including by affecting the activity of catalysts, the stability of enzyme production, reaction stereoselectivity, and the equilibrium constant [22]. The carbonyl reductase in ZJPH806 was expected to be a heat-resistant enzyme with good performance in biological reduction involving 2-chloro-1-(2,4-dichlorophenyl) ethanol in the interest of achieving a technical breakthrough in the heat-resistant transformation of *Acinetobacter* sp. Therefore, the effect of different reaction temperatures ranging from 25 to 50 °C on yield and enantioselectivity was subsequently investigated, and the results displayed in Figure 6. As expected, the yield decreased slightly at higher reaction temperatures (40–50 °C). The maximum yield of 59.3% was obtained at 40 °C with a product ee greater than 99.9% at 2 g/L for 2-chloro-1-(2,4-dichlorophenyl) ethanone.

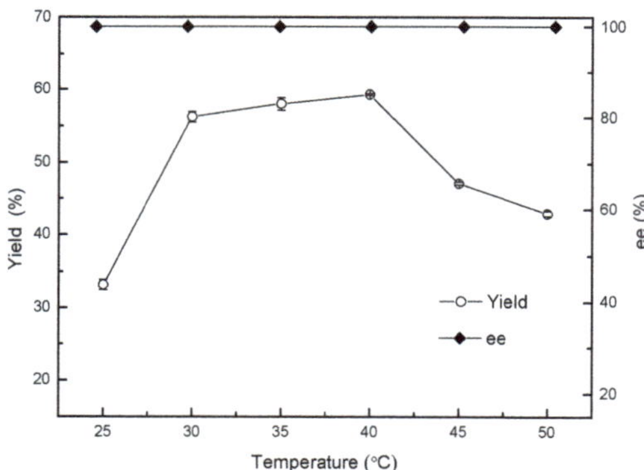

Figure 6. Effect of temperature on the asymmetric reduction.

2.7. Effects of Cell Concentration, Substrate Concentration, and Reaction Time on Biocatalytic Reduction

In biocatalytic reactions, increasing cell concentration can make a great difference on the product yield [20]. At a certain point, with an increase in cell concentration, the amount of enzyme provided also increases, which is conducive to the progress of the reaction and the improvement of the yield. However, cell oxygen consumption also increases simultaneously, with the metabolites produced by cell metabolism likewise increasing within a short period, leading conversely to a decrease in the yield. Therefore, addition of the appropriate cell concentration plays an important role in improving the yield of the reaction. Thus, cell concentrations varying from 50 to 300 g/L (wet cell weight) were designed in order to ascertain the optimum concentration for the bioreduction of 2-chloro-1-(2,4-dichlorophenyl) ethanone to (*R*)-2-chloro-1-(2,4-dichlorophenyl) ethanol. The results revealed that the maximal yield for (*R*)-form ethanol reached 76.6% at 150 g/L wet cell weight (approximately 26.7 g/L dry cell concentration) with >99.9% of product ee. However, a further increase in the wet cell concentration resulted in a significant decrease in the yield. The ee value of the product remained optimal (Figure 7a).

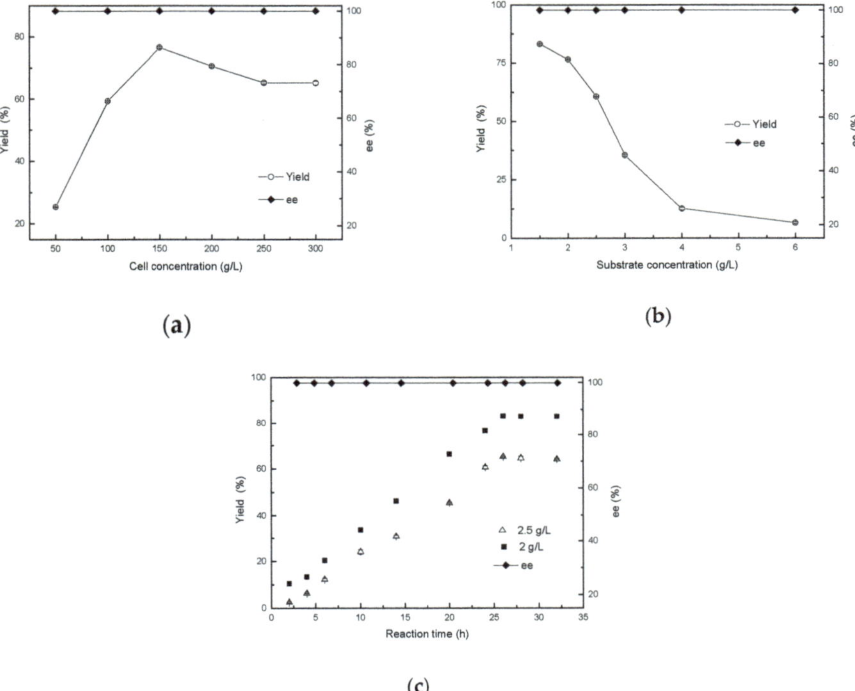

Figure 7. Effects of cell concentration (**a**), substrate concentration (**b**), and reaction time (**c**) on the asymmetric reduction.

In order to evaluate the ZJPH1806 cells mediated biological capacity, 2-chloro-1-(2,4-dichlorophenyl) ethanone was used with different concentrations of 1.5 to 6.0 g/L to examined the effect of substrate concentration on the synthesis of (*R*)-2-chloro-1-(2,4-dichlorophenyl) ethanol. Figure 7b demonstrates the results, with 83.2% yield being achieved within 24 h at 1.5 g/L of 2-chloro-1-(2,4-dichlorophenyl) ethanone concentration. This paper attempted to increase the substrate concentration, causing a reduction in the yield. This may be due to the existence of toxicity of substrates and products to cells and enzymes [23]. When the substrate concentration was increased to 2 g/L, the yield reached 83.2%when the reaction time was extended to 26 h, while the yield was only 65.12% at 2.5 g/L of substrate concentration (Figure 7c).

3. Discussion

Biocatalysts have been gradually introduced into the medical field. The biocatalytic method has attracted significant attention due to its mild reaction conditions, fewer by-products, and reduced pollution from biodegradability [24]. Zhu's report used alcohol dehydrogenase from *Pyrococcus furiosus* (PFADH), with a substrate concentration of 250 mM 2-chloro-1-(2,4-dichlorophenyl) ethanone, the yield up to 92%, and 99% ee [25]. This route was dependent on NADH addition. Compared with the use of purified enzymes, whole-cell biocatalysis does not require additional cofactors and is in favor of maintaining the stability of intracellular enzymes [26]. In this study, whole cells were used as biocatalysts to produce (*R*)-2-chloro-1-(2,4-dichlorophenyl) ethanol, the key chiral intermediate in chiral miconazole production. The strain ZJPH1806 was obtained by screening soil samples with perfect enantioselectity. Based on its morphology, physiological-biochemical identification and 16*S* rDNA sequence analysis results identified the strain as an *Acinetobacter* sp. The key factors affecting the yield and ee value were optimized, with the results demonstrating that the yield reached 83.2% in 26 h at 2 g/L of substrate concentration, which significantly improved the yield prior to optimization.

Some reports concern the synthesis of single enantiomers by using *Acinetobacter* sp. as a biocatalyst and proved that *Acinetobacter* sp. has a good potential in biocatalysis applications. *Acinetobacter* sp. SC13874 contained ketoreductase I, ketoreductase II, and ketoreductase III [21]. Among them, purified ketoreductase III was able to asymmetrically catalyze the transformation of ethyl 3,5-diketo-6-benzyloxyhexanoate to syn-(3*R*, 5*S*)-dihydroxy ester with absolute stereoselectivity and a yield of 99.3%. A diketoreductase (DKR), cloned from *Acinetobacter baylyi* ATCC 33305, was also used to reduce these two carbonyl groups via the addition of NADH with more than 99.5% ee. In 0.1 M potassium phosphate buffer, the ketoreductase III showed the pH optimum centered at pH 7.0, and the DKR was pH 6.0 [27,28]. The pH of the reduction of ketones catalyzed differs from that of *Acinetobacter* sp. ZJPH1806 under alkaline conditions. It is observed that *Acinetobacter* sp. ZJPH1806 possesses a wide catalytic range for asymmetric reduction because the yield was not significantly influenced by the pH and ionic strength of the phosphate buffer under the tested range. Furthermore, within the tested range (40–50 °C), the species also maintained high catalytic capacity. For industrial biocatalysis, novel biocatalysts with eminent biochemical properties such as thermostability and pH stability have, therefore, become increasingly attractive [29].

4. Materials and Methods

4.1. Materials

2-Chloro-1-(2,4-dichlorophenyl) ethanone (≥99% purity) and (*R*)-2-chloro-1-(2,4-dichlorophenyl) ethanol (≥98% purity) were purchased from Hangzhou Xinhai Biological Co., Ltd., China. Soil samples were collected from different areas of China, such as Fujian, Shanxi, and Zhejiang Province. *Acinetobacter* sp. ZJPH1806 was isolated from a soil sample collected in Jinghua, Zhejiang Province, China. All other chemicals were obtained commercially and were of analytical grade or better.

4.2. Screening and Cultivation of the Microorganisms

One gram of soil sample was added to 100 mL of normal saline (0.85%, *w/v*) and vibrated for 30 min. Subsequently, 1 mL of the turbid bacterial solution was added to a seeding solution and cultured it for 24 h. The seeding solution contained 15 g/L glucose, 20 g/L peptone, 10 g/L yeast extract, 2 g/L $(NH_4)_2SO_4$, 2 g/L KH_2PO_4, 1 g/L NaCl, and 0.5 g/L $MgSO_4{\cdot}7H_2O$ in pH 6.5. 2-Chloro-1-(2,4-dichlorophenyl) ethanone was used as the sole carbon source in the screening medium to select the best microorganisms and was incubated at 30 °C and 200 rpm for 5–7 days. The screening medium contained 2 g/L $(NH_4)_2SO_4$, 2 g/L KH_2PO_4, 1 g/L NaCl, 0.5 g/L $MgSO_4{\cdot}7H_2O$, and 2 g/L 2-chloro-1-(2,4-dichlorophenyl) ethanone. The enriched culture was obtained and diluted to 10^{-4}, 10^{-5}, and 10^{-6}, respectively. The diluents (200 μL) were respectively plated over the screening medium agar (2 g/L $(NH_4)_2SO_4$, 2 g/L KH_2PO_4, 1 g/L NaCl, 0.5 g/L $MgSO_4{\cdot}7H_2O$, 2 g/L 2-chloro-1-(2,4-dichlorophenyl) ethanone, and 20 g/L agar, pH 6.5), and then incubated at 30 °C and 200 rpm for 5–7 days. Following maturation, the pure colonies were obtained by continuous streaking of single colonies onto agar plates with all nutrition culture mediums (15 g/L glucose, 20 g/L peptone, 10 g/L yeast extract, 2 g/L $(NH_4)_2SO_4$, 2 g/L KH_2PO_4, 1 g/L NaCl, 0.5 g/L $MgSO_4{\cdot}7H_2O$, and 20 g/L agar, pH 6.5) at 30 °C for 2–3 days.

The single colonies were inoculated into 100 mL seed medium of 250 mL shaken flasks and incubated at 30 °C, 200 rpm for 12 h. Then, 10 mL of inoculum was transferred into 100 mL fermentation medium of 250 mL Erlenmeyer flasks (15 g/L glucose, 20 g/L peptone, 10 g/L yeast extract, 2 g/L $(NH_4)_2SO_4$, 2 g/L KH_2PO_4, 1 g/L NaCl, 0.5 g/L $MgSO_4{\cdot}7H_2O$, pH 6.5), and further incubated at 30 °C and 200 rpm for 24 h.

4.3. Phenotypic Characterization of Isolate ZJPH1806

The morphological characterization of strain ZJPH1806 was observed following the incubation on the nutrient agar plates. Physiological-biochemical identification was performed by a standardized method employing the VITEK 2 compact GN system (bioMérieux, France).

4.4. 16S rDNA Sequence Determination and Phylogenetic Analysis

The taxonomic identification of strain ZJPH1806 was performed through partial 16*S* rDNA sequence determination using a SK8255 column bacterial genomic DNA extraction kit, and 16*S* rDNA was amplified by PCR. The extracted whole cell genomes were amplified by PCR with the universal primer 27F: (5′-AGTTTGATCMTGGCTCAG-3′) and 1429R: (5′-GGTTACCTTGTTACGACTT-3′). The PCR amplification conditions were as follows: 94 °C for 4 min, repeated for 30 cycles of 45 s at 9 °C, 55 °C for 45 s, 72 °C for 1 min, and a final extension step of 72 °C for 10 min. The resulting approximately 1467 bp PCR products were purified and sequenced by Sangon Biotech Co., Ltd. For the homologous analysis, related sequences with high degrees of similarity were found through the GenBank database (National Center for Biotechnology Information, NCBI). A neighbor-joining phylogenetic tree with a nucleotide distance model and a bootstrap analysis for the evolution of the phylogenetic topology was constructed with MEGA Version 6.0.

4.5. The Growth Curve of Acinetobacter sp. ZJPH1806

The cells were cultured in seeding liquid at 200 rpm and 30 °C for 12 h; then, they were inoculated into an optimized fermentation medium (27.63 g/L glucose, 57.35 g/L corn steep liquor, and 0.9 g/L KH_2PO_4, pH 8.0), and their biomass and enzymatic activity were measured.

4.6. Asymmetric Bioreduction Process

After culture fermentation, the bioreduction was performed in 50 mL Erlenmeyer flasks with 10 mL of phosphate buffer [0.1 M, pH 6.5, DMSO (10%, *v*/*v*)], 100 g/L cells (wet cell weight), 2 g/L 2-chloro-1-(2,4-dichlorophenyl) ethanone, and 100 g/L glucose. The samples at conditions of asymmetric reduction were vibrated at 200 rpm and 30 °C for 24 h. The product was extracted with ethyl acetate for 30 min. All bioconverted samples were carried out in triplicate for detection. The yield and product ee were assayed by HPLC analysis.

4.7. Analytical Methods

Biomass was assayed by densitometry. Cell growth was monitored on a spectrophotometer (model UV-16-1; Shimadzu, Japan) at a wavelength of 600 nm. One unit of enzyme activity was defined as the amount of enzyme needed to produce 1 μmol of (*R*)-2-chloro-1-(2,4-dichlorophenyl) ethanol per minute under the assay conditions.

The NMR spectra of the substrate and product were detected by a method using a Bruker spectrometer (Ascend^{TM}; Bruker Ltd., Germany). The product was recovered by ethyl acetate extraction and purified by column chromatography on silica gel. (*R*)-2-Chloro-1-(2,4-dichlorophenyl) ethanol: ^{1}H NMR (600 MHz, DMSO): δ7.66–7.57 (m, 2H), 7.48 (d, J = 8.4 Hz, 1H), 6.10 (d, J = 4.9 Hz, 1H), 5.08 (dt, J = 7.8, 4.0 Hz, 1H), 3.78 (dd, J = 11.3, 3.8 Hz, 1H), 3.64 (dd, J = 11.3, 6.9 Hz, 1H). ^{13}C NMR (151 MHz, DMSO): δ138.39 (s), 132.90 (s), 131.97 (s), 129.87 (s), 128.44 (s), 127.44 (s), 68.75 (s), 48.48 (s). 2-Chloro-1-(2,4-dichlorophenyl) ethanone: ^{1}H NMR (600 MHz, DMSO) δ7.82 (dd, J = 3.2, 5.2 Hz, 2H), 7.61 (dd, J = 8.4, 2.0 Hz, 1H), 5.05 (s, 2H). ^{13}C NMR (151 MHz, DMSO) δ192.66 (s), 136.97 (s), 134.21 (s), 131.71 (s), 131.06 (s), 130.21 (s), 127.56 (s), 49.18 (s), 40.08–39.94 (m), 39.92 (s), 39.71 (d, J = 21.0 Hz), 39.51–39.35 (m), 39.35–39.30 (m), 39.22 (s), 39.08 (s).

The obtained substrate and product were subjected to HPLC analysis. HPLC analyses for 2-chloro-1-(2,4-dichlorophenyl) ethanone and (*R*)-2-chloro-1-(2,4-dichlorophenyl) ethanol were performed on an LC-20A system (Shimadzu) combined with an SPD-20A UV detector (Shimadzu);

The conditions were: Column of Daicel Chiralpak®OB-H (250 mm × 4.6 mm i.d., particle size 5 μm, Daicel Chemical Ltd., Tokyo, Japan), mobile phase of isopropanol-n-hexane buffer (3:97, *v*/*v*), flow rate of 1.0 mL min^{-1}, and in the wavelength of 220 nm.

$$\text{Yield }(\%) = \frac{Cp}{C0} \times 100\%$$

In the formula, C_p is the concentration of the resultant (*R*)-2-chloro-1-(2,4-dichlorophenyl) ethanol, and C_0 is the initial concentration of 2-chloro-1-(2,4-dichlorophenyl) ethanone.

$$\text{ee} = \frac{C_S - C_R}{C_S + C_R} \times 100\%$$

In the formula, C_S is the resultant concentration of (*S*)-2-chloro-1-(2,4-dichlorophenyl) ethanol; C_R is the resultant concentration of (*R*)-2-chloro-1-(2,4-dichlorophenyl) ethanol.

4.8. *The Effects of Key Variables on the Asymmetric Reduction*

In the quest for enhancing the ability of asymmetric reductions, the cosubstrates, glycerol addition, buffer pH, temperature, cell concentration, substrate concentration, and reaction time were optimized. The yield and product ee were assayed by HPLC analysis.

4.8.1. Effect of Different Cosubstrates on the Asymmetric Reduction

The results are shown in Table 2. The reaction conditions were: 100 g/L wet cell weight, 10 mL of phosphate buffer [0.1 M, pH 6.5, DMSO (10%, *v*/*v*)], 2 g/L of 2-chloro-1-(2,4-dichlorophenyl) ethanone, and different cosubstrates at a fraction of 10% (*w*/*v*, or *v*/*v*), at 30 °C and 200 rpm for 24 h.

4.8.2. Effect of Different Glycerol Addition on the Asymmetric Reduction

The optimized results of glycerol addition are revealed in Figure 5. The reaction conditions are: 100 g/L *Acinetobacter* sp. ZJPH1806 cells, 10 mL of phosphate buffer [0.1 M, pH 6.5, DMSO (10%, *v*/*v*)], and 2 g/L of 2-chloro-1-(2,4-dichlorophenyl) ethanone, using glycerol as the cosubstrate, with a proportion of 5% to 25% (*w*/*v*), at 30 °C and 200 rpm for 24 h.

4.8.3. Effects of Buffer pH and Ionic Strength on the Asymmetric Reduction

In the phosphate buffer system, pH and ionic strength are optimized and the results are shown in Table 3. The reaction conditions are: 100 g/L *Acinetobacter* sp. ZJPH1806 cells, 10 mL of phosphate buffer [DMSO (10%, *v*/*v*)], 2 g/L of 2-chloro-1-(2,4-dichlorophenyl) ethanone, and 20% (*w*/*v*) glycerol, at 200 rpm and 30 °C, with a 24 h reduction time.

4.8.4. Effect of Temperature on the Asymmetric Reduction

The optimized results of temperature are outlined in Figure 6. Reaction conditions: 100 g/L *Acinetobacter* sp. ZJPH1806 cells, 10 mL of phosphate buffer [0.1 M, pH 7.6, DMSO (10%, *v*/*v*)], 2 g/L of 2-chloro-1-(2,4-dichlorophenyl) ethanone, and 20% (*w*/*v*) glycerol, at 200 rpm, and at different temperatures (from 25 °C to 50 °C) for 24 h.

4.8.5. Optimization of Cell Concentration, Substrate Concentration, and Reaction Time on the Asymmetric Reduction

Cell concentration, substrate concentration, and reaction time are optimized, and the results are shown in Figure 7. The reaction conditions are: 10 mL of phosphate buffer [0.1 M, pH 7.6, DMSO (10%, *v*/*v*)], 20% (*w*/*v*) glycerol, and 2 g/L of 2-chloro-1-(2,4-dichlorophenyl) ethanone (a) at 200 rpm and 40 °C for 24 h (a,b).

5. Conclusions

In the present study, a biological method used to produce (*R*)-2-chloro-1-(2,4-dichlorophenyl) ethanol with *Acinetobacter* sp. ZJPH1806 was successfully developed. The findings were that of achieving the desired product in a single step by the whole-cell method. This study also provides a useful strategy to develop efficient and economical approach to cofactor regeneration. Glycerol was selected as an economical and environmentally friendly cosubstrate. Therefore, the reaction did not depend on any additional coenzyme, and the cost of the biological process was reduced. Meanwhile, the reaction was in an aqueous medium at mild conditions and the subsequent treatment was simple. The results indicated that *Acinetobacter* sp. ZJPH1806 containing a thermophilic carbonyl reductase is a promising biocatalyst for the production of the chiral intermediates and possessing a wide pH range for asymmetric reduction. Although biosynthesis efficiency of (*R*)-2-chloro-1-(2,4-dichlorophenyl) ethanol was considerably improved in the current research, similarly, the problem of substrate inhibition is a limiting factor. Thus, the research on medium engineering for biocatalysis or the immobilization of the enzyme-containing cells are important subjects worthy of further study to improve the efficiency of this biological process.

Supplementary Materials: The following are available online at http://www.mdpi.com/xxx/s1, Figure S1: The results of the bioconverted sample by HPLC analysis, Figure S2: The colony morphology of ZJPH1806, Figure S3: The phylogenetic tree of *Acinetobacter* sp. ZJPH1806, Figure S4: The ^{1}H NMR spectrum of the substrate.

Author Contributions: Conceptualization, P.W.; methodology, P.W.; software, Y.M.; validation, Y.M.; investigation, Y.M.; resources, P.W.; data curation, Y.M.; writing—original draft preparation, Y.M.; writing—review and editing, Y.M. and P.W.; visualization, Y.H. and Y.L.; project administration, P.W.

Funding: This work was funded by the National Natural Science Foundation of China (No. 21676250), and the Zhejiang Provincial Natural Science Foundation of China (LY16B060010).

Conflicts of Interest: The authors declare no conflict of interest. The funders had no role in the design of the study; in the collection, analyses, or interpretation of data; in the writing of the manuscript, or in the decision to publish the results.

References

1. Giraud, F.; Loge, C.; Pagniez, F.; Crepin, D.; Pape, P.L.; Borgne, M.L. Design, synthesis, and evaluation of 1-(*N*-benzylamino)-2-phenyl-3-(1*H*-1,2,4-triazol-1-yl)propan-2-ols as antifungal agents. *Bioorg. Med. Chem. Lett.* **2008**, *18*, 1820–1824. [CrossRef] [PubMed]
2. Mangas-Sanchez, J.; Busto, E.; Gotor-Fernandez, V.; Malpartida, F.; Gotor, V. Asymmetric chemoenzymatic synthesis of miconazole and econazole enantiomers. The importance of chirality in their biological evaluation. *J. Org. Chem.* **2011**, *76*, 2115–2122. [CrossRef] [PubMed]
3. Barasch, A.; Griffin, A.V. Miconazole revisited: New evidence of antifungal efficacy from laboratory and clinical trials. *Future Microbiol.* **2008**, *3*, 265–269. [CrossRef] [PubMed]
4. Fonseca, T.S.; Lima, L.D.; Oliveira, M.C.F.; Lemos, T.L.G.; Zampieri, D.; Molinari, F.; Mattos, M.C. Chemoenzymatic synthesis of luliconazole mediated by lipases. *Eur. J. Org. Chem.* **2018**, *2018*, 2110–2116. [CrossRef]
5. Zhang, X. Industrial Production Method of Chiral 2-Chloro-1-(2,4-dichlorophenyl)ethanol. C.N. Patent 106008166 A, 12 October 2016.
6. Luo, J.Y.; Xu, X.B.; Guo, P. New Synthetic Method for Key Chiral Intermediates of Luliconazole. C.N. Patent 108299156 A, 20 January 2018.
7. Shang, Y.P.; Chen, Q.; Kong, X.D.; Zhang, Y.J.; Xu, J.H.; Yu, H.L. Efficient synthesis of (*R*)-2-chloro-1-(2,4-dichlorophenyl) ethanol with a ketoreductase from *Scheffersomyces stipitis* CBS 6045. *Adv. Synth. Catal.* **2017**, *359*, 426–431. [CrossRef]
8. Grynkiewicz, G.; Borowiecki, P. New applications of biotechnology in the field of pharmaceutical syntheses. *Przem. Chem.* **2019**, *98*, 434–441.
9. Xie, Y.; Xu, J.H.; Xu, Y. Isolation of a *Bacillus* strain producing ketone reductase with high substrate tolerance. *Bioresour. Technol.* **2010**, *101*, 1054–1059. [CrossRef]

10. Lin, H.; Chen, Y.Z.; Xu, X.Y.; Xia, S.W.; Wang, L.X. Preparation of key intermediates of adrenergic receptor agonists: Highly enantioselective production of (*R*)-α-halohydrins with *Saccharomyces cerevisiae* CGMCC 2.396. *J. Mol. Catal. B Enzym.* **2009**, *57*, 1–5. [CrossRef]
11. Xie, Y.; Xu, J.H.; Lu, W.Y.; Lin, G.Q. Adzuki bean: A new resource of biocatalyst for asymmetric reduction of aromatic ketones with high stereoselectivity and substrate tolerance. *Bioresour. Technol.* **2009**, *100*, 2463–2468. [CrossRef]
12. Tang, Y.P.; Ding, G.F.; Yu, F.M.; Yang, Z.S.; Huang, F.F. A Method for Biosynthesis of Intermediates of Miconazole. C.N. Patent 108396040 A, 14 August 2018.
13. Ferreira-Leitão, V.S.; Cammarota, M.C.; Aguieiras, E.C.G.; de Sá, L.R.V.; Fernandez-Lafuente, R.; Freire, D.M.G. The protagonism of biocatalysis in green chemistry and its environmental benefits. *Catalysts* **2017**, *7*, 9.
14. Hollmann, F.; Arends, I.W.C.E.; Holtmann, D. Enzymatic reductions for the chemist. *Green Chem.* **2011**, *13*, 2285–2313. [CrossRef]
15. Sahin, E.; Serencam, H.; Dertli, E. Whole cell application of *Lactobacillus paracasei* BD101 to produce enantiomerically pure (*S*)-cyclohexyl(phenyl) methanol. *Chirality* **2019**, *31*, 211–218. [CrossRef]
16. Vitale, P.; Perna, F.M.; Agrimi, G.; Scilimati, A.; Salomone, A.; Cardellicchioe, C.; Capriati, V. Asymmetric chemoenzymatic synthesis of 1,3-diols and 2,4-disubstituted aryloxetanes by using whole cell biocatalysts. *Org. Biomol. Chem.* **2016**, *14*, 11438–11445. [CrossRef]
17. Kara, S.; Schrittwieser, J.H.; Hollmann, F.; Ansorge-Schumacher, M.B. Recent trends and novel concepts in cofactor-dependent biotransformations. *Appl. Microbiol. Biotechnol.* **2014**, *98*, 1517–1529. [CrossRef]
18. Gu, Y.L.; Jerome, F. Glycerol as a sustainable solvent for green chemistry. *Green Chem.* **2010**, *12*, 1127–1138. [CrossRef]
19. Asami, K.; Machida, T.; Jung, S.; Hanaya, K.; Shoji, M.; Sugai, T. Synthesis of (*R*)-bambuterol based on asymmetric reduction of 1-[3,5-bis(dimethylcarbamoyloxy)phenyl]-2-chloroethanone with incubated whole cells of *Williopsis californica* JCM 3600. *J. Mol. Catal. B Enzym.* **2013**, *97*, 106–109. [CrossRef]
20. Sun, J.; Huang, J.; Ding, X.Z.; Wang, P. Efficient enantioselective biocatalytic production of a chiral intermediate of sitagliptin by a newly filamentous fungus isolate. *Appl. Biochem. Biotechnol.* **2016**, *180*, 695–706. [CrossRef]
21. Guo, Z.W.; Chen, Y.J.; Goswami, A.; Hanson, R.L.; Patel, R.N. Synthesis of ethyl and *t*-butyl (3*R*,5*S*)-dihydroxy-6-benzyloxy hexanoates via diastereo- and enantioselective microbial reduction. *Terahedron Asymmetry* **2006**, *17*, 1589–1602. [CrossRef]
22. Zhang, Y.J.; Chen, J.L.; Chen, C.S.; Wu, S.J. Isolation of a *Bacillus Aryabhattai* strain for the resolution of (*R*,*S*)-ethyl indoline-2-carboxylate to produce (*S*)-indoline-2-carboxylic acid. *Catalysts* **2019**, *9*, 206. [CrossRef]
23. Wang, S.S.; Xu, Y.; Zhang, R.Z.; Zhang, B.T.; Xiao, R. Improvement of (*R*)-carbonyl reductase-mediated biosynthesis of (*R*)-1-phenyl-1,2-ethanediol by a novel dual-cosubstrate-coupled system for NADH recycling. *Process Biochem.* **2012**, *47*, 1060–1065. [CrossRef]
24. Simon, R.C.; Mutti, F.G.; Kroutil, W. Biocatalytic synthesis of enantiopure building blocks for pharmaceuticals. *Drug. Discov. Today Technol.* **2013**, *10*, e37–e44. [CrossRef]
25. Zhu, D.M.; Hyatt, B.A.; Hua, L. Enzymatic hydrogen transfer reduction of α-chloro aromatic ketones catalyzed by a hyperthermophilic alcohol dehydrogenase. *J. Mol. Catal. B Enzym.* **2009**, *56*, 272–276. [CrossRef]
26. Wachtmeister, J.; Rother, D. Recent advances in whole cell biocatalysis techniques bridging from investigative to industrial scale. *Curr. Opin. Biotechnol.* **2016**, *42*, 169–177. [CrossRef]
27. Goldberg, S.; Guo, Z.W.; Chen, S.; Goswami, A.; Patel, R.N. Synthesis of ethyl-(3*R*,5*S*)-dihydroxy-6-benzyloxyhexanoates via diastereo- and enantioselective microbial reduction: Cloning and expression of ketoreductase III from *Acinetobacter* sp. SC 13874. *Enzym. Microb. Technol.* **2008**, *43*, 544–549. [CrossRef]
28. Wu, X.R.; Liu, N.; He, Y.M.; Chen, Y.J. Cloning, expression, and characterization of a novel diketoreductase from *Acinetobacter baylyi*. *Acta Biochim. Biophys. Sin.* **2009**, *41*, 163–170. [CrossRef]
29. Atalah, J.; Cáceres-Moreno, P.; Espina, G.; Blamey, J.M. Thermophiles and the applications of their enzymes as new biocatalysts. *Bioresour. Technol.* **2019**, *280*, 478–488. [CrossRef]

Article

Enhancing Enzymatic Properties of Endoglucanase I Enzyme from *Trichoderma Reesei* via Swapping from Cellobiohydrolase I Enzyme

Aslı Yenenler [1,2,*], Hasan Kurt [3] and Osman Uğur Sezerman [2]

1 Molecular Biology, Genetics and Bioengineering, Faculty of Engineering and Natural Sciences, Sabanci University, Tuzla, Istanbul 34956, Turkey
2 Department of Biostatistics and Medical Informatics, School of Medicine, Acibadem Mehmet Ali Aydinlar University, Atasehir, Istanbul 34752, Turkey; ugur.sezerman@acibadem.edu.tr
3 School of Engineering and Natural Sciences, Istanbul Medipol University, Beykoz, Istanbul 34810, Turkey; hasankurt@medipol.edu.tr
* Correspondence: asliyenenler@sabanciuniv.edu

Received: 4 January 2019; Accepted: 23 January 2019; Published: 1 February 2019

Abstract: Utilizing plant-based materials as a biofuel source is an increasingly popular attempt to redesign the global energy cycle. This endeavour underlines the potential of cellulase enzymes for green energy production and requires the structural and functional engineering of natural enzymes to enhance their utilization. In this work, we aimed to engineer enzymatic and functional properties of Endoglucanase I (EGI) by swapping the Ala43-Gly83 region of Cellobiohydrolase I (CBHI) from *Trichoderma reesei*. Herein, we report the enhanced enzymatic activity and improved thermal stability of the engineered enzyme, called EGI_swapped, compared to EGI. The difference in the enzymatic activity profile of EGI_swapped and the EGI enzymes became more pronounced upon increasing metal-ion concentrations in the reaction media. Notably, the engineered enzyme retained a considerable level of enzymatic activity after thermal incubation for 90 min at 70 °C while EGI completely lost its enzymatic activity. Circular Dichroism spectroscopy studies revealed distinctive conformational and thermal susceptibility differences between EGI_swapped and EGI enzymes, confirming the improved structural integrity of the swapped enzyme. This study highlights the importance of swapping the metal-ion coordination region in the engineering of EGI enzyme for enhanced structural and thermal stability.

Keywords: Endoglucanese I; Cellbiohydrolase I; *Trichoderma reesei*; Swapping; Protein Engineering; divalent metal ion

---

1. Introduction

Lignocellulose is considered the most abundant and sustainable energy resource in the world [1,2]. Due to the growing demand on green alternative energy sources, a substantial global effort is being exerted to redesign the global carbon cycle [3] in an effort to provide sustainabile fuels from plant-based cellulosic materials [4]. Plant-based cellulosic materials as a biofuel resource offer significant advantages over fossil fuel sources, e.g., decreasing global emission, providing a long-term rural development [5,6], the particular challenges still have to be surmounted in the deconstruction processes of cellulose to cellulosic biomass. The naturally endowed complexity of cellulose requires the synergistic action of cellulase enzymes for the effective breaking down of cellulosic biomass [7,8]. Apart from efforts in different pre-treatment techniques for a more efficient breaking down of cellulose, researchers focused on the development of new cellulase enzymes with improved efficiency via protein engineering [9,10].

In protein engineering, the combinations of experimental and computational methods are frequently employed to study the structure-function relationships of enzymes and to reveal the impacts of structural variations on their stability. Domain swapping is one of the prominent tools in the protein engineering arsenal that aims to provide the alterations on the scaffold of proteins for attaining the improvements in existing catalytic and enzymatic activities [11]. So far, numerous studies were reported on domain swapping (DS), ranging from cysteine proteinase inhibitors [12], exploring SH2 and SH3 domains [13], L-histidonol dehydrogenase proteins [14], and prion proteins [15].

Naturally occurring cellulase enzymes usually underperform in biotechnological and industrial applications, and hence this limitation intrigues researchers to develop new cellulase enzymes that can efficiently perform under less restricted conditions [16]. Understanding the molecular basis of how enzymes maintain their stability at harsh conditions is a recurring topic biochemistry [17]. Thermostability could be affected by numerous factors; e.g., hydrogen bond networks, hydrophobic and packing interactions, and charge clusters [18]. The presence of metal ions could directly alter thermostability [18–20] via the enhancement of hydrogen networking and packing interactions with the aliphatic residues' tendency to interact with metal ions via unpaired electrons [21].

The aim of this study is to create a suitable environment for metal-ion coordination in EGI enzyme via swapping a region from CBHI enzyme, being isolated from *Trichoderma reseei* [22,23]. The findings indicate that swapping the metal-ion coordination region from the CBHI enhanced structural integrity and resulted in a novel endoglucanase enzyme, EGI_swapped, which displayed unique structural and functional properties and is a promising catalyst candidate for industrial applications. To the best of our knowledge, the work presented here is the first study in terms of engineering the structural and functional properties of an EGI enzyme from *Trichoderma reesei* via swapping in the literature.

2. Results & Discussion

2.1. Construction of EGI_Swapped and Cloning Approach

In this study, we aim to alter the existing structural features of EGI enzyme via domain swapping from CBHI enzyme. These enzymes are both isolated from *Trichoderma reesei* by sharing the certain level of structural similarity but differentiated in the mode of action (Figure S1). First, we perform the structural alignment between EGI and CBHI (PDB id: 1EG1 and 1DY4) to find appropriate regions for domain swapping. We find that choosing the Ala43-Gly83 region (CBHI numbering) is promising due to its closeness to the Co^{2+} metal ion in the crystal structure of CBHI, which does not exist in the crystal structure of EGI [24] (see Figure S2). Among many metal ions, Co^{2+} ion is reported as a cellulase activator [25].

In the crystal structure of CBHI, two Co^{2+} ions are reported as one is linked to His206 and Glu239, and another one is linked to Glu295 and Glu325. We just focused on a Co^{2+} ion linked to His206 and Glu239 [24], based on the structural alignment results. Around this particular Co^{2+} ion, there is a network of negatively charged and polar residues [24], e.g., Asp63, Asn64, and Glu65, which can provide support for the coordination of Co^{2+}. Before deciding on a swapped region, we measure the distances of the OD1 atom of Asp63, ND2 atom of Asn64 and OE2 atom of Glu65 to Co^{2+} ion, in CBHI, as ~5.6 Å, ~11 Å, and ~9 Å, respectively. Even these distances are out of range for minimum direct metal-ion binding to residues (Asp/Glu) [26], yet these residues are in the range of coordination as they are close enough to create a "high hydrophobicity" contrast [26,27]. Mainly, metal ions are classified into three according to their polarizability; (i) hard, (ii) soft, and (iii) borderline. Hard means a "low-polarizability" atom or ion that is deformed only by difficulty, and soft means a "high-polarizability" atom or ion that is easily deformed [27,28]. The Co^{2+} ion is reported as borderline based on its polarizability. Due to the nature of not being easily deformed, these particular distances would be enough to keep Co^{2+} ion in true geometry with His residue [28,29].

Specifically, a fixed aromatic residue or negatively charged one provides an appropriate environment for effective metal ion binding by creating new H-bond networking or contributing to the existing one in advance [21]. Hence, swapping of Ala43-Gly83 region of CBHI to EGI has satisfied these constraints by covering the residues of Asp63 to Glu65, close to fixed Histidine in EGI's location in order to create a favorable neighbour pocket for Co^{2+} ion coordination. With Ala to Glu replacement 64th position (EGI numbering) upon swapping, this favorable environment has been also supported in EGI_swapped for Co^{2+} ion coordination, in terms of donating more electrons and thus creating an advanced H-bond network, with respect to its native counterpart (see Figure S3). In Figure 1, the domain swapping is described by indicating Co-linked His and Glu residues.

Stahlberg et al. reported that divalent metal ions, as Ni^{2+}, Co^{2+}, Ca^{2+}, Mn^{2+}, and Mg^{2+}, seem to be required by crystallization [24]. There are several CBHI structures containing the different ions, e.g., Sm^{3+} and Gd^{3+}, than divalent ones [30,31]. Before deciding on the swapped region for Co^{2+} coordination, the crystal structures displaying more than 95% sequence similarity and also the high level of structural similarity within our reference structure (CBHI enzyme) are carefully examined. Among 26 reported crystal structures, only 7 ones contain different metal ions than Co^{2+}, e.g., Sm^{3+} in PDB id: 4P1H, Gd^{3+} in PDB id: 5TC9, and Ca^{2+} in PDB id: 1CEL, 2CEL, 3CEL and 4CEL [30–33]. To make a true decision for swapping, we also perform the structural alignments between EGI (PDB id: 1EG1) with them. These alignment results suggest that the ion coordination regions of CBHI's structures are not well aligned in EGI's structure with the existence of structural breaks, and thus they are not suitable to perform swapping (Figure S4). Actually, Ca^{2+}, Gd^{3+}, and Sm^{3+} ions are coordinated by CBHI in the region, covering Glu295 and Glu325, as similar to another Co^{2+} ion in our reference CBHI enzyme. This particular region is too far from our swapped-region, Ala43-Gly83, as displayed in Figure S4.

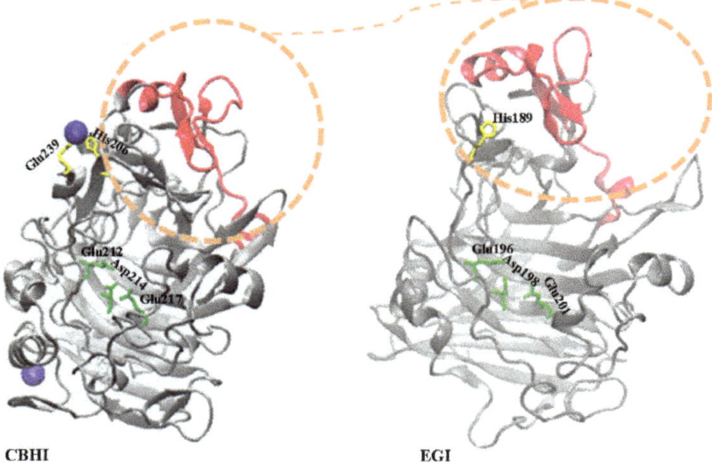

Figure 1. The rationality of the swapping approach. The crystal structures of CBHI (PDB id: 1DY4) and EGI (PDB id: 1EG1) are drawn by colouring the swapped-domain in red. The catalytic residues of CBHI, Glu212, Asp214 and Glu217, and EGI, Glu196, Asp198 and Glu201, are depicted in 'licorice' format and coloured as 'green'. The His206 and Glu239 residues directly linked to Co^{2+} in CBHI's structure is drawn in 'licorice' format and colored as 'yellow'. To perform distance measurement and visualization of the structure, the Visual molecular dynamic is used [34].

It is stated in the literature that tunnel forming loops existed in the structure of CBHI is not observed in EGI [24,35] due to the loss of particular genes via horizontal gene transfer during the evolutionary process [36]. Even a certain level of structural similarity has been preserved between the two, this particular gene loss results in the differentiation in their mode of action as depicted in Figure S1. Instead of swapping just a few residues in close contact, the swapping of the whole region eliminates the possibility of generation of the low- or non-functional enzyme due to misfolding or unfolding (Figure S2). Further, this swapped region is rich, for Gly content that provides additional conformational flexibility in the swapped enzyme to adopt Co^{2+} coordination (Figure S3).

Before transforming the swapped gene into *P.pastoris* expression vector, the positive clones were double-checked by *EcoR* I single digestion (Figure S5a,b). As displayed in Figure S3, there is a high level of similarity and identity between EGI_swapped and EGI enzymes, 94.3% and 96.4% respectively. In parallel to prediction (Table S1), the molecular weight of the swapped enzyme is reported around ~50 kDa, including a 6xHis tag in the expression vector used for the purification step, see Figure S6.

2.2. Enzymatic Activities

First, we find the optimal pH-buffer combinations for both EGI_swapped and EGI enzymes. The optimal pH-buffer combination is found by scanning pH 4–5 and pH 6–7 ranges in 50 mM sodium acetate and potassium phosphate buffers toward 0.5% w/v CMC at 50 °C for 3 h, respectively. For EGI_swapped and EGI enzymes, the highest enzymatic activity is recorded at 50 mM sodium acetate buffer pH 4.7 as in line with the reported PI value (Table S1). Then, the rest of all enzyme activity and thermal stability tests are performed in 50 mM sodium acetate pH 4.7 buffer. Just for the circular dichroism (CD) spectroscopy measurement, the concentration of buffer is decreased to 20 mM at pH 4.7.

As depicted in Figure 1, the swapped region is not close to the catalytic triad. This particular distance from the swapped region to the catalytic triad creates an expectation that alterations in enzyme activity are most probably due to a change in the thermal stability of EGI_swapped. Among 1 h results, the first notable difference in enzymatic activities is captured at the highest temperature, 50 °C (see Figure S9). Upon extending the reaction time to 3 h, the notable difference is again recorded at 50 °C (Figure 2). Indeed, the difference in activities of EGI and EGI_swapped become more visible, around ~20% difference, at 6 h (Figure 2). These results point out the alterations in thermal stability of EGI_swapped with respect to the EGI enzyme.

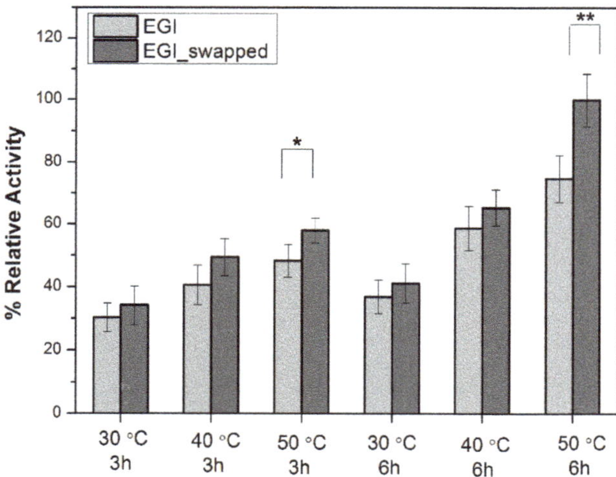

Figure 2. The enzyme activity profiles of EGI and EGI_swapped enzymes toward CMC for 3 h and 6 h at 30–50 °C (/10 °C). Data are express as mean ± SD, n = 3. *p*-values between 0.1 and 0.05, and *p*-value smaller than 0.05 are indicated with '*' and '**' on plots, respectively.

Moreover, we aim to investigate either there is any impact of Co^{2+} addition on enzymatic activities of EGI and EGI_swapped, or not. To reveal this, we performed activity tests on enzymes toward 0.5% w/v CMC by following two paths; (i) pre-incubating enzymes with Co^{2+} (Figure 3) and (ii) pre-incubating CMC with Co^{2+} (Figure S7) prior to the hydrolysis reaction. We report that the relative activities of EGI are 48% without any Co^{2+}, and 47% for 0.05 mM and 0.1 mM Co^{2+} ion additions. As Co^{2+} ion addition increases to 0.2 mM and 0.5 mM, its enzymatic activity dramatically decreases to 28% and 24%, respectively. This particular trend could be explained as the accumulated Co^{2+} ions randomly bind to the tertiary structure of EGI due to the lack of a favorable environment for ion binding. This result is in line with the literature search that there is no reported crystal structure in PDB, displaying more than a 70% sequence similarity with EGI (PDB id: 1EG1) to be considered as a homolog, with the presence of Co^{2+} or any relevant metal ions [35].

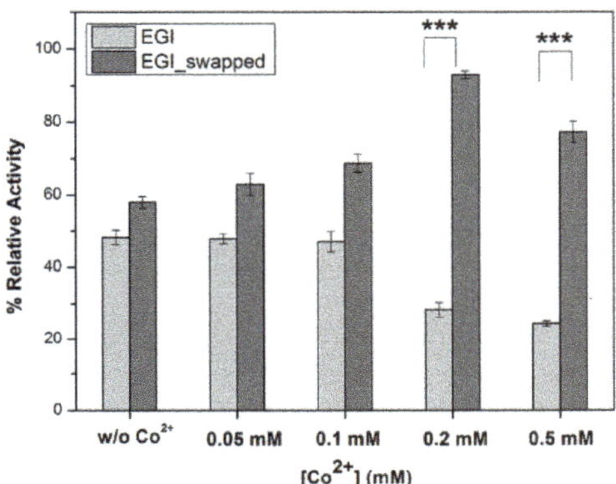

Figure 3. The effects of Co^{2+}-ion pre-incubation with enzymes prior to hydrolysis reaction on their enzymatic activity profiles. Through enzyme assays toward 0.5% w/v CMC, the impacts of Co^{2+} ion pre-incubation with enzymes were investigated for different Co^{2+} ion concentrations, 0 mM, 0.05 mM, 0.1 mM, 0.2 mM and 0.5 mM, in final. The highest data among all activity tests, EGI_swapped at 50 °C for 6 h, is taken as 100% to calculate the rest. Data are express as mean ± SD, n = 3. To indicate the significant differences in enzymatic activities of EGI and EGI_swapped enzymes, p-value smaller than 0.001 are indicated with '***' on plots.

For EGI_swapped, we report that the relative activities of EGI_swapped are 58% without any Co^{2+}, 63% with 0.05 mM Co^{2+} and 69% with 0.1 mM Co^{2+} (Figure 3). The most notable result is recorded as 92% with 0.2 mM and 77% with 0.5 mM Co^{2+} ion concentrations (Figure 3). Herein, we conclude that Co^{2+} ion is coordinated better in EGI_swapped with 0.2 mM concentration. Due to the addition of more Co^{2+} ion, its enzymatic activity decreases to 77% but still it is better than the enzymatic activity of EGI_swapped with no Co^{2+} ion addition, at 57%. When we compare the enzymatic activities of EGI_swapped with respect to native, Co^{2+} ion coordination is significantly different in EGI_swapped with a p-value smaller than 0.001 for 0.2 mM and 0.5 mM Co^{2+} ion concentrations, see Figure 3.

As previously stated, two different paths [(i) and (ii)] are followed to investigate whether Co^{2+} ions inhibit enzymatic activity by binding to alternative sites in their tertiary structure or if they occupy any side of the CMC, long polymeric chain, to prevent the proper attack of enzymes to CMC. Based on the results of path (ii) in Figure S7, we conclude that Co^{2+} ion incubating with CMC does not alter the results.

2.3. *Thermal Stabilities of EGI and EGI_Swapped*

Thermal stabilities of EGI and EGI_swapped enzymes are evaluated by measuring their remaining activities toward 0.5% w/v CMC at 30 °C for 1 h after subjecting them to thermal incubation at 30–70 °C/10 °C for 30 min, 60 min, and 90 min. An enzymatic activity without any thermal incubation 30°C for 1 h is taken as a reference, 100%, to calculate the rest. In parallel to the enzymatic activity profile in Figure 2, the notable differences in enzymatic activities are captured upon the increase in incubation temperature and extension in incubation time.

As displayed in Figure 4, the first notable difference among 30 min incubation results is reported at 70 °C, e.g., EGI_swapped displays ~1.5-fold better activity than EGI. Among 60 min incubation results, the notable differences are at 60 °C and 70 °C, e.g., EGI_swapped displays ~1.6-fold and ~2.8-fold higher activity than EGI at 60 °C and 70 °C, respectively. Upon extending the incubation time to 90 min, ~1.8-fold at 50 °C and ~2-fold at 60 °C higher enzymatic activities are reported for EGI_swapped with respect to EGI. Among all data sets, the most striking thermal stability result is reported for 90 min thermal incubation at 70 °C. Notably, swapped enzyme still shows enzymatic activity after thermal incubation for 90 min at 70 °C, e.g., 14% relative activity, while the enzymatic activity of EGI is completely lost. These thermal stability assay results together with the enzyme activity assay results performed with and without Co^{2+} ions (Figures 2 and 3) prove to us that EGI_swapped is a more promising catalyst than its native counterpart by meeting the demands of the harsh conditions of biotechnological application.

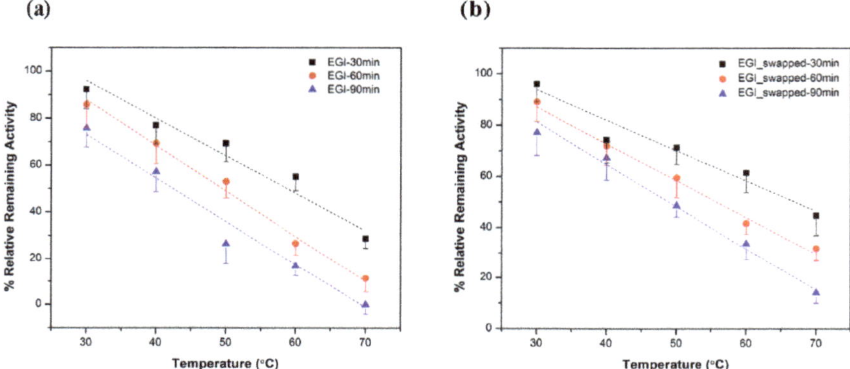

Figure 4. The thermal stability results; (**a**) EGI and (**b**) EGI_swapped enzymes toward CMC. The remaining relative activities of enzymes toward 0.5% w/v CMC were measured after incubating enzymes at different temperatures (30–70 °C/10 °C) for 30, 60 and 90 minutes. For each EGI and EGI_swapped enzymes, the enzymatic activities at 30 °C without any thermal incubation were set to 100% in order to calculate the rest. Data are express as mean ± SD, n = 3.

The thermal stability of the enzyme is controlled by numerous factors; e.g., hydrogen bond networks, hydrophobic and packing interactions, and charge clusters [18]. It is stated for metal ions that they can provide a geometrically defined scaffold, evolving into the reorganization of the local interactions [26]. In some cases, this defined scaffold can modulate the active site of enzymes through long-range interactions to some extent that are dependent on the position of metal in the protein, its interaction with solvent, and the accessibility of counter-ions, etc. [26] Based on the findings in Figure 4, we could conclude that swapping the Ala43-Gly83 region of CBHI to EGI has resulted in an alteration to the enzyme's structure; e.g., the enhancement of hydrogen networking and packing interactions, [18–20]. Hence, we evidently end up with a novel endoglucanase enzyme and displayed altered thermal stability with respect to its native counterpart. The existence of local reorganizations are expected in proteins' scaffold upon the metal ion binding [26]. Consequently, observing the significant differences in thermal stabilities of EGI_swapped and EGI would imply the existence of these local reorganizations in the scaffold of EGI_swapped.

2.4. *Analysis of Coordinated Co^{2+} Ions in the Structures of EGI and EGI_Swapped Enzymes*

As already demonstrated, EGI_swapped displays altered enzymatic activity and thermal stability with respect to its native counterpart, see Figures 2–4. These findings strongly suggest that the favorable environment for Co^{2+} ion coordination is created in EGI_swapped but still additional findings are required for further demonstrations. We perform ICP-OES measurements with EGI_swapped and EGI enzymes. Immediately after 20 min heat denaturation of EGI_swapped and EGI_swapped at 95 °C, the amount of released Co^{2+} ion is measured as 2.2 μM and 8 μM in EGI_swapped and EGI enzymes, respectively. Herein, ~3.6-fold stronger Co^{2+} ion coordination in EGI_swapped with respect to EGI could be accounted as another strong implication for the existence of Co^{2+} ion coordination in the EGI_swapped enzyme. In ICP-OES analysis, we detected a minute amount of Co^{2+} ions in expression buffer and thus free Co^{2+} ions alternatively bind to EGI's structure and they are released upon heating at 95 °C.

Stahlberg et al. suggested that divalent metal ions are required for the crystallization of the CBHI (PDB id: 1DY4) enzyme [24]. This suggestion seems to be suffered since there are several CBHI structures in literature that are crystallized with metal ion different than divalent ones; e.g., 4P1H and 5TC9 with Sm^{3+} and Gd^{3+} [30–33] (see Figure S4). It has been already suggested in the literature that there is a selective binding mode of metal ions to proteins, and it is controlled by several factors, e.g., metal ion radii, number of unpaired electrons, coordination number and preferred coordination geometry of metal ion, the availability of His, and Glu and Asp residues in proteins' structure, etc. [27]. With the well-combination of all these factors, proteins bind to one appropriate metal ion much more tightly than others [26]. To investigate whether EGI_swapped displays any specific binding to other divalent atoms (Ca^{2+}, Mg^{2+}, and Zn^{2+}) or not, we measured the amount of released Ca^{2+}, Mg^{2+} and Zn^{2+} ions upon the thermal denaturation of EGI_swapped and EGI enzymes at 95°C for 20 min through ICP-OES method. We report the ratio of released ions from EGI_swapped to EGI as ~1.0 for Ca^{2+}, ~1.1 for both Mg^{2+} and Zn^{2+}. Hence, these particular differences in the ratio of released Co^{2+} and other divalent atoms from EGI_swapped to EGI enzymes, ~3.6 and ~1.0/~1.1, could be accounted as strong evidence for specificity toward Co^{2+}, but not Ca^{2+}, Mg^{2+}, and Zn^{2+}. For Sm^{3+} and Gd^{3+} ions, there is no proper detector in our ICP-OES instrument. That is why we could not report the amount of released Sm^{3+} and Gd^{3+} ions from EGI_swapped and EGI enzymes.

There is no set of rules for metal ions' binding to proteins as similar to their binding to inorganic molecules, several constrains should be satisfied for metal ion binding to protein molecules. For instance, the existence of true geometry in the protein structure, the presence of His and Asp/Glu residues in proper geometry, and the overcoming size and shape limitations are several constraints that originate from the proteins' structures [26,27]. The polarizability, radii of metal ion, and the number of unpaired electrons are other constraints' coming from metal ions' site [26,27]. Upon the meeting of these constraints both from proteins' and metal ions' sites, the specific metal binding to protein has occurred. In the light of enzyme assay, thermal stability and ICP-OES measurement results, we draw a conclusion that swapping of Ala43-Gly83 region of CBHI to EGI enzyme enabled EGI_swapped enzyme to provide the several constrains mentioned above that are actually lacking in EGI enzyme (see Figures 2 and 3).

2.5. The Change in Secondary Structures Upon Swapping

The altered enzymatic activity and thermal stability results in the swapped enzyme have already pointed out that domain swapping should transform the secondary structure of the EGI significantly. Now we use Circular Dichroism (CD) spectroscopy approach in order to evaluate the possible structural variations of EGI_swapped and EGI enzymes. The temperature sweep of CD spectra between 25 °C and 70 °C reveals distinct conformational and thermal susceptibility between the secondary structure of EGI and EGI_swapped enzymes, see Figure 5a,b. EGI enzyme shows antiparallel β-sheet features in the spectra region of 210–240 nm with a distinctive peak ca. 215 nm. In contrast, EGI_swapped demonstrates the superposition of an antiparallel β-sheet at ca. 215 nm and α-helix like features at ca. 225 nm [37]. Peculiarly, temperature dependent CD study reveals that the structure of EGI enzyme significantly changes with increasing temperature. Although the maxima of ca. 215 is increased slightly, the structural changes exhibit a significant broadening at ca. 230 nm. These structural changes correlate with the temperature dependent activity assays of EGI enzyme. Increasing temperature induces critical alterations in the secondary structure of EGI enzyme, and prolonged temperature exposures could render the enzyme completely inert as shown in Figure 4a.

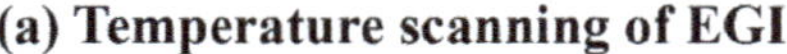

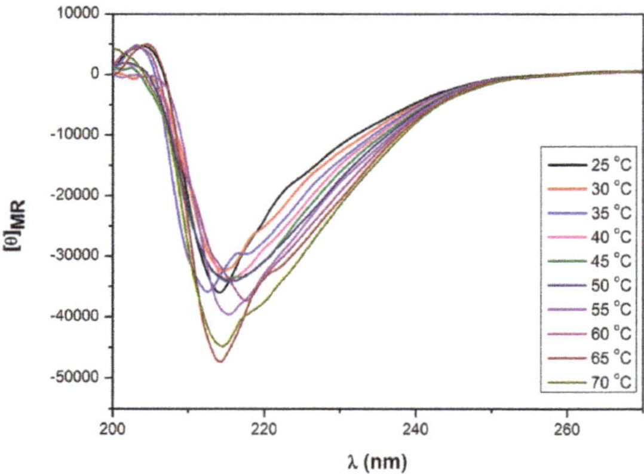

(b) Temperature scanning of EGI_swapped

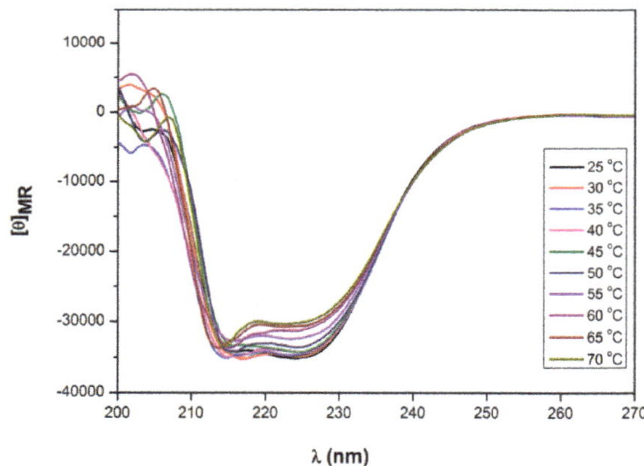

Figure 5. Temperature scanning of EGI (**a**) and EGI_swapped enzymes (**b**). Through Circular Dichroism, the change in the structure of endoglucanase I enzyme upon domain swapping has been displayed. The enzyme samples were prepared in 20 mM sodium acetate buffer (pH 4.7) that was corrected for the backgrounds. The changes in secondary structures of native and swapped enzymes upon an increase in temperature (25–70 °C/5 °C) were captured. The secondary structure changes of native and domain-swapped enzymes are carefully tracked within 210–220 nm and 215–230 nm, respectively. All results are displayed in mean residue ellipticity ([θ]).

Unlike to EGI, the EGI_swapped enzyme exhibits subtle changes in temperature sweep. Antiparallel β-sheet dominant CD feature ca. 215 nm does not show changes in increased temperature. Meanwhile, α-helix associated CD absorption at ca. 225 remains highly stable until the temperature of 50 °C then is slightly reduced but remains prominent as it correlates with the temperature dependent activity assays of EGI_swapped, see Figure 4b. Here, we deduce that the temperature-dependent activity variation between EGI and EGI_swapped enzymes originates from the structural changes and confirms that domain swapping has altered the conformation of the enzyme, unequivocally leading to a novel endoglucanase enzyme, EGI_swapped.

Furthermore, we also investigate the changes in secondary structures of EGI and EGI_swapped enzymes upon thermal denaturation, for the duration of an hour at 95 °C. Conforming to the thermal stability profiles (Figure 4), EGI_swapped contains a considerable level of its secondary structure while antiparallel β-sheet structures of EGI is drastically degraded (Figure 6). Evidently, swapping of the Ala43-Gly83 region of CBHI to EGI enzymes improves the structural integrity and thermal stability significantly (Figure S10).

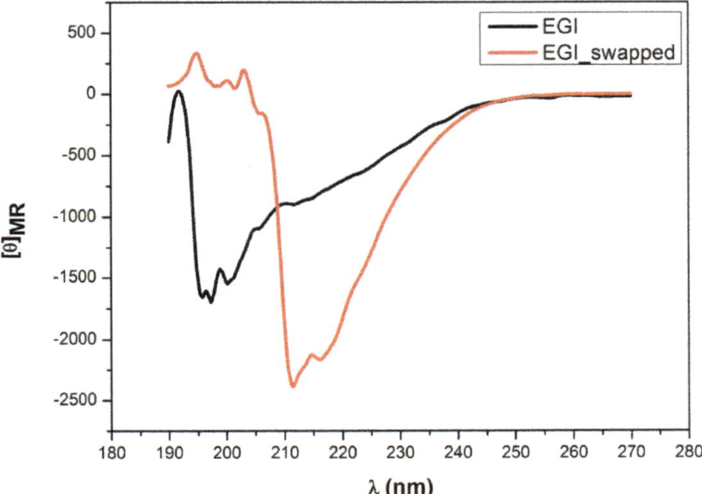

Figure 6. The CD spectroscopy measurement of EGI and EGI_swapped enzymes after 1 h denaturation at 95 °C. Through Circular Dichroism, the change in structures of EGI and EGI_swapped was carefully traced between 195 nm and 230 nm after 1-h heat denaturation at 95 °C and 3-h incubation at room temperature. The CD measurement was performed at room temperature with samples prepared in 20 mM sodium acetate buffer (pH 4.7) that was corrected for the backgrounds. For EGI_swapped, the peaks at 211 nm and 216 nm are indicated. All results are displayed in mean residue ellipticity ([θ]).

To elucidate the impact of Co^{2+} presence on secondary structures enzymes, we track the secondary structure of these enzymes at various Co^{2+} concentrations (Figure 7). In the case of the EGI enzyme, increasing Co^{2+} concentration results in narrowing and the red-shift of the antiparallel β-sheet dominant CD profile. Upon the addition of 0.05 mM Co^{2+}, two negative maxima in the CD pattern emerges at a wavelength of 216 nm and 220 nm, as is similar to EGI_swapped but in a narrower range (Figure 7a). With the addition of higher concentrations of Co^{2+} to EGI enzyme, the sharper peaks are observed. The sharpened peaks remain to be linked with the presence of Co^{2+} while they are not correlated with the concentration of Co^{2+} suggesting nonspecific interactions with Co^{2+}. The enzyme assay results performed with various Co^{2} ion concentrations (Figure 3a) and the release of ~2.2 μM Co^{2+} ion from the EGI enzyme upon heat denaturation are also verified by these non-specific interactions. Contrary to EGI, the impact of Co^{2+} is more stable and obvious in EGI_swapped. Increasing Co^{2+} concentrations exhibit the correlation with a peak intensity ca. 211 nm. The newly formed peak could be interpreted as the higher amount of stable antiparallel β-sheet formation, contributing the improved enzymatic activity and higher thermal stability, as previously showed in Figure 4.

(a) The effects of Co2+ addition on EGI

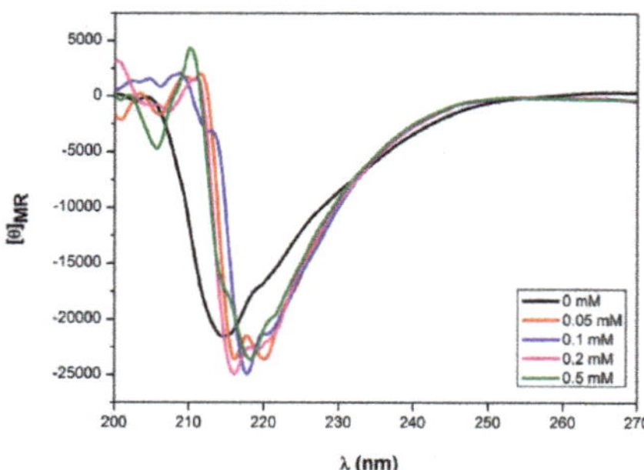

(b) The effects of Co2+ addition on EGI_swapped

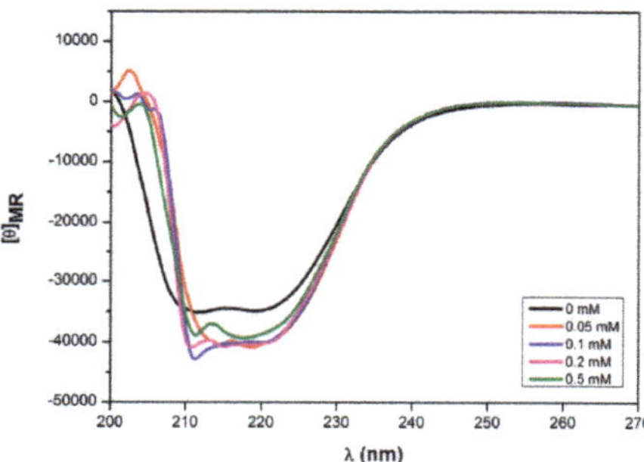

Figure 7. The effects of Co^{2+} ion addition on secondary structures of EGI and EGI_swapped enzymes. Through Circular Dichroism, the differences in secondary structures of EGI (**a**) and EGI_swapped (**b**) upon Co^{2+} ion addition have been captured within 210–220 nm and 215–230 nm, respectively. Protein samples were prepared in 50 mM sodium acetate buffer (pH 4.7) that was corrected for background at room temperature. All results are displayed in mean residue ellipticity ([θ]).

3. Material & Methods

3.1. The Construction of EGI_Swapped and Cloning Approach

A foresight for swapped domain between EGI (pdb id: 1EG1) [35] and CBHI (PDB id: 1DY4) [24] was provided with the structural alignment, performed via TM-Align web-server [38]. Based on the structural alignment results, EGI_swapped was constructed, and the finalized nucleotide sequence of the swapped enzyme was synthesized by Microsynth AG® (Balgach, Switzerland) in the pUC57 plasmid.

To clone this swapped gene into *pPICZαA* vector, its initial amplification was performed with PCR by using forward (F_{EGI}) and reverse (R_{EGI}) primers. The reaction conditions were set to 96 °C for 120 s, 94 °C for 30 s, 55 °C for 30 s, 72 °C for 4 min with 35 cycles and 72 °C for 7 min. After digesting the amplified gene and *pPICZαA* vector with *EcoRI* and *XbaI* restriction enzymes (Thermo Scientific, USA), the rapid ligation kit (Thermo Scientific, Waltham, MA, USA) was used to ligate the digested EGI_swapped gene into the digested sites of a *pPICZαA* expression vector (Invitrogen, San Diego, CA, USA). Finally, 20 μL of ligation mixture was transformed into *Escherichia Coli* XL-1 blue cells, cultured on low-salt Luria-Bertani (LB) broth containing 25 $\mu g \cdot mL^{-1}$ of Zeocin [16]. The selection of positive colonies from LB plates was done with colony PCR by priming the AOX site of *pPICZαA* vector and they were isolated with QIAprep miniprep spin kit (QIAGEN). For double verification, these isolated plasmids were subjected to single digestion with *EcoRI* restriction enzyme. Prior to transformation of selected plasmids to *Pichia pastoris* cells, they were linearized with *SacI* to enable its integration to the AOX1 chromosomal locus of *P. pastoris* KM71H (AOX: ARG4, arg4) (Invitrogen). The competent *P. pastoris* KM71H cells with slow methanol utilizing phenotype were prepared according to a procedure of Wu and Letchworth [39]. Approximately, ~2 µg of linearized plasmid was transformed into KM71H *P. pastoris* cells by using 1.5 kV charging voltage, 25 F capacitance and 200 Ω resistance. The transformation product was spread onto YPD plates containing 100 $\mu g \cdot mL^{-1}$ of Zeocin. After 48 h incubation at 30 °C, the positive ones were also selected with colony PCR from YPD-zeo plates by priming the AOX site of *pPICZαA* vector. Eventually, they were transferred to YPD Broth medium at 30 °C for 48 h with 250 rpm shakings.

3.2. *P. pastoris Expression & Purification*

Previously, EGI enzyme was cloned into *pPICZαA* vector and expressed in *P.pastoris* [40]. By following a similar procedure, EGI and EGI_swapped genes in *pPICZαA* vector were initially inoculated into 50 mL YPD-Broth medium for 48 h at 30 °C with 250 rpm shaking. As the optical density of cultures was reached to ~30 arbitrary units at 600 nm, the cells were collected by centrifugation (4200 rpm for 10 min at 4 °C). The pellets were suspended into expression buffer, containing 100 mM potassium phosphate buffer (pH 6.0), 1.34% Yeast Nitrogen Base (YNB), 4×10^{-5} % biotin and 2% (v/v) methanol in final. They were expressed at 30 °C for 72 h with 250 rpm shaking by inducing expression cultures with 2% v/v methanol in final, in every 24 h. At the end of the expression, proteins were collected by centrifugation at 4200 rpm for 25 min at 4 °C. After concentration of collected products with Sartocon Micro and Ultrafiltration system with 10 and 100 kDa cutoffs (Sartorius Stedim), they were purified with nickel-coated beads. During binding and elution steps, 20 mM sodium phosphate buffers with 50 mM imidazole and 500 mM imidazole were used.

3.3. *Gel Electrophoresis*

Followed by purification, the presence of swapped enzyme was checked with sodium dodecyl sulphate polyacrylamide gel electrophoresis (SDS-PAGE), prepared as 5% (w/v) stacking and 12% (w/v) separation gel. Enzyme samples were boiled at 95 °C for 5 min before loading them to SDS-PAGE. The gel was stained with Coomassie Brillant Blue Dye R-250 (Thermo Scientific, Waltham, MA, USA).

3.4. *Bioinformatic Analysis of EGI and EGI_Swapped*

Several bioinformatics tools were used to indicate changes in EGI enzyme upon swapping. First, the missing parts of EGI, deposited up to Thr371 (PDB id: 1EG1) [35], were predicted with I-TASSER [41,42] before performing its structural alignment with CBHI enzyme (PDB id: 1DY4) [24] via TM-Align web-server [38]. For visualization purpose, Visual Molecular Dynamics (VMD) was used [34]. By providing default parameters, the multiple alignments between EGI_swapped and EGI enzymes was performed [43]. The change in molecular weight and PI value of enzyme was predicted via ExPaSy-PI/MW tool [44,45].

3.5. *Enzyme Assays*

First of all, the purification buffer of enzymes was exchanged with reaction buffer through Sartocon Micro and Ultrafiltration system with 10 kDa cutoffs (Sartorius Stedim, Göttingen, Germany), and enzyme concentrations were determined via Bradford Assay [46]. The activities of these enzymes were determined by the 3,5-dinitrosalicyclic acid (DNS) against Carboxymethyl Cellulase (CMC) (Megazyme®, Bray, Ireland) as described by Gusakov et al [47]. The first step was to check the optimal buffer and pH combinations for both. For this purpose, activity tests were performed in 50 mM sodium acetate buffer (pH 4–5) and in 50 mM potassium phosphate buffer (pH 6–7) by narrowing pH ranges as 4, 4.3, 4.7, and 5 for sodium acetate buffer and 6, 6.3, 6.7, and 7 for potassium phosphate buffer. The activities of enzymes were measured against 0.5% w/v CMC, prepared separately for each buffer-pH combination, for 3 h at 50 °C by keeping the amount of enzyme as 20 µg in final.

After determination of optimal pH-buffer combination for enzymes, the following enzyme assays were performed in 50 mM sodium acetate buffer (pH 4.7) for 1 h, 3 h and 6 h at 30–50 °C (/10 °C) against 0.5% w/v CMC. Again, the amount of enzymes was kept as 20 µg in final. Prior to all assays, enzymes and substrate were separately pre-incubated at the designed temperature for 10 min. To terminate the reaction, DNS reagent was added to the reaction mixture that was boiled at 95 °C for 5 min. Upon cooling down the reaction mixture to room temperature, the absorbance of the reduced sugar was measured at a wavelength of 540 nm. The amount of reduced sugar was calculated from standard calibration curve after subtracting A_{540} background of enzymes and substrate. To test the reproducibility of data, three samples from two independent protein expression and purification batches of EGI and EGI_swapped enzymes, were taken. Since the reaction volume and the amount of enzyme were kept constant in all reactions, mean values were represented as relative values, obtained by setting the highest value to 100% for calculating the rest, with standard deviations below 10%.

Moreover, the enzyme activity tests were performed with the presence of Co^{2+} ions at several final concentrations, 0 mM, 0.05 mM, 0.1 mM, 0.2 mM and 0.5 mM. Two different paths were followed to reveal the impacts of Co^{2+} ions on activity; (i) incubating EGI_swapped and EGI enzymes with Co^{2+} ions or (ii) incubating 0.5% w/v CMC with Co^{2+} ions for 10 min at room temperature just before performing enzymatic assays described by Gusakov et al [47]. As outlined in above, the reaction was terminated by adding DNS reagent to reaction mixture that was further boiled at 95 °C for 5 min. The absorbance of reduced sugar was measured at a wavelength of 540 nm after cooling the reaction mixture to room temperature. Followed by proper subtraction of enzymes' and substrate' absorbance at 540 nm (A_{540}), the amount of reduced sugar was calculated from the calibration curve. Three samples from two independent protein expressions and purification batches were taken to test the reproducibility of data. To follow the impact of Co^{2+} ion addition on enzymes' activities, the highest data among all activity test results, e.g., EGI_swapped at 50°C for 6 h in our case, was taken as 100% to calculate the rest.

3.6. Thermal Stability Assay

To delve into the effects of swapping on thermal stability, EGI_swapped and EGI enzymes were initially incubated at 30–70 °C (/10 °C) for 30 min, 60 min and 90 min in 50 mM sodium acetate buffer (pH 4.7) by keeping the amount of enzymes as 20 µg in final. Followed by thermal incubation, they were chilled on ice for 20 min to allow structural recovery. Then, their remaining enzymatic activities were determined by DNS against 0.5% w/v CMC at 30 °C for 1 h in the same way as described above [47]. Similarly, the absorbance of reduced sugar was measured at a wavelength of 540 nm upon cooling of the reaction mixture to room temperature, and the amount of reduced sugar was calculated from standard calibration curve after subtracting A_{540} background of enzymes and substrate. To test the reproducibility of data, three samples from two independent expressions and purification batches of swapped and native enzymes were taken. The relative remaining activities were calculated by setting the enzymatic activity value of each enzyme, at 30 °C for 1 h, without any thermal incubation to 100%.

As mentioned above, all enzyme activity and thermal stability tests are performed by the 3,5-Dinitrosalicylic acid (DNS) method against 0.5% w/v CMC as outlined by Gusakov et al [47]. Among all available methods, colorimetric methods are most commonly used for measuring the amount of reduced sugar released during the enzymatic reaction [48]. 3,5-Dinitrosalicylic acid (DNS) is a recently used colorimetric assay [49]. Compared to the Nelson–Somogyi method, it is reported as being faster, less toxic and more convenient compared to the Nelson–Somogyi method [50]. To obtain the realistic results in DNS assay, we aimed to improve the sensitivity and precision of the method, DNS assay for our case as much as possible. For this particular purpose, we initially satisfied it by working on the linearization of enzyme assay with trying different enzyme-substrate combinations, incubation time and temperatures. The reproducibility of our data was also tested before reporting by increasing the number of samples coming from two independent purification and expression batches.

3.7. Measurement of Co(II) Ion Concentration in EGI and EGI_Swapped Enzymes

To assess whether suitable environment for Co^{2+} ion coordination was successfully integrated into EGI_swapped or not, we performed inductively coupled optical emission spectrometry measurement (ICP-OES) (Vista-Pro Axial; Varian Pty Ltd., Mulgrave, Australia). Followed by expression and purification steps, enzyme samples of 0.5 mg/ml concentration were boiled at 95 °C for 20 min, and the amount of released Co^{2+} upon denaturation was detected with an ICP-OES measurement. In addition to the Co^{2+} ion, the amount of released Ca^{2+}, Mg^{2+} and Zn^{2+} ions were also measured through ICP-OES method after 20 min of thermal denaturation at 95 °C.

3.8. Circular Dichroism (CD) Spectroscopy Measurements

CD spectrums of enzymes were collected by using a J-815 Circular Dichroism Spectropolarimeter (Jasco International Co., Tokyo, Japon) under a N_2 stream purge equipped with a thermostatically controlled cuvette. The measurement was performed with a quartz cuvette (Hellma Analytics, 1.0 mm path length) and the scanning speed was 100 nm/min [51]. Temperature scanning was performed from 25 °C to 70 °C as 5 °C increments. In each temperature, ten scans were averaged to obtain the full spectra of samples prepared in 20 mM of sodium acetate buffer (pH 4.7) that were corrected for the background.

To explore more about the effects of Co^{2+} addition on the structures of EGI and EGI_swapped enzymes, Co^{2+} ions at various concentrations (0.05 mM, 0.1 mM, 0.2 mM and 0.5 mM) were added to the enzyme samples. Prior to the collection of full spectra, they were incubated for 10 min at room temperature, in 20 mM sodium acetate buffer (pH 4.7). The full spectra of samples were collected at room temperature with scanning speeds as 100 nm/min with ten scans that were averaged to obtain the full spectra of samples.

Further, the effects of thermal incubation on enzymes' structures were investigated with CD spectroscopy measurement after they were subjected to thermal incubation at 95 °C.

Enzymes were subjected to thermal incubation at 95 °C for 1 h and then were incubated at room temperature for 3 h. Again; CD spectra samples were collected as ten averages with scanning speed as 100 nm/min at room temperature. There was no Co^{2+} ion addition on enzyme samples that subjected to thermal denaturation.

By using the following equation, the mean residue ellipticity was calculated;

$$[\theta] = \frac{\theta}{10 \cdot c \cdot l} \quad (1)$$

where θ is the observed ellipticity mmDeg, c is the molar concentration, l is the cell length in centimetres.

4. Conclusions

In this study, we present how swapping is a strong protein-engineering tool in order to create an endoglucanase enzyme, called as EGI_swapped, with novel structural and functional properties. Basically, we aimed to create the favorable environment for providing a Co^{2+} ion coordination in the endoglucanase enzyme by swapping the Ala43-Gly83 region of CBHI to EGI.

In the first part of this study, we described the impacts of swapping on enzyme activities with improvements in EGI_swapped. The difference in enzymatic activities of EGI_swapped and EGI become more prominent upon increasing Co^{2+} ion concentrations. This particular result could be considered as a strong implication for the existence of Co^{2+} ion coordination in EGI_swapped. As being different than EGI, EGI_swapped displays improved thermal stability that becomes more apparent with the extension of reaction time and temperature. Specifically, the EGI_swapped enzyme shows quantifiable activity against CMC while EGI enzyme completely loses its activity after 90 min thermal incubation at 70 °C. ICP-OES measurements suggest that the existence of Co^{2+} ion in EGI_swapped is significantly different than EGI; e.g.~3.6-fold stronger Co^{2+} ion coordination in EGI_swapped with respect to the EGI enzyme. This particular difference could provide a basis to explain improvements in enzymatic activity and thermal stability of EGI_swapped with respect to its native counterpart.

In the second part of the study, we identify the structural differences in EGI_swapped and native enzymes via Circular Dichroism (CD) spectroscopy. Swapping the Ala43-Gly83 region of CBHI to EGI leads to a transformation of the secondary structure, and results in distinct conformational and thermal susceptibility. In EGI_swapped enzyme, the newly formed peaks are observed upon Co^{2+} ion addition via CD spectroscopy. Unlike EGI, the considerable level of secondary structure is reported in CD spectroscopy of EGI_swapped after heat denaturation at 95 °C for 1 h. This result verifies the thermal stability results by indicating that swapping improves the structural integrity of the enzyme and contributes significantly to its thermal stability without disrupting its activity. Indeed, we end up with a novel endoglucanase enzyme with more promising enzymatic and functional properties, in terms of meeting the demands of harsh conditions of industrial applications, compared to its native counterpart.

To the best of our knowledge, this study is the first attempt in literature to alter the functional and structural properties of endoglucanase enzyme from *Trichoderma reesei* via swapping from cellobiohydrolase I enzyme. The computational and experimental tools of protein-engineering approaches have been perfectly combined to end up with a creation of new endoglucanase enzyme with altered thermal and structural properties. In literature, this study can be considered as the first successful attempt to understand the impacts of divalent metal ion coordination in endoglucanase enzyme from *Trichoderma reseei*. We believe that this study will be used as a basis and motivation for others, aiming to fulfill the gaps in cellulase research about the structural and functional importance of metal ion coordination.

Supplementary Materials: The following are available online at http://www.mdpi.com/2073-4344/9/2/130/s1, Figure S1: The depiction of cellobiohydrolase and endoglucanase mode of action; Figure S2: The structural alignment results between EGI and CBHI enzymes (a) TM-alignment results between enzymes are displayed. ':' donates aligned residue pairs whose distances are smaller than 5 Å. The aligned region, Ala43-Gly83 (CBHI numbering) is indicated with *. The structural break happened around His189 (EGI numbering) is indicated with **. (b) The structural presentation of alignment results are provided in 'cartoon' format for overall, and 'surf' format just for swapped region. His206 in CBHI and His189 in EGI are drawn in 'licorice' format as yellow and red, respectively. In 'surf' presentation, the end points of swapped domain (Ala43 and Gly 83 as CBHI numbering) are indicated; Figure S3: The multiple protein alignment results between EGI_swapped and EGI; Figure S4: The crystal structures of cellobiohydrolase I enzymes. These cellobiohydrolase I enzymes have displayed more than 95% sequence similarity to CBHI (pdb id: 1DY4). These structures are crystallized with metal ions different than Co2+ ion. On the left side, the crystal structures are displayed with metal ions. Structures and metal ions are drawn in 'cartoon' and 'VDW' format, respectively, with Visual Molecular dynamics. The corresponding parts of our swapped region, Ala43-Gly83, in cellobiohydrolase enzymes are displayed by indicating the distances of Ala43 and Gly83 to metal ions. On the right side, the structural alignment results of cellobiohydrolase enzymes, pdb id: 1CEL, 4P1H and 5TC9, with EGI enzyme (pdb id: 1EG1) are displayed by indicating Glu residues involved in metal ions coordination; Figure S5: The cloning results of EGI_swapped; (a) the colony PCR results and (b) the double verification of isolated plasmid with EcoR I, and (c) Raw files of colony PCR results and single digest verification with EcoR I. After selection of positive clones through colony PCR, the isolated plasmids (last four ones) were double checked with single EcoR I digestion as displayed in (b). The colony PCR's products and linearized plasmid were visualized in 1.2% w/v Agarose gel by ethidium bromide staining. M indicates a GeneRuler Ladder Mix (Fermentas); Figure S6. SDS-PAGE results of EGI_swapped; Figure S7: The effects of Co2+-ion pre-incubation with CMC prior to hydrolysis reaction on their enzymatic activity profiles; Figure S8. The construction of pH profiles for native and domain swapped enzymes; Figure S9: The enzyme activity profiles of EGI and EGI_swapped enzymes toward CMC for 1 h at 30–50 °C(/10 °C); Figure S10: The temperature path of EGI_swapped and EGI enzymes.

Author Contributions: O.U.S. initiated the research. A.Y. and O.U.S. designed the study. A.Y. performed and analyzed all computational and experimental work excluding circular dichroism (CD) spectroscopy study under the supervision of O.U.S. H.K. performed the CD spectroscopy measurements. H.K. and A.Y. analyzed the CD results. A.Y. wrote the manuscript, and H.K. and O.U.S. revised the manuscript. All authors read and approved the manuscript.

Acknowledgments: The Scientific and Technological Research Council of Turkey supported this work with grant number 112T901. The authors have special thanks to Yusuf Tutuş (Sabanci University) for contribution to ICP-OES readings.

Conflicts of Interest: The authors declare that they have no competing interest.

Abbreviations

Cellobiohydrolase I	CBHI
CD	Circular Dichroism
Carboxymethyl cellulase	CMC
3,5-Dinitrosalicylic acid	DNS
Endoglucanese I	EGI
Endoglucanese I swapped	EGI_swapped
PDB	Protein Data Bank

References

1. Lin, L.L.; Yan, R.; Liu, Y.Q.; Jiang, W.J. In-depth investigation of enzymatic hydrolysis of biomass wastes based on three major components: Cellulose, hemicellulose and lignin. *Bioresour. Technol.* **2010**, *101*, 8217–8223. [CrossRef] [PubMed]
2. Zhang, Y.H.; Lynd, L.R. Toward an aggregated understanding of enzymatic hydrolysis of cellulose: Noncomplexed cellulase systems. *Biotechnol. Bioeng.* **2004**, *88*, 797–824. [CrossRef] [PubMed]
3. Pérez, J.; Muñoz-Dorado, J.; de la Rubia, T.; Martínez, J. Biodegradation and biological treatments of cellulose, hemicellulose and lignin: An overview. *Int. Microbial.* **2002**, *5*, 53–63. [CrossRef] [PubMed]
4. Eibinger, M.; Sattelkow, J.; Ganner, T.; Plank, H.; Nidetzky, B. Single-molecule study of oxidative enzymatic deconstruction of cellulose. *Nat. Commun.* **2017**, *8*, 894. [CrossRef] [PubMed]
5. Himmel, M.E.; Ding, S.Y.; Johnson, D.K.; Adney, W.S.; Nimlos, M.R.; Brady, J.W.; Foust, T.D. Biomass recalcitrance: Engineering plants and enzymes for biofuels production. *Science* **2007**, *315*, 804–807. [CrossRef]
6. Donohoe, B.S.; Resch, M.G. Mechanisms employed by cellulase systems to gain access through the complex architecture of lignocellulosic substrates. *Curr. Opin. Chem. Biol.* **2015**, *29*, 100–107. [CrossRef] [PubMed]
7. Percival Zhang, Y.H.; Himmel, M.E.; Mielenz, J.R. Outlook for cellulase improvement: Screening and selection strategies. *Biotechnol. Adv.* **2006**, *24*, 452–481. [CrossRef] [PubMed]
8. Lynd, L.R.; Weimer, P.J.; van Zyl, W.H.; Pretorius, I.S. Microbial cellulose utilization: Fundamentals and biotechnology. *Microbiol. Mol. Biol. Rev.* **2002**, *66*, 506–577. [CrossRef] [PubMed]
9. Wilson, D.B. Cellulases and biofuels. *Curr. Opin. Biotechnol.* **2009**, *20*, 295–299. [CrossRef] [PubMed]
10. Bayer, E.A.; Chanzy, H.; Lamed, R.; Shoham, Y. Cellulose, cellulases and cellulosomes. *Curr. Opin. Struct. Biol.* **1998**, *8*, 548–557. [CrossRef]
11. Bennett, M.; Choe, S.; Eisenberg, D. Domain swapping: Entangling alliances between proteins. *Proc. Natl. Acad. Sci. USA* **1994**, *91*, 3127–3131. [CrossRef] [PubMed]
12. Janowski, R.; Abrahamson, M.; Grubb, A.; Jaskolski, M. Domain swapping in N-truncated human cystation. *Chem. J. Mol. Biol.* **2004**, *341*, 151–160. [CrossRef] [PubMed]
13. Schering, N.; Casale, E.; Caccia, P.; Giordano, P.; Battistini, C. Dimer formation through domain swapping in the crystal structure of the Grb2-SH2-Ac-pYVNV complex. *Biochemistry* **2000**, *39*, 13376–13382. [CrossRef]
14. Barbosa, J.; Sivarman, J.; Li, Y.; Larocque, R.; Matte, A. Mechanism of action and NAD+ binding mode revealed by the crystal structure of L-histidinol dehydrogenase. *Proc. Natl. Acad. Sci. USA* **2002**, *99*, 1859–1864. [CrossRef] [PubMed]
15. Knaus, K.; Morillas, M.; Swietnicki, W.; Malone, M.; Surewicz, WK.; Yee, V.C. Crystal structure of the human prion protein reveals a mechanism for oligomrerization. *Nat. Struct. Biol.* **2001**, *8*, 770–774. [CrossRef] [PubMed]
16. Yenenler, A.; Sezerman, O.U. Design and characterizations of two novel cellulases through single-gene shuffling of Cel12A (EG3) gene from Trichoderma reseei. *Protein Eng. Des. Sel.* **2016**, *29*, 219–229. [CrossRef] [PubMed]
17. Timucin, E.; Sezerman, O.U. The conserved lid tryptophan, W211, potentiates thermostability and thermoactivity in bacterial thermoalkalophilic lipases. *PLoS ONE* **2013**, *8*, e85186. [CrossRef]

18. Jaenicke, R.; Bohm, G. The stability of proteins in extreme environments. *Curr. Opin. Struct. Biol.* **1998**, *8*, 738–748. [CrossRef]
19. Tanaka, S.; Igarashi, S.; Ferri, S.; Sode, K. Increasing stability of water-soluble PQQ glucose dehydrogenase by increasing hydrophobic interaction at dimeric interface. *BMC Biochem.* **2005**, *6*, 1. [CrossRef]
20. Kirino, H.; Aoki, M.; Aoshima, M.; Hayashi, Y.; Ohba, M. Hydrophobic interaction at the subunit interface contributes to the thermostability of 3-isopropylmalate dehydrogenase from an extreme thermophile, Thermus thermophilus. *Eur. J. Biochem.* **1994**, *220*, 275–281. [CrossRef]
21. Keeling, A.A. *Understanding Humic Substances: Advanced Methods, Properties and Applications*; Wiley: Cambridge, UK, 2000.
22. Hakansson, U.; Fagerstam, L.G.; Pettersson, G.; Andersson, L. 1,4-β-glucan glucanohydrolase from the cellulolytic fungus Trichoderma viride QM 9414. *Biochem. J.* **1979**, *179*, 141–149. [CrossRef] [PubMed]
23. Shoemaker, S.; Schweickart, V.; Ladner, M.; Gelfand, D.; Kwok, S.; Myambo, K.; Innis, M. Molecular cloning of exo-cellobiohydrolase from Trichoderma reesei strain L27. *Nat. Biotechnol.* **1983**, *1*, 691–696. [CrossRef]
24. Stahlberg, J.; Henriksson, H.; Divne, C.; Isaksson, R.; Pettersson, G.; Johansson, G.; Jones, T.A. Structural basis for enantiomer binding and separation of a common beta-blocker: Crystal structure of cellobiohydrolase Cel7A with bound (S)-propranolol at 1.9 A resolution. *J. Mol. Biol.* **2001**, *305*, 79–93. [CrossRef]
25. Cassia Pereira, J.; Giese, E.C.; Souza Moretti, M.M.; Santos Gomes, A.C.; Perrone, O.M.; Boscolo, M.; Silva, R.; Gomes, E.; Bocchini Martins, D.A. Effect of Metal Ions, Chemical Agents and Organic Compounds on Lignocellulolytic Enzymes Activities. In *Enzyme Inhibitors and Activators*; Senturk, M., Ed.; IntechOpen: London, UK, 2017.
26. Glusker, J.P. Structural aspects of metal liganding to functional groups in proteins. *Adv. Protein Chem.* **1991**, *42*, 1–76.
27. Sigel, R.K.; Pyle, A.M. Alternative roles for metal ions in enzyme catalysis and the implications for ribozyme chemistry. *Chem. Rev.* **2007**, *107*, 97–113. [CrossRef] [PubMed]
28. Vedani, A. YETI: An interactive molecular mechanics program for small-molecule protein complexes. *J. Comput. Chem.* **1986**, *9*, 269–280. [CrossRef]
29. Vedani, A.; Huhta, D.W. A new force field for modeling metalloproteins. *J. Am. Chem. Soc.* **1990**, *112*, 4759–4767. [CrossRef]
30. Bodenheimer, A.M.; Meilleur, F. Crystal structures of wild-type Trichoderma reesei Cel7A catalytic domain in open and closed states. *FEBS Lett.* **2016**, *590*, 4429–4438. [CrossRef] [PubMed]
31. Bodenheimer, A.M.; Cuneo, M.J.; Swartz, P.D.; Myles, D.A.; Meilleur, F. Crystal Structure of Wild Type Hypocrea Jecorina Cel7a in a Monoclinic Crystal Form. Available online: https://www.rcsb.org/structure/4p1h (accessed on 29 January 2019).
32. Divne, C.; Stahlberg, J.; Reinikainen, T.; Ruohonen, L.; Pettersson, G.; Knowles, J.K.; Teeri, T.T.; Jones, T.A. The three-dimensional crystal structure of the catalytic core of cellobiohydrolase I from Trichoderma reesei. *Science* **1994**, *265*, 524–528. [CrossRef]
33. Stahlberg, J.; Divne, C.; Koivula, A.; Piens, K.; Claeyssens, M.; Teeri, T.T.; Jones, T.A. Activity studies and crystal structures of catalytically deficient mutants of cellobiohydrolase I from Trichoderma reesei. *J. Mol. Biol.* **1996**, *264*, 337–349. [CrossRef]
34. Humphrey, W.; Dalke, A.; Schulten, K. VMD: Visual molecular dynamics. *J. Mol. Graph.* **1996**, *14*, 33–38. [CrossRef]
35. Kleywegt, G.J.; Zou, J.Y.; Divne, C.; Davies, G.J.; Sinning, I.; Stahlberg, J.; Reinikainen, T.; Srisodsuk, M.; Teeri, T.T.; Jones, T.A. The crystal structure of the catalytic core domain of endoglucanase I from Trichoderma reesei at 3.6 A resolution, and a comparison with related enzymes. *J. Mol. Biol.* **1997**, *272*, 383–397. [CrossRef] [PubMed]
36. Rouvinen, J.; Bergfors, T.; Teeri, T.; Knowles, J.K.; Jones, T.A. Three-dimensional structure of Pcellobiohydrolase II from Trichoderma reesei. *Science* **1990**, *249*, 380–386. [CrossRef] [PubMed]
37. Greenfield, N.J. Using circular dichroism spectra to estimate protein secondary structure. *Nat. Protoc.* **2007**, *1*, 2876–2890. [CrossRef] [PubMed]
38. Zhang, Y.; Skolnick, J. TM-align: A protein structure alignment algorithm based on the TM-score. *Nucleic Acids Res.* **2005**, *33*, 2302–2309. [CrossRef] [PubMed]
39. Wu, S.; Letchworth, G.J. High efficiency transformation by electroporation of Pichia pastoris pretreated with lithium acetate and dithiothreitol. *Biotechniques* **2004**, *36*, 152–154. [CrossRef] [PubMed]

40. Akcapinar, G.B.; Gul, O.; Sezerman, U. Effect of codon optimization on the expression of Trichoderma reesei endoglucanase 1 in Pichia pastoris. *Biotechnol. Prog.* **2011**, *27*, 1257–1263. [CrossRef] [PubMed]
41. Zhang, Y. I-TASSER server for protein 3D structure prediction. *BMC Bioinform.* **2008**, *9*, 40. [CrossRef] [PubMed]
42. Yang, J.; Zhang, Y. I-TASSER server: New development for protein structure and function predictions. *Nucleic Acids Res.* **2015**, *43*, W174–W181. [CrossRef]
43. Rice, P.; Longden, I.; Bleasby, A. EMBOSS: The European Molecular Biology Open Software Suite. *Trends Genet.* **2000**, *16*, 276–277. [CrossRef]
44. Gasteiger, E.; Gattiker, A.; Hoogland, C.; Ivanyi, I.; Appel, R.D.; Bairoch, A. ExPASy: The proteomics server for in-depth protein knowledge and analysis. *Nucleic Acids Res.* **2003**, *31*, 3784–3788. [CrossRef] [PubMed]
45. Wilkins, M.R.; Gasteiger, E.; Bairoch, A.; Sanchez, J.C.; Williams, K.L.; Appel, R.D.; Hochstrasser, D.F. Protein identification and analysis tools in the ExPASy server. *Methods Mol. Biol.* **1999**, *112*, 531–552. [PubMed]
46. Bradford, M.M. A rapid and sensitive method for the quantitation of microgram quantities of protein utilizing the principle of protein-dye binding. *Anal. Biochem.* **1976**, *72*, 248–254. [CrossRef]
47. Gusakov, A.V.; Kondratyeva, E.G.; Sinitsyn, A.P. Comparison of two methods for assaying reducing sugars in the determination of carbohydrase activities. *Int. J. Anal. Chem.* **2011**, *2011*, 283658. [CrossRef] [PubMed]
48. Bhat, M.K.; Hazlewood, G.P. *Enzymology and Other Characteristics of Cellulases and Xylanases*; CABI Publishing: Oxford, UK, 2001; pp. 11–60.
49. Chen, X.; Li, W.; Ji, P.; Zhao, Y.; Hua, C.; Han, C. Engineering the conserved and noncatalytic residues of a thermostable beta-1,4-endoglucanase to improve specific activity and thermostability. *Sci. Rep.* **2018**, *8*, 2954. [CrossRef] [PubMed]
50. Bailey, M.J.; Biely, P.; Poutanen, K. Interlaboratory testing of methods for assay of xylanase activity. *J. Biotechnol.* **1992**, *23*, 257–270. [CrossRef]
51. Kurt, H.; Yuce, M.; Hussain, B.; Budak, H. Dual-excitation upconverting nanoparticle and quantum dot aptasensor for multiplexed food pathogen detection. *Biosens. Bioelectron.* **2016**, *81*, 280–286. [CrossRef] [PubMed]

Review

Bioproduction of Isoprenoids and Other Secondary Metabolites Using Methanotrophic Bacteria as an Alternative Microbial Cell Factory Option: Current Stage and Future Aspects

Young Chan Jeon, Anh Duc Nguyen and Eun Yeol Lee *

Department of Chemical Engineering, Kyung Hee University, Yongin-si, Gyeonggi-do 17104, Korea; wjsdud6709@gmail.com (Y.C.J.); nguyenducanh01051991@gmail.com (A.D.N.)
* Correspondence: eunylee@khu.ac.kr; Tel.: +82-31-201-3839

Received: 28 September 2019; Accepted: 21 October 2019; Published: 24 October 2019

Abstract: Methane is a promising carbon feedstock for industrial biomanufacturing because of its low price and high abundance. Recent advances in metabolic engineering and systems biology in methanotrophs have made it possible to produce a variety of value-added compounds from methane, including secondary metabolites. Isoprenoids are one of the largest family of secondary metabolites and have many useful industrial applications. In this review, we highlight the current efforts invested to methanotrophs for the production of isoprenoids and other secondary metabolites, including riboflavin and ectoine. The future outlook for improving secondary metabolites production (especially of isoprenoids) using metabolic engineering of methanotrophs is also discussed.

Keywords: methane; methanotrophs; secondary metabolites; isoprenoid; metabolic engineering

---

1. Introduction

Methane is the major component of natural gas and a huge amount of methane is available due to advances in natural and shale gas production technology [1]. Unfortunately, atmospheric methane is 20 times as potent as carbon dioxide as a greenhouse gas that needs to be mitigated. Of the total methane emissions, anthropogenic emissions (combustion of fossil fuels, livestock industry, landfills, and biomass burning) account for 63%, and natural sources (wetlands, oceans, rivers, lakes, and permafrost) account for 37%. Anthropogenic emissions have steadily increased, with an increase in methane gas production as relevant industry develops [2,3]. Currently, most methane is used to produce electricity in gas turbine boilers and is supplied to homes for heating and cooking [4].

One challenging technical issue on methane utilization is gas-to-liquid (GTL) conversion because methane gas is not easily transported [5]. Various chemical GTL conversion technologies have been developed to date, but there are some disadvantages. The chemical gas-to-liquid conversion technology consists of a methane reforming process to produce syngas, a Fischer-Tropsch process to convert syngas into hydrocarbons, and an upgrading and separation process [6]. Generally, these processes require high temperature up to 800 °C and high-pressure conditions. Therefore, chemical gas-to-liquid conversion requires high operating costs. Biological gas-to-liquid conversion technology is being developed to address these issues [5]. Biological gas-to-liquid conversion technology uses methanotrophs, which can use methane as a sole carbon source. This gas-to-liquid bioconversion using methanotrophs shows higher efficiency than the chemical conversion [7]. In addition, bioprocess proceeds under normal temperature and pressure conditions. By applying metabolic engineering and systems biology approaches, currently, engineered methanotrophs have been deployed for the production of several products, including secondary metabolites [8–10]. However, there are still some technical issues in the gas-to-liquid bioconversion using methanotrophs, such as limited mass transfer rates of methane, low volumetric productivity, and instability of recombinant strains [11].

Secondary metabolites produced by plants and microorganisms have many useful biological activities. Among the many secondary metabolites, isoprenoids are the largest family of natural products, with more than 50,000 members identified [12–14]. Many isoprenoids have been used in many applications such as pharmaceuticals, nutraceuticals, fragrances, as well as the chemical industry. Additionally, due to the methyl-branched and cyclized hydrocarbon alkene structure, some isoprenoids have attracted more attention for their potential use as advanced biofuels. Plants are the major sources of isoprenoids. However, there are still many limitations for the production of high quantities of natural isoprenoids, including slow growth and tissue-specific biosynthesis, and difficulties in harvesting and extraction [12–14]. In contrast, a promising alternative method for isoprenoid production is the reconstruction of isoprenoid biosynthesis pathways from plants into microbes. The application of this approach has been extensively studied on model organisms like *Escherichia coli* or *Saccharomyces cerevisiae* for a variety of isoprenoids [12,14]. *E. coli* and *S. cerevisiae* are promising hosts since they possess simple genetic backgrounds, high growth rates as well as well-developed genetic tools. For example, high production of isoprenoid-based biofuels has been achieved in the metabolic engineering of microbes via endogenous or heterologous biosynthetic pathways in these hosts. However, the vast majority of these microbe-based isoprenoid production cases rely on sugar-based metabolism, which is likely to increase in the price of isoprenoids produced in large quantity.

The use of alternative carbon sources is attractive in industrial biotechnology for isoprenoid production. Furthermore, different type of pathway regulation in different microbes utilizing unusual carbon substrate like methane can play a vital role in metabolic engineering of various microbes including model organisms. Thus, metabolic engineering of methane-utilizing methanotrophs has recently attracted much attention, due to not only cheaper price of methane but also the novelty and potential of secondary metabolites biosynthetic capability of methanotrophs. Some secondary metabolites were naturally produced in methanotrophs, such as carotenoids and ectoine, which are responsible for the pigmentation and osmotic stress in a high salt environment, respectively [8,10]. Recently, some rational metabolic engineering strategies have been used in methanotrophs as a microbial cell factory platform for the bioconversion of methane to a variety of value-added products [9]. Based on these backgrounds, in this study, the biosynthesis pathways and characteristics of isoprenoids and other secondary metabolites in methanotrophs are reviewed. The current state of the production of various isoprenoids, riboflavin, and ectoine using methanotrophs is overviewed. In later sections of this review, we discuss the potential of methanotrophs and strategies for improving the production of isoprenoids through metabolic engineering of methanotrophs.

2. Methane Metabolism in Methanotrophs

Methanotrophs play important roles in global methane cycling and in the degradation of harmful substances. They use methane as a sole carbon source and were isolated in several conditions where methane and oxygen are both present. Freshwater and sediments are sinks of atmospheric methane. A variety of different methanotrophs inhabit these environments, and *Methylomonas, Methylobacter, Methylosarcina, Methylococcus*, and *Methylosoma* are predominant. Landfills and rice field soils are major sources of atmospheric methane. Studies have reported that landfills and rice field soils contain a variety of methanotrophs, such as *Methylomonas, Methylobacter, Methylosarcina, Methylomicrobium, Methylococcus, Methylocaldum, Methylocystis*, and *Methylosinus*. The distribution and abundance of methanotrophs in these methane-generating environments are affected by oxygen availability. Methanotrophs are also found in various extreme environments such as hot springs, the Antarctic, peatbogs, and deep-sea environments [15].

Methanotrophs are bacteria that have the ability to assimilate methane. This has always been considered a very unique microbial function. Over the past decade, knowledge of methane metabolism has been broadened by the help of systems biology approaches [16]. However, it also highlighted a large amount of fundamental knowledge gaps of methanotrophs that must be addressed to exploit the full potential of methanotrophs [17]. Oxidation of methane is carried out by methane monooxygenase (MMO), a membrane-associated particulate methane monooxygenase (pMMO) and soluble methane monooxygenase (sMMO) [18–20]. After that, methanol dehydrogenase (MDH) oxidized methanol into formaldehyde, which is then assimilated through C1 metabolism [18–20]. While calcium-dependent MDH encoded by *mxaFI* has been well-investigated, a novel MDH (i.e., lanthanide-dependent MDH encoded by *xoxF*) has also been discovered [21]. The role of *xoxF* in methane and methanol oxidation and the divergence and wide distribution of these enzymes among bacterial taxa demonstrated the importance of *xoxF* enzymes environmentally, and furthermore, *xoxF* may be ancestral to *mxaF*.

There are two types of methanotrophs—γ-proteobacteria (Type I) and α-proteobacteria (Type II). The distinction between these groups is based on the process of assimilating formaldehyde, the organization and arrangement of cell membranes, and related characteristics such as cell morphology [22,23]. Typically, type I methanotrophs assimilate formaldehyde through the RuMP pathway (including Type X *Methylocaldum* and *Methylococcus*). They also form a flat, disc-shaped inner cytoplasm and have a TCA cycle lacking α-ketoglurarate dehydrogenase activity [24]. However, there have been many novel findings related to the methane metabolism of this group that have been reported recently. For example, ^{13}C carbon tracings were used to discover a complete oxidative TCA cycle operating during methanotrophic growth in *Methylomicrobium buryatense* 5GB1 [25]. Others demonstrated that the Embden Meyerhof pathway (EMP) is the main route for C1 assimilation in *Methylomicrobium alcaliphilum* 20Z and *Methylomonas* sp. LW13 [26,27].

The assimilation of formaldehyde in type II methanotroph is through the serine pathway. First, formaldehyde is oxidized to formate, subsequently incorporating into biomass via tetrahydrofolate-linked pathways and the serine cycle. The serine cycle first converts formaldehyde and CO_2 into C3 and C4 compounds, converts them to acetyl-CoA, and sends them to the TCA cycle (carbon dioxide accounts for about 50% of the total carbon) [18]. It has been hypothesized that an efficient pathway for C1 assimilation requires the coupling of the serine cycle with a glyoxylate regeneration pathway such as a glyoxylate shunt or an ethylmalonyl-coA (EMC) pathway. A flux analysis based on ^{13}C metabolomics technology was used to identify that the EMC pathway is essential for glyoxylate regeneration in *M. trichosporium* OB3b [28]. Unlike type I methanotrophs, Type II methanotrophs possess α-ketoglurarate dehydrogenase activity, but the TCA cycle does not work perfectly because of its low activity. Type II methanotrophs also forms inner cytoplasm, which is similar to cytoplasm, except in *Methylocella* [22,23].

3. Genetic Tool Development for Methanotrophs

Genetic tool development is basic requiment for enabling the potential of industrial methane biocatalysis. To date, many attempts for the development of genetic tools for methantrophs have been conducted [8,29]. The introduction of replicable plasmids containing RK2/RP4 origin into methanotrophs via conjugation using *Escherichia coli* S17-1 as a donor strain was the most common method for foreign gene transfer and expression in methanotrophs [8,17,18,29]. In addition, integration and deletion vectors were also used in wide range of methanotrophs by marker exchange via homologous recombination methods [8,28]. Along with this, counter-selectable marker-based techniques for mutagenesis was developed for methanotrophs in which *sacB* or *pheS* gene were employed as the most common markers [9,30].

Electroporation-based genetic manipulation methods have been developed in both type I and type II methanotrophs [9,31]. Recently, one research group from the National Renewable Energy Laboratory has successfully developed a dual-plasmid, broad-host-range CRISPR/Cas9 gene-editing system for *Methylococcus capsulatus* (Bath) [32]. The DNA targeting and double-strand DNA cleavage by the deployment of heterologous Cas9 endonuclease and synthetic single-guide RNA were applied to the manipulation of cell vitality, fluorescent protein conversion, and protein function disruption by point mutation. This is a proof-of-concept study of novel genetic toolbox development for methanotrophic genetic manipulation. The advances in genetic toolbox development will provide efficient methods to study metabolic features of secondary metabolites in methanotrophs and enable genetic editing for methanotrophs for production of high value-added secondary metabolites.

4. Isoprenoid Production in Methanotrophs

4.1. Isoprenoid Biosynthesis Via MEP Pathway

The basic isoprenoid building blocks isopentenyl pyrophosphate (IPP) and dimethylallyl pyrophosphate (DMAPP) are synthesized through one of two distinct biosynthetic routes—the mevalonic acid (MVA) pathway or the 2-C-methyl-D-erythritol 4-phosphate (MEP) pathway. Over the last few decades, the MVA pathway was considered to be the unique pathway for IPP and DMAPP biosynthesis [33]. However, in the early 1990s, an alternative pathway named the MEP pathway was discovered in bacteria for the biosynthesis of IPP and DMAPP (Figure 1). While archaea and most eukaryotes use the MVA pathway for IPP and DMAPP synthesis, the bacteria (including key pathogens such as all Gram-negative bacteria and mycobacteria) use the MEP pathway [34].

The MEP pathway consists of seven enzymatic steps which initiate with the condensation of pyruvate and D-glyceraldehyde 3-phosphate (G3P) to form 1-deoxy-D-xylulose 5-phosphate (DXP) by 1-deoxy-D-xylulose-5-phosphate synthase (DXS). Next, DXP is converted to MEP by 1-deoxy-D-xylulose-5-phosphate reductoisomerase (DXR) and is subsequently converted to methylerythritol 2,4-cyclodiphosphate (MEcDP) through three consecutive enzymatic steps, cytidylation (CTP dependent), phosphorylation (ATP-dependent), and cyclization encoded by *ispD, ispE,* and *ispF*. In the next step, MEcDP is converted to hydroxymethylbutenyl diphosphate (HMBDP) catalyzed by HMBDP synthase (HDS) encoded by the *ispG* gene. In the final step, HMDBP is reduced to IPP and DMAPP by the action of the *ispH* gene product, HMBDP reductase (HDR). IPP and DMAPP are isomerized by isopentenyl diphosphate isomerase (IDI) [35]. The regulatory network of the MEP pathway consists of different regulatory mechanisms involving intermediate metabolites and enzymes. Methylerythritol cyclodiphosphate (MEcDP), the fifth intermediate of this pathway, was recently reported as a key metabolite and may play a critical role in MEP pathway regulation. MEcDP may connect the MEP pathway with other cellular metabolism in making precursors for isoprenoids [35]. An efficient fermentation-based process for isoprenoid production has been achieved in engineered *E. coli* by the heterologous expression of the MVA pathway or upregulation of the native MEP pathway. Among them, the expression of the heterologous MVA pathway is the superior route for enhancing isoprenoid production compared to the MEP pathway. However, the MEP pathway possesses many advantages for isoprenoid production, including higher theoretical yield and better balance than that of the MVA pathway [36]. Specifically, the theoretical maximum IPP yield on glucose via the MEP pathway is 0.83 C-mol IPP/C-mol glucose and is higher than that via the MVA pathway, which is 0.56 C-mol IPP/C-mol glucose IPP [37].

Methanotrophs currently play an important role in a wide range of biotechnological applications with their ability to utilize single carbon (C1) feedstock, i.e., methane and methanol, to produce various high-value compounds. Our genome analysis indicated the existence of a set of genes encoding the methylerythritol 4-phosphate (MEP) pathway in both type I and type II methanotrophs [24] (Table 1). Interestingly, many strains belong to type I methanotrophs such as *Methylomonas* sp. DH-1, *M. alcaliphilum* 20Z, and *M. buryatense* 5GB1, and possess two genes encoding DXS, which catalyzes the first step of the MEP pathway. On the other hand, some strains belonging to type II methanotrophs such as *M. trichosporium* OB3b consisted of two copies of DXR, which converted DXP to MEP in the second step of MEP pathway (Table 1). In contrast to *E. coli* and *Bacillus subtilis*, IDI, which isomerized IPP and DMAPP in the last step of MEP, did not exist in either type I or type II methanotrophs. These results suggest that the regulatory mechanism of the MEP pathway in methanotrophs might be different compared to other microbes, which need to be elucidated in further studies. In agreement with this suggestion, our comparative transcriptome analysis of *Methylomonas* sp. DH-1 showed the extreme upregulation of MEP pathway encoding genes when shifting substrates from methane to methanol [38]. In addition, the pools of pyruvate and G3P, precursors for the MEP pathway, were approximately 150 and 30-fold higher than that of the acetyl-CoA pool, suggesting that the MEP pathway in methanotrophs is a suitable route to produce isoprenoid-related products [39].

Table 1. The existence of genes encoding the methylerythritol 4-phosphate (MEP) pathway in methanotrophs and other microbes.

	dxs1	*dxs2*	*dxr1*	*dxr2*	*ispD*	*ispE*	*ispF*	*ispG*	*ispH*	*idi*
Methylomonas sp. DH-1	x	x	x		x	x	x	x	x	
Methylomicrobium alcaliphilum 20Z	x	x	x		x	x	x	x	x	
Methylomicrobium buryatense 5GB1	x	x	x		x	x	x	x	x	
Methylomonas methanica MC09	x	x	x		x	x	x	x	x	
Methylosinus trichosporium OB3b	x		x	x	x	x	x	x	x	
Escherichia coli K12 MG1655	x		x		x	x	x	x	x	x
Bacillus subtilis	x		x		x	x	x	x	x	x

4.2. C5: Isoprene

Isoprene, a volatile 5-carbon hydrocarbon, is an important platform chemical for the production of polyisoprene (for tires and the rubber industry), elastomers (for sporting goods), and isoprenoids (for adhesives and pharmaceuticals) [40,41]. Isoprene is produced by direct separation from the oil cracking fraction and dehydration of C5 isoalkanes and isoalkenes. However, seven gallons of crude oil are needed to make one gallon of isoprene. In other words, demand is higher than supply, and studies are being conducted to biologically synthesize isoprene to solve this problem. Biologically, isoprene is produced by various organisms such as microorganisms, plants, and animals through the MVA and MEP pathways [42]. Currently, algae and cyanobacteria are used as biomass for biological isoprene production. These have the advantage of being renewable resources, but more than half of the mass is composed of oxygen, which results in low conversion efficiency. This is because the oxygen content of the isoprene is lower than the oxygen content of the cellulosic biomass, so oxygen must be removed as waste [43]. Therefore, it is still not suitable for commercial production.

To solve this problem, studies are being conducted to synthesize isoprene from methane gas through methanotrophs. Methanotrophs have genes encoding the enzymatic components of the MEP pathway. Therefore, if methanotrophs have an isoprene synthase, it can synthesize isoprene from both methane and methanol (Figure 1). Recently, recombinant technology of methanotrophs has been developed, and the isoprene synthase (*ispS*) derived from sweet potato was expressed in *M. alcaliphilum* 20Z to synthesize 50 μg/mL of isoprene from methanol (Table 2) [44].

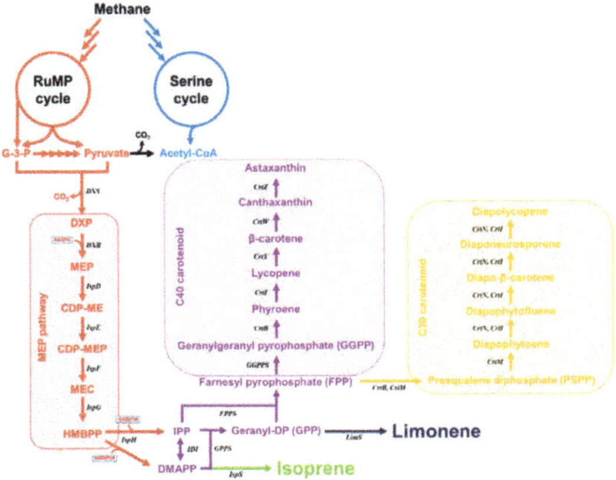

Figure 1. Biosynthetic pathway of isoprenoids in methanotrophs.

Table 2. Production processes and productivity of secondary metabolites by methanotrophs.

Product	Strain	C1 Feedstock	Process	Productivity	Reference
Isoprene	*Methylomicrobium alcaliphilum* 20Z	Methanol	Expression of the isoprene synthase (*ispS*)	50 µg/mL	[44]
Limonene	*Methylomonas* sp. 16a	Methane	Expression of the limonene synthase (*limS*)	0.5 ppm	[45]
Astaxanthin	*Methylomonas* sp. 16a	Methane	Expression of the geranylgeranyl diphosphate synthase (*crtE*), phytoene synthase (*crtB*), phytoene desaturase (*crtI*), lycopene cyclase (*crtY*), beta-carotene ketolase (*crtW*), and beta-carotene hydroxylase (*crtZ*) genes	2.4 mg/gDCW	[46]
Astaxanthin	*Methylomonas* sp. 16a	Methane	Use of transposon promoter-probe vectors for enhancing C40 carotenoid	2 mg/gDCW	[47]
Vitamin B2	*Methylocystis* sp. Strain M	Methane	Exponential phase	0.05 µM	[48]
			Late exponential phase	0.056 µM	
			Stationary phase	0.102 µM	
			Iron deficient condition	0.145 µM	
Ectoine	*Methylomicrobium alcaliphilum* 20Z	Methane	Stirred tank reactors (non-sterile conditions) varies depending on the NaCl concentration and the stirring rates.	16.5–37.4 mg/g cell	[49]
			bio-milking process	70.4 mg/g cell	[50]

4.3. C10: Limonene

Among the components contained in the epidermis of citrus fruits such as mandarins and oranges, limonene is an essential component of plants. Its chemical formula is $C_{10}H_{16}$, and it is naturally produced from monoterpene. (S)-limonene is an (S)-menthol precursor, which is the main component of mint. (R)-limonene is widely used in antimicrobials, deodorants, food, pharmaceuticals, and cosmetics and has recently been used as an alternative detergent. In recent studies, the efficacy of (R)-limonene's anticancer properties has been determined, and research has been actively conducted in the medical field [51]. This limonene is produced by modifying the precursor geranyl pyrophosphate (GPP) by limonene synthase to form aromatic rings. Therefore, methanotrophs with the MEP pathway, which is the pathway for producing GPP, also have the potential to produce limonene (Figure 1). The heterologous expression of *Mentha spicata* limonene synthase in *Methylomonas* sp. 16a resulted in limonene production with a titer of 0.5 ppm (Table 2) [45].

4.4. C30 and C40: Carotenoids

Carotenoids are isoprenoid compounds that are often used as feed or food additives because of their unique color. In addition, since they have a long double bond chain and a ketone group or a hydroxyl group, they have strong antioxidative effects [52,53]. Currently, most carotenoids are produced through chemical synthesis, but there is a growing interest in natural synthesis as well as in the use of medicines and health foods.

Carotenoids are classified into C30 and C40 carotenoids depending on the number of carbon atoms. The most commonly known C40 carotenoids are lycopene, β-carotene, canthaxanthin, and astaxanthin. Many studies have been conducted on ways to biosynthesize these compounds. Studies have also been carried out to biosynthesize C40 carotenoids in methanotrophs (Figure 1). Farnesyl pyrophosphate, which is a precursor of carotenoids, was prepared through the MEP pathway of methanotrophs, and the gene of the subsequent process was expressed. Rick et al. synthesized canthaxanthin, adonixanthin, and astaxanthin by expressing *crtE*, *crtB*, *crtI*, *crtY*, *crtW*, and *crtZ* genes in *Methylomonas* sp. 16a. The content of astaxanthin was increased by controlling the number of copies of *crtW* and *crtZ* genes to synthesize up to 2.4 mg/g DCW (Table 2) [46,47]. Additionally, lycopene was naturally produced in *Methylomonas* sp. ZR1, as confirmed by HPLC-DAD and HPLC-MS/MS analysis methods [54].

C30 carotenoids are less studied than C40 carotenoids. However, it is known that C30 carotenoids such as diapolycopene, diapolycopen-dial, and diaponeurosporene have higher antioxidative effects than tocopherol [55]. Therefore, C30 carotenoids can sufficiently replace C40 carotenoids. C30 carotenoid biosynthesis related genes *crtN*, *ald*, and *crtNb* are contained in *Methylomonas* sp. [24]. It was confirmed that C30 carotenoid was synthesized via mutagenesis analysis. A colorless mutant was isolated through deletion of a promoter upstream of the C30 encoding gene cluster [46]. Methanotrophs can naturally synthesize C30 carotenoids so that high productivity could be achieved by applying the recent studies on metabolic engineering and systems biology of methanotrophs.

4.5. Advances in Engineering Central Carbon Metabolism to Improve Isoprenoid Production in Methanotrophs

As noted above, the MEP pathway serves as a good candidate to enhance isoprenoid production in methanotrophs by metabolic engineering. The MEP pathway uses glyceraldehyde 3-phosphate and pyruvate in the same molar ratio. However, the EMP pathway that is the main route for C1 assimilation in type I methanotrophs with 75% pyruvate produces an imbalanced ratio of G3P and pyruvate, which is unfavorable for the efficiency of the MEP pathway. Engineering the feeding module of these precursors via the Entner-Doudoroff (ED) pathway suggests the possibility to achieve the balancing of precursor pool, which subsequently improves the isoprenoid titer. The isoprene titer was increased 3.1-fold compared to the strain that used the EMP pathway only by overexpressing the ED and the pentose phosphate pathways [56]. Reconstruction of the MVA pathway in type I methanotrophs is another promising strategy for isoprenoid production if the intracellular concentration of acetyl-CoA (which is a critical branch-point metabolite in central metabolism) can be enhanced. An effective approach for metabolic engineering to improve acetyl-coA pools in methanotrophs could increase carbon flux through the phosphoketolase (PKT) pathway (Figure 2). The overexpression of this pathway in type I methanotrophs has been successfully used to increase the yield of acetyl-CoA. High carbon flux through the PKT pathway was achieved by the overexpression of two PKT isoforms in *M. buryatense*; among them, *ptkB* overexpression showed a two-fold increase in intracellular acetyl-CoA [57]. Therefore, the integration of PKT and MVA pathways in type I methanotrophs could improve isoprenoid production in type I methanotrophs. Furthermore, since glycogen and extracellular polymeric substance are the main carbon sink in Group I strains, the elimination of these processes could also increase the pool of secondary metabolites.

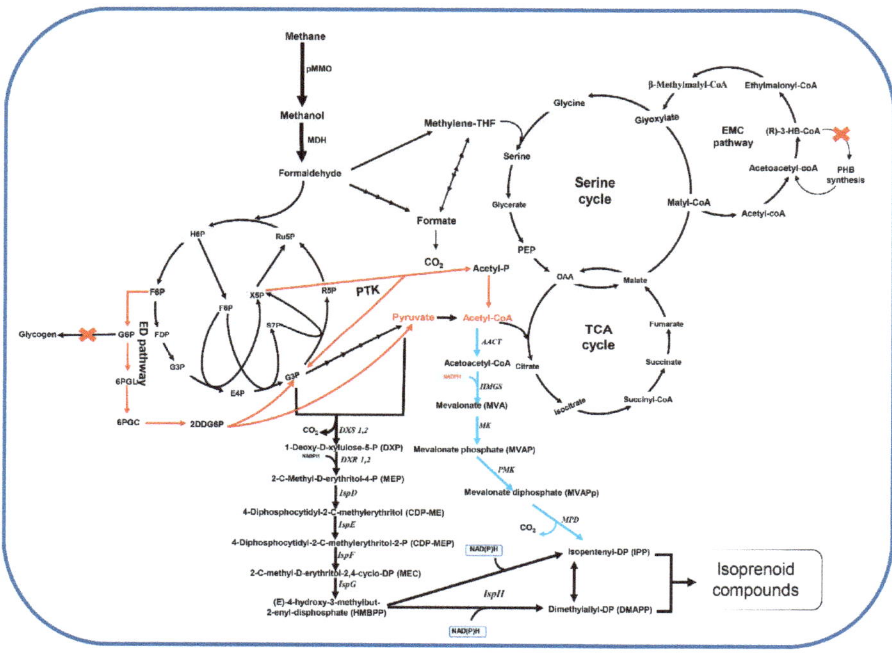

Figure 2. Central metabolism in methanotrophs and proposed engineering strategies for isoprenoid production. Red arrows: predicted overexpression targets to improve carbon conversion efficiency for production of isoprenoid via MEP and mevalonic acid (MVA) pathway; red X: suggested knockout targets for improving flux into isoprenoids; blue arrows: heterologous expression of MVA pathway in methanotrophs.

The serine cycle is the main route for the C1 assimilation pathway in type II methanotrophs, in which formaldehyde was converted to formate and further incorporated into the serine cycle to produce acetyl-CoA. It should be noted that the production of acetyl-CoA from the serine cycle enhances the overall carbon conversion efficiency compared to RuMP-based C1 assimilation without loss of CO_2 [17]. Type II methanotrophs possess higher flux to acetyl-coA production than type I methanotrophs and could be increased by eliminating polyhydroxybutyrate accumulation. Therefore, isoprenoids could also be produced by using both the heterologous expression of the MVA pathway and the endogenous MEP pathway. During initiation of isoprenoid biosynthesis via the MVA pathway, two molecules of acetyl-CoA are condensed to form acetoacetyl-CoA by an acetoacetyl-CoA thiolase. However, the heterologous expression of this enzyme has often been difficult to express in non-native hosts. Interestingly, this enzyme is also already presented in the primary metabolism of type II methanotrophs via the EMC pathway. This feature could be an advantage for engineering isoprenoid biosynthesis in type II methanotrophs (Figure 2). By applying this approach, reconstruction of the MVA pathway in a model methylotroph, *Methylobacterium extorquens* AM1, resulted in humulene concentrations of up to 1.65 g/L from methanol [58].

5. Non-Isoprenoid Secondary Metabolite Production from Methane Using Methanotrophs

5.1. Vitamin B2 (Riboflavin)

Riboflavin, a water-soluble vitamin, is biosynthesized by various microorganisms and most plants. However, it is not biosynthesized in vertebrates, including humans, and should be replenished by external food. It is a precursor to flavin adenine dinucleotide (FAD) and flavin mononucleotide (FMN), the coenzyme necessary for the oxidation-reduction in all organisms and an essential nutrient for humans [59]. Riboflavin is produced in an annual capacity of 9000 tons and can be produced by chemical synthesis and biological synthesis. The chemical synthesis method is accomplished through a complicated process from glucose via ribose, rivamine, etc. Also, the production cost is high since the ribose is used as a precursor. Biological synthesis methods have been developed to replace the chemical method. The biological synthesis method involving the isolation of the microorganisms that naturally produce riboflavin is promising. There is also a method of producing riboflavin that involves improving the microorganism by a genetic engineering method or a chemical-physical method. The microorganisms that produce riboflavin include yeasts belonging to the genus *Saccharomyces* sp. and *Candida* sp., bacteria belonging to the genus *Clostridium*, *Bacillus*, and *Corynebacterium*, and fungi belonging to the genus *Eremothecium* and *Ashbya* [59,60].

Methanotrophs also biosynthesize riboflavin and release it in the form of riboflavin and flavin mononucleotide (FMN) (Figure 3). Both Type I and II methanotrophs can release the flavin to the medium. *Methylocystis* sp. strain M releases the flavin to the medium at 0.05 μM in the exponential phase, 0.056 μM in the late exponential phase, and 0.102 μM in the stationary phase. In the case of iron deficiency, up to 0.145 μM of flavin is released into the medium (Table 2). It is predicted that the flavin mononucleotide (FMN) will chelate the iron and act as an iron shuttle for cell use and respiration [48].

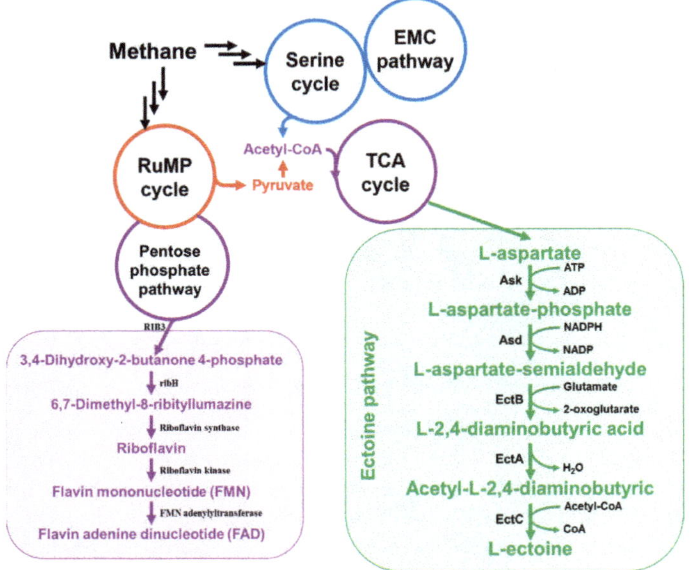

Figure 3. Biosynthetic pathway of riboflavin and ectoine in methanotrophs.

5.2. Ectoine

Ectoine (1,4,5,6-tetrahydro-2-methyl-4-pyrimidine carboxylic acid) is a substance produced by microorganisms to withstand osmotic stress in a high salt environment. It prevents water loss, regulates cell collapse, and does not interfere with cell metabolism. Ectoine can be used for cosmetic and medical purposes. At present, ectoine has a demand of 15,000 tonnes per year, the market price is $1000 USD per kg, and it is commercially produced using *Halomonas elongate* [61].

The ectoine biosynthetic pathways have been investigated in a variety of halophilic bacteria. First, the aspartate commonly used in the synthesis of lysine, threonine and methionine is phosphorylated by aspartokinase (*ask*) to synthesize aspartyl phosphate. It is then converted to aspartate semialdehyde by aspartate semialdehyde dehydrogenase, which is used as a precursor to the ectoine synthesis pathway. Ectoine is synthesized by three enzymes, diaminobutyric acid transaminase (EctB), diaminobutyric acid acetyltransferase (EctA), and ectoine synthase (EctC). The synthesis of ectoine starts with conversion of aspartate semialdehyde to L-2,4-diaminobutyric acid (DABA) by EctB. L-2,4-diaminobutyric acid (DABA) is converted to acetyl-L-2,4-diaminobutyric acid by EctA. Acetyl-L-2,4-diaminobutyric acid is synthesized as ectoine by EctC. The genes coding this synthetic pathway have been studied in several halophilic bacteria and have been named *ectABC*, which is an ectoine synthesis cluster [62].

Studies on the synthesis route of ectoine were also conducted for methanotrophs. *M. alcaliphilum* 20Z, which is known to be a halophilic methanotroph, has an ectoine synthesis pathway. This pathway includes *ask*, L-2,4-diaminobutyric acid transaminase (*ectB*), L-2,4-diaminobutyric acid acetyltransferase (*ectA*) and L-ectoine synthase (*ectC*) (Figure 3), and it exists as one operon. Studies have been conducted to find factors that regulate the ectoine biosynthesis of *M. alcaliphilum* 20Z. As a result, the promoters *ectAp1* and *ectAp2* of the *ectABAask* operon were identified. This is similar to the σ^{70}-dependent promoters of *E. coli*. In addition, a MarR-like transcriptional regulator, EctR1, was identified above the gene cluster, which acts as a negative regulator of the *ectABCask* operon. This confirmed that the purified EctR1 specifically binds to the *ectAp1* promoter [63,64].

Studies on the production of ectoine through *M. alcaliphilum* 20Z are also underway. According to Cantera et al., the production of ectoine by *M. alcaliphilum* 20Z in stirred tank reactors (non-sterile conditions) varies depending on the NaCl concentration and the stirring rates [49]. When the concentration of NaCl in the medium increased from 3% to 6%, the intracellular ectoine content increased to 37.4 mg/g cell from 16.5 mg/g cell (Table 2). A high agitation rate stresses the cells and lowers the yield of ectoine. In addition, when the concentration of copper ion in the medium was increased from 0.05 mM to 25 mM, the amount of ectoine released from the cell increased by about 20%, although it did not affect the production of the intracellular ectoine [49].

A bio-milking process for the commercial production of ectoine was used to increase the production of ectoine from *M. alcaliphilum* 20Z. This is a method of continuously producing ectoine by sequentially in the culture medium. *M. alcaliphilum* 20Z was cultured in 6% NaCl and then alternately cultured in 0% controlling the NaCl concentration NaCl. As a result, it was observed that *M. alcaliphilum* 20Z released the accumulated ectoine at a low osmotic pressure and absorbed the released ectoine at high osmotic pressure. It was confirmed that the ectoine concentration in the cell was maintained at 70.4 mg/g cell in 6% NaCl and that 70% of the ectoine was released in 0% NaCl (Table 2). This shows that it is possible to continuously produce ectoine from methane using methanotrophs [50].

6. Future Aspects for Secondary Metabolite Production from Methane Using Methanotrophs

Methanotrophs might have advantages for secondary metabolite production, especially isoprenoid production, because the tolerance and the storage capacity are major limiting steps for isoprenoid production in other microbes. That is, one of the potentials of methanotrophs is the ability to accumulate higher amounts of isoprenoid on the basis of cell mass because methanotrophs need isoprenoids as a modulator of the oxidative stress or reactive oxygen species generated from oxygen split for C–H bond activation during methane oxidation. In addition, the carbon conversion efficiency (CCE) and yield of the secondary metabolites produced from methane are much higher than those of other substrates, highlighting the potential of secondary metabolite production in methanotrophs. For example, in comparison the theoretical yield for *E. coli* vs. methanotroph–based catalysts for isoprene production, the theoretical yield in methanotrophs is 71% (w/w) and is much higher than that of *E. coli*, which has a theoretical yield of 38% [18].

However, some critical issues need to be overcome to improve the productivity of both secondary metabolites and other target compounds. First, transferring metabolic engineering strategies of *E. coli* and yeast for increased isoprenoid production can be successfully applied to methanotrophs, although productivity enhancement of isoprenoids in methanotrophs is still limited by an incomplete understanding of flux control in the isoprenoid biosynthetic pathway. Isoprenoid production can be increased by optimal gene combination and dosage in methanotrophs. The DXS, DXR, and HDR are expected to act as the major flux control in MEP pathway, thus overexpression of these genes could be possible solution to increase isoprenoids productivity in methanotrophs.

For an efficient production of isoprenoids using *E. coli*, MVA pathway has been reconstructed. In addition, in order to circumvent energetically expensive MVA pathway consuming three ATP for the biosynthesis of an IPP molecule, the archaeal MVA pathway has been reconstructed in *E. coli* to save one mole ATP. However, MVA pathway reconstruction in methanotrophs might be not a good approach because the cellular amount of acetyl-CoA in methanotrophs is very low compared to pyruvate and GAP. Thus, metabolic engineering of the natural MEP pathway might be right option for constructing methanotrophic cell factory for isoprenoid production.

To provide potential metabolic engineering strategies to improve productivity of isoprenoids and other secondary metabolites in methanotrophs, a more deeper understanding on relevant secondary metabolism and biosynthetic regulatory network in methanotrophs is absolutely needed. For example, multi-omics studies together with construction of genome-scale model could be applied to achieve a system-level understanding of cellular behavior for accumulation of isoprenoids and other secondary metabolites, including specific and global gene expression regulation and control of their carbon flux. In particular, a systems biology–based understanding of the interrelationship between methane oxidation process and isoprenoids biosynthesis will be one of the key issues on isoprenoids productivity enhancement.

M. alcaliphilum 20Z strain, one of the platform strains used in methane bioconversion, is capable of growing under high salt conditions of up to 8.7%. Thus, *M. alcaliphilum* 20Z is expected to regulate the synthesis of relevant secondary metabolite, for example, ectoine, with high salt-induced stress. Thus, in-depth understanding of stress-regulatory network can play a key role in the isoprenoid production from methane using methanotrophs. Of course, the first step for this approach will be identification of appropriate stress factor.

The biosynthetic pathways and characteristics of isoprenoids in methanotrophs have not yet fully discovered. This means there might be novel genes, enzymes, and biosynthetic route in methanotrophs that can be used for the generation of new classes of isoprenoids. However, the amounts of secondary metabolites are low—not enough to be isolated efficiently for structure analysis and function characterization. Therefore, the development of precursor-overproducing platform methanotrophic strain—for example, MEP-related intermediates-overproducing strains—can be a promising approach for the production of novel isoprenoids that is too low to be isolated and characterized.

Funding: This research was supported by the C1 Gas Refinery Program through the National Research Foundation of Korea (NRF) funded by the Ministry of Science and ICT (2015M3D3A1A01064882).

Conflicts of Interest: The authors declare no conflict of interest.

References

1. Hwang, I.Y.; Hur, D.H.; Lee, J.H.; Park, C.; Chang, I.S.; Lee, J.W.; Lee, E.Y. Batch conversion of methane to methanol using *Methylosinus trichosporium* OB3b as biocatalyst. *J. Microbiol. Biotechnol* **2015**, *25*, 375–380. [CrossRef] [PubMed]

2. Abbasi, T.; Tauseef, S.; Abbasi, S. Anaerobic digestion for global warming control and energy generation-An overview. *Renew. Sustain. Energy Rev.* **2012**, *16*, 3228–3242. [CrossRef]
3. Kirschke, S.; Bousquet, P.; Ciais, P.; Saunois, M.; Canadell, J.G.; Dlugokencky, E.J.; Bergamaschi, P.; Bergmann, D.; Blake, D.R.; Bruhwiler, L. Three decades of global methane sources and sinks. *Nat. Geosci.* **2013**, *6*, 813. [CrossRef]
4. Nwaoha, C.; Wood, D.A. A review of the utilization and monetization of Nigeria's natural gas resources: Current realities. *J. Nat. Gas Sci. Eng.* **2014**, *18*, 412–432. [CrossRef]
5. Conrado, R.J.; Gonzalez, R. Chemistry. Envisioning the bioconversion of methane to liquid fuels. *Science* **2014**, *343*, 621–623. [CrossRef]
6. Haynes, C.A.; Gonzalez, R. Rethinking biological activation of methane and conversion to liquid fuels. *Nat. Chem. Biol.* **2014**, *10*, 331. [CrossRef]
7. Han, J.; Ahn, C.; Mahanty, B.; Kim, C. Partial oxidative conversion of methane to methanol through selective inhibition of methanol dehydrogenase in methanotrophic consortium from landfill cover soil. *Appl. Biochem. Biotechnol.* **2013**, *171*, 1487–1499. [CrossRef]
8. Hwang, I.Y.; Nguyen, A.D.; Nguyen, T.T.; Nguyen, L.T.; Lee, O.K.; Lee, E.Y. Biological conversion of methane to chemicals and fuels: Technical challenges and issues. *Appl. Microbiol. Biotechnol.* **2018**, *102*, 3071–3080. [CrossRef]
9. Nguyen, A.D.; Hwang, I.Y.; Lee, O.K.; Kim, D.; Kalyuzhnaya, M.G.; Mariyana, R.; Hadiyati, S.; Kim, M.S.; Lee, E.Y. Systematic metabolic engineering of *Methylomicrobium alcaliphilum* 20Z for 2, 3-butanediol production from methane. *Metab. Eng.* **2018**, *47*, 323–333. [CrossRef]
10. Khmelenina, V.N.; Rozova, N.; But, C.Y.; Mustakhimov, I.I.; Reshetnikov, A.S.; Beschastnyi, A.P.; Trotsenko, Y.A. Biosynthesis of secondary metabolites in methanotrophs: biochemical and genetic aspects (review). *Appl. Biochem. Microbiol.* **2015**, *51*, 140–150. [CrossRef]
11. Fei, Q.; Guarnieri, M.T.; Tao, L.; Laurens, L.M.; Dowe, N.; Pienkos, P.T. Bioconversion of natural gas to liquid fuel: opportunities and challenges. *Biotechnol Adv* **2014**, *32*, 596–614. [CrossRef] [PubMed]
12. Wang, C.; Zada, B.; Wei, G.; Kim, S.W. Metabolic engineering and synthetic biology approaches driving isoprenoid production in *Escherichia coli*. *Bioresour. Technol.* **2017**, *241*, 430–438. [CrossRef]
13. Tippmann, S.; Chen, Y.; Siewers, V.; Nielsen, J. From flavors and pharmaceuticals to advanced biofuels: production of isoprenoids in *Saccharomyces cerevisiae*. *Biotechnol. J.* **2013**, *8*, 1435–1444. [CrossRef]
14. Li, Y.; Pfeifer, B.A. Heterologous production of plant-derived isoprenoid products in microbes and the application of metabolic engineering and synthetic biology. *Curr. Opin. Plant Biol.* **2014**, *19*, 8–13. [CrossRef] [PubMed]
15. Murrell, J.C.; McDonald, I.R.; Bourne, D.G. Molecular methods for the study of methanotroph ecology. *FEMS Microbiol. Ecol.* **1998**, *27*, 103–114. [CrossRef]
16. Akberdin, I.R.; Thompson, M.; Kalyuzhnaya, M.G. Systems Biology and Metabolic Modeling of C1-Metabolism. In *Methane Biocatalysis: Paving the Way to Sustainability*; Springer: Cham, Switzerland, 2018; pp. 99–115.
17. Kalyuzhnaya, M.G.; Puri, A.W.; Lidstrom, M.E. Metabolic engineering in methanotrophic bacteria. *Metab. Eng.* **2015**, *29*, 142–152. [CrossRef] [PubMed]
18. Lee, O.K.; Hur, D.H.; Nguyen, D.T.N.; Lee, E.Y. Metabolic engineering of methanotrophs and its application to production of chemicals and biofuels from methane. *Biofuel Bioprod. Biorefin.* **2016**, *10*, 848–863. [CrossRef]
19. Hakemian, A.S.; Rosenzweig, A.C. The biochemistry of methane oxidation. *Annu. Rev. Biochem.* **2007**, *76*, 223–241. [CrossRef]
20. Hanson, R.S.; Hanson, T.E. Methanotrophic bacteria. *Microbiol. Rev.* **1996**, *60*, 439–471.
21. Chistoserdova, L.; Kalyuzhnaya, M.G. Current trends in methylotrophy. *Trends Microbiol.* **2018**, *26*, 703–714. [CrossRef]
22. Dedysh, S.N.; Dunfield, P.F. Facultative and obligate methanotrophs: how to identify and differentiate them. In *Methods in Enzymology*; Elsevier: Amsterdam, The Netherlands, 2011; Volume 495, pp. 31–44.
23. Trotsenko, Y.A.; Murrell, J.C. Metabolic aspects of aerobic obligate methanotrophy? *Adv. Appl. Microbiol.* **2011**, *63*, 183.
24. Nguyen, A.D.; Hwang, I.Y.; Lee, O.K.; Hur, D.H.; Jeon, Y.C.; Hadiyati, S.; Kim, M.; Yoon, S.H.; Jeong, H.; Lee, E.Y. Functional analysis of *Methylomonas* sp. DH-1 genome as a promising biocatalyst for bioconversion of methane to valuable chemicals. *Catalysts* **2018**, *8*, 117.

25. Fu, Y.; Li, Y.; Lidstrom, M. The oxidative TCA cycle operates during methanotrophic growth of the Type I methanotroph *Methylomicrobium buryatense* 5GB1. *Metab. Eng.* **2017**, *42*, 43–51. [CrossRef] [PubMed]
26. Kalyuzhnaya, M.G.; Yang, S.; Rozova, O.N.; Smalley, N.E.; Clubb, J.; Lamb, A.; Gowda, G.N.; Raftery, D.; Fu, Y.; Bringel, F. Highly efficient methane biocatalysis revealed in a methanotrophic bacterium. *Nat. Commun.* **2013**, *4*, 2785. [CrossRef] [PubMed]
27. Nguyen, A.D.; Park, J.Y.; Hwang, I.Y.; Hamilton, R.; Kalyuzhnaya, M.G.; Kim, D.; Lee, E.Y. Genome-scale evaluation of core one-carbon metabolism in gammaproteobacterial methanotrophs grown on methane and methanol. *Metab. Eng.* **2019**. [CrossRef] [PubMed]
28. Kalyuzhanaya, M.G.; Yang, S.; Matsen, J.B.; Konopka, M.; Green-Saxena, A.; Clubb, J.; Sadilek, M.; Orphan, V.J.; Beck, D. Global molecular analyses of methane metabolism in methanotrophic alphaproteobacterium, *Methylosinus trichosporium* OB3b. Part II. Metabolomics and ^{13}C-labeling study. *Front. Microbiol.* **2013**, *4*, 70.
29. Lee, O.K.; Nguyen, D.T.; Lee, E.Y. Metabolic Engineering of Methanotrophs for the Production of Chemicals and Fuels. In *Methanotrophs*; Springer: Cham, Switzerland, 2019; pp. 163–203.
30. Ishikawa, M.; Yokoe, S.; Kato, S.; Hori, K. Efficient counterselection for Methylococcus capsulatus (Bath) by using a mutated pheS gene. *Appl. Environ. Microbiol.* **2018**, *84*, 1875. [CrossRef] [PubMed]
31. Ro, S.Y.; Rosenzweig, A.C. Recent advances in the genetic manipulation of Methylosinus trichosporium OB3b. In *Methods in Enzymology*; Elsevier: Amsterdam, The Netherlands, 2018; Volume 605, pp. 335–349.
32. Tapscott, T.; Guarnieri, M.T.; Henard, C.A. Development of a CRISPR/Cas9 system for Methylococcus capsulatus in vivo gene editing. *Appl Environ Microbiol* **2019**, *85*, 340. [CrossRef]
33. Zhao, L.; Chang, W.; Xiao, Y.; Liu, H.; Liu, P. Methylerythritol phosphate pathway of isoprenoid biosynthesis. *Annu. Rev. Biochem.* **2013**, *82*, 497–530. [CrossRef]
34. Odom, A.R. Five questions about non-mevalonate isoprenoid biosynthesis. *PLoS Pathog.* **2011**, *7*, e1002323. [CrossRef]
35. Banerjee, A.; Sharkey, T. Methylerythritol 4-phosphate (MEP) pathway metabolic regulation. *Nat. Prod. Rep.* **2014**, *31*, 1043–1055. [CrossRef]
36. Ajikumar, P.K.; Xiao, W.H.; Tyo, K.E.; Wang, Y.; Simeon, F.; Leonard, E.; Mucha, O.; Phon, T.H.; Pfeifer, B.; Stephanopoulos, G. Isoprenoid pathway optimization for Taxol precursor overproduction in *Escherichia coli*. *Science* **2010**, *330*, 70–74. [CrossRef]
37. Niu, F.; Lu, Q.; Bu, Y.; Liu, J. Metabolic engineering for the microbial production of isoprenoids: Carotenoids and isoprenoid-based biofuels. *Synth. Syst. Biotechnol.* **2017**, *2*, 167–175. [CrossRef] [PubMed]
38. Nguyen, A.D.; Kim, D.; Lee, E.Y. A comparative transcriptome analysis of the novel obligate methanotroph *Methylomonas* sp. DH-1 reveals key differences in transcriptional responses in C1 and secondary metabolite pathways during growth on methane and methanol. *BMC Genomics* **2019**, *20*, 130.
39. Akberdin, I.R.; Thompson, M.; Hamilton, R.; Desai, N.; Alexander, D.; Henard, C.A.; Guarnieri, M.T.; Kalyuzhnaya, M.G. Methane utilization in *Methylomicrobium alcaliphilum* 20Z R: A systems approach. *Sci. Rep.* **2018**, *8*, 2512. [CrossRef] [PubMed]
40. Kuzuyama, T. Mevalonate and nonmevalonate pathways for the biosynthesis of isoprene units. *Biosci. Biotechnol. Biochem.* **2002**, *66*, 1619–1627. [CrossRef]
41. Clement, N.D.; Routaboul, L.; Grotevendt, A.; Jackstell, R.; Beller, M. Development of palladium–carbene catalysts for telomerization and dimerization of 1, 3-dienes: From basic research to industrial applications. *Chem. Eur. J.* **2008**, *14*, 7408–7420.
42. Julsing, M.K.; Rijpkema, M.; Woerdenbag, H.J.; Quax, W.J.; Kayser, O. Functional analysis of genes involved in the biosynthesis of isoprene in *Bacillus subtilis*. *Appl. Microbiol. Biotechnol.* **2007**, *75*, 1377–1384. [CrossRef] [PubMed]
43. Whited, G.M.; Feher, F.J.; Benko, D.A.; Cervin, M.A.; Chotani, G.K.; McAuliffe, J.C.; LaDuca, R.J.; Ben-Shoshan, E.; Sanford, K.J. Technology update: Development of a gas-phase bioprocess for isoprene-monomer production using metabolic pathway engineering. *Ind. Biotechnol.* **2010**, *6*, 152–163. [CrossRef]
44. Song, J.; Cho, K.K.; Lee, K.S.; La, Y.H.; Kalyuzhnaya, M. Method for Producing Isoprene Using Recombinant Halophilic Methanotroph. U.S. Patent US20170211100A1, 27 July 2017.
45. Dicosimo, D.J.; Koffas, M.; Odom, J.M.; Wang, S. Production of cyclic terpenoids. U.S. Patent US6818424B2, 2004.
46. Rick, W.Y.; Yao, H.; Stead, K.; Wang, T.; Tao, L.; Cheng, Q.; Sharpe, P.L.; Suh, W.; Nagel, E.; Arcilla, D. Construction of the astaxanthin biosynthetic pathway in a methanotrophic bacterium *Methylomonas* sp. strain 16a. *J. Ind. Microbiol. Biotechnol.* **2007**, *34*, 289.

47. Sharpe, P.L.; DiCosimo, D.; Bosak, M.D.; Knoke, K.; Tao, L.; Cheng, Q.; Rick, W.Y. Use of transposon promoter-probe vectors in the metabolic engineering of the obligate methanotroph *Methylomonas* sp. strain 16a for enhanced C40 carotenoid synthesis. *Appl. Environ. Microbiol.* **2007**, *73*, 1721–1728. [CrossRef]
48. Balasubramanian, R.; Levinson, B.T.; Rosenzweig, A.C. Secretion of flavins by three species of methanotrophic bacteria. *Appl. Environ. Microbiol.* **2010**, *76*, 7356–7358. [CrossRef]
49. Cantera, S.; Lebrero, R.; Rodríguez, E.; García-Encina, P.A.; Muñoz, R. Continuous abatement of methane coupled with ectoine production by *Methylomicrobium alcaliphilum* 20Z in stirred tank reactors: A step further towards greenhouse gas biorefineries. *J. Clean. Prod.* **2017**, *152*, 134–141. [CrossRef]
50. Cantera Ruiz de Pellón, S.; Lebrero Fernández, R.; Rodríguez, S.; García Encina, P.A.; Muñoz Torre, R. Ectoine bio-milking in methanotrophs: A step further towards methane-based bio-refineries into high added-value products. *Chem. Eng. J.* **2017**, *328*, 44–48.
51. Sun, J. D-Limonene: Safety and clinical applications. *Altern. Med. Rev.* **2007**, *12*, 259. [PubMed]
52. Noviendri, D.; Hasrini, R.F.; Octavianti, F. Carotenoids: Sources, medicinal properties and their application in food and nutraceutical industry. *J. Med. Plant. Res.* **2011**, *5*, 7119–7131.
53. Umeno, D.; Tobias, A.V.; Arnold, F.H. Diversifying carotenoid biosynthetic pathways by directed evolution. *Microbiol. Mol. Biol. Rev.* **2005**, *69*, 51–78. [CrossRef] [PubMed]
54. Guo, W.; Li, D.; He, R.; Wu, M.; Chen, W.; Gao, F.; Zhang, Z.; Yao, Y.; Yu, L.; Chen, S. Synthesizing value-added products from methane by a new *Methylomonas*. *J. Appl. Microbiol.* **2017**, *123*, 1214–1227. [CrossRef] [PubMed]
55. Kim, S.H.; Kim, M.S.; Lee, B.Y.; Lee, P.C. Generation of structurally novel short carotenoids and study of their biological activity. *Sci. Rep.* **2016**, *6*, 21987. [CrossRef] [PubMed]
56. Liu, H.; Sun, Y.; Ramos, K.R.M.; Nisola, G.M.; Valdehuesa, K.N.G.; Lee, W.; Park, S.J.; Chung, W. Combination of Entner-Doudoroff pathway with MEP increases isoprene production in engineered *Escherichia coli*. *PLoS ONE* **2013**, *8*, e83290. [CrossRef]
57. Henard, C.A.; Smith, H.K.; Guarnieri, M.T. Phosphoketolase overexpression increases biomass and lipid yield from methane in an obligate methanotrophic biocatalyst. *Metab. Eng.* **2017**, *41*, 152–158. [CrossRef]
58. Sonntag, F.; Kroner, C.; Lubuta, P.; Peyraud, R.; Horst, A.; Buchhaupt, M.; Schrader, J. Engineering *Methylobacterium extorquens* for de novo synthesis of the sesquiterpenoid α-humulene from methanol. *Metab. Eng.* **2015**, *32*, 82–94. [CrossRef]
59. Schwechheimer, S.K.; Park, E.Y.; Revuelta, J.L.; Becker, J.; Wittmann, C. Biotechnology of riboflavin. *Appl. Microbiol. Biotechnol.* **2016**, *100*, 2107–2119. [CrossRef]
60. Kato, T.; Park, E.Y. Riboflavin production by *Ashbya gossypii*. *Biotechnol. Lett.* **2012**, *34*, 611–618. [CrossRef]
61. Strong, P.; Kalyuzhnaya, M.; Silverman, J.; Clarke, W. A methanotroph-based biorefinery: Potential scenarios for generating multiple products from a single fermentation. *Bioresour. Technol.* **2016**, *215*, 314–323. [CrossRef] [PubMed]
62. Kuhlmann, A.U.; Bremer, E. Osmotically regulated synthesis of the compatible solute ectoine in *Bacillus pasteurii* and related *Bacillus* spp. *Appl. Environ. Microbiol.* **2002**, *68*, 772–783. [CrossRef] [PubMed]
63. Reshetnikov, A.S.; Khmelenina, V.N.; Trotsenko, Y.A. Characterization of the ectoine biosynthesis genes of haloalkalotolerant obligate methanotroph "*Methylomicrobium alcaliphilum* 20Z". *Arch. Microbiol.* **2006**, *184*, 286–297. [CrossRef] [PubMed]
64. Mustakhimov, I.I.; Reshetnikov, A.S.; Glukhov, A.S.; Khmelenina, V.N.; Kalyuzhnaya, M.G.; Trotsenko, Y.A. Identification and characterization of EctR1, a new transcriptional regulator of the ectoine biosynthesis genes in the halotolerant methanotroph *Methylomicrobium alcaliphilum* 20Z. *J. Bacteriol.* **2010**, *192*, 410–417. [CrossRef]

Review

Microbial Phosphotriesterase: Structure, Function, and Biotechnological Applications

Wahhida Latip [1,*], Victor Feizal Knight [1], Norhana Abdul Halim [2], Keat Khim Ong [2], Noor Azilah Mohd Kassim [2], Wan Md Zin Wan Yunus [3], Siti Aminah Mohd Noor [2,*] and Mohd Shukuri Mohamad Ali [4]

1. Center for Chemical Defence, National Defence University of Malaysia, Kem Perdana Sungai Besi, Kuala Lumpur 57000, Malaysia
2. Center for Defence Foundation Studies, National Defence University of Malaysia, Kem Perdana Sungai Besi, Kuala Lumpur 57000, Malaysia
3. Center for Tropicalisation, National Defence University of Malaysia, Kem Perdana Sungai Besi, Kuala Lumpur 57000, Malaysia
4. Enzyme and Microbial Technology Research Center, Department of Biochemistry, Faculty of Biotechnology and Biomolecular sciences, Universiti Putra Malaysia, Serdang 43400, Malaysia

* Correspondence: wahhidalatip@gmail.com (W.L.); s.aminah@upnm.edu.my (S.A.M.N.); Tel.: +60-390513400 (ext. 4455) (W.L. & S.A.M.N.)

Received: 24 June 2019; Accepted: 18 July 2019; Published: 7 August 2019

Abstract: The role of phosphotriesterase as an enzyme which is able to hydrolyze organophosphate compounds cannot be disputed. Contamination by organophosphate (OP) compounds in the environment is alarming, and even more worrying is the toxicity of this compound, which affects the nervous system. Thus, it is important to find a safer way to detoxify, detect and recuperate from the toxicity effects of this compound. Phosphotriesterases (PTEs) are mostly isolated from soil bacteria and are classified as metalloenzymes or metal-dependent enzymes that contain bimetals at the active site. There are three separate pockets to accommodate the substrate into the active site of each PTE. This enzyme generally shows a high catalytic activity towards phosphotriesters. These microbial enzymes are robust and easy to manipulate. Currently, PTEs are widely studied for the detection, detoxification, and enzyme therapies for OP compound poisoning incidents. The discovery and understanding of PTEs would pave ways for greener approaches in biotechnological applications and to solve environmental issues relating to OP contamination.

Keywords: phosphotriestease; structure; function or reaction; biotechnological application

1. Introduction

Organophosphate (OP) compounds—otherwise known as phosphotriesters—are a class of highly neurotoxic synthetic compounds that are widely used as agricultural insecticides, petroleum additives, plasticizers, refrigerants, dyes, and even as chemical warfare agents [1]. Contamination by these compounds may occur because of terrorist attacks, unsafe disposal methods, and spillage from poor handling by agricultural workers [2]. OP compounds are known to inhibit mammalian acetylcholinesterases (AChE) [3–5]. Acetylcholinesterase (AChE) is an enzyme that hydrolyzes the neurotransmitter acetylcholine (ACh) and serves as a regulator of neurotransmission [3]. The most terrifying fact of its action is that the inhibition process is naturally irreversible, whereby the enzyme can no longer hydrolyze acetylcholine resulting in paralysis of muscles, seizures, difficulty in breathing and death by suffocation. In 2009, the World Health Organization reported that over 3 million people were poisoned by OP pesticides, and 200,000 had died and furthermore these numbers are increasing yearly. Moreover, OP compounds have been reported to be involved in gene mutations, DNA damage, cancerogenesis, and endocrine disorders.

OP compounds can be briefly divided into three major groups, which are the phosphotriesters, thiophosphotriesters, and phosphorothiolesters [6]. Phosphotriesters such as paraoxon contain a phosphate in the center with an o-linked group Figure 1A. Similarly, in thiophosphotriesters (methyl parathion) and phosphorothiolesters (malathion), the phosphate group is still in the center but the phosphoryl oxygen is replaced by sulfur and one or more of the ester oxygens are replaced by sulfur Figure 1B,C. To date these compounds are still widely used especially in agricultural industries [7]. Moreover, OPs compounds are easily drained into ground water that in turn will pollute water supplies and surrounding land areas [8]. Due to its high toxicity, numerous efforts to detect the presence of OP compounds and to detoxify their effects on humans and the environment have been made.

Many alternative technologies have been studied and developed to resolve this problem, using various chemical, physical, and biological methods. Unfortunately, most of these methods are expensive and may not suitable or even safe for the decontamination of human beings [7]. They may be very harsh and may not be very specific [3]. Currently, using enzymes is one of the more promising and safer ways to solve this problem. Organophosphate hydrolysis enzymes (OPHs) are enzymes that can hydrolyze the OP compounds to become nontoxic agents [9]. These enzymes have been reported to be found in bacteria as well as in cephalopods (squids) and mammals [10,11]. The OP compounds have been reported as being used as food by several bacteria species where they would cleave the phosphorester bond, and thereby reduce their toxicity [12,13]. A few OPHs have been isolated recently and are classified according to their protein homology. Phosphotriesterases (PTEs) are a group of these OPHs and are classified under the aryldialkylphosphatase super family [14]. This enzyme has been extensively studied in molecular biology and with much emphasis on their applications.

In this review, the 3D structure, mechanism of hydrolysis, and substrate specificity of PTEs are described. The biotechnological application of this enzyme in bioremediation, biosensor development and human therapy is also discussed.

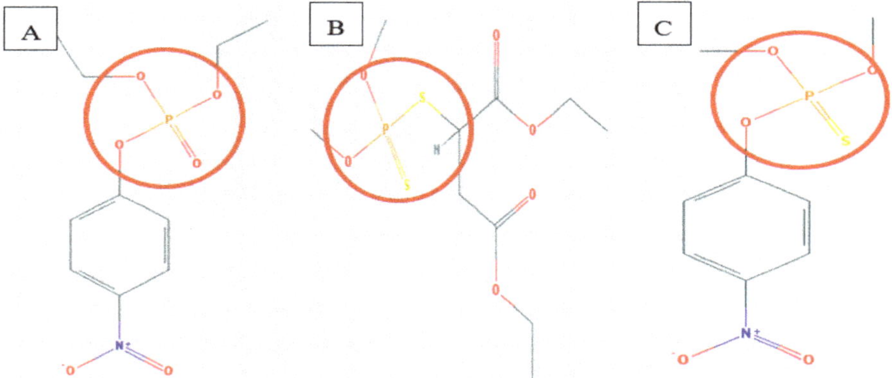

Figure 1. Three major groups of organophosphate compound. (**A**) Phosphotriester (paraoxon), (**B**) thiophosphotriester (methyl parathion), and (**C**) phosphorothiolester (Malathion). The red circles embody the differences between the OP-S compounds. The structures were adopted from Pubchem.

2. Structure and Function

2.1. Bacterial Phosphotriesterase 3D Structures

A number of bacteria that are capable of degrading OP pesticides have been identified from soil samples from different parts of the world [15]. These include *Pseudomonas pseudoalcaligenes*, *Pseudomonas diminuta*, *Agrobacterium radiobacter*, *Alteromonas haloplanktis* and *Saccharomyces cerevisiae*, *Pseudomonas monteilii*, and *Geobacillus stearothermophilus* [13,16,17]. *Flavobacterium* sp. ATCC27551 is the first OP degrading bacteria and was isolated in the Philippines in 1973 [1]. The gene encoding the phosphotriesterase (PTE) enzyme was isolated and expressed in several expression hosts including *E. coli*, yeasts, insect cells and in in vitro compartmentalization [13,14,18,19]. Some of the genes showed a conserved sequence homology between two species such as with *Pseudomonas diminuta* and *Flavobacterium* sp. ATCC27551 even though the plasmid containing the PTE genes were discovered from among different plasmids [20]. *Pseudomonas diminuta*, *Flavobacterium* sp., and *Alteromonas* sp. possess the best characterized PTE enzymes isolated from bacteria [19,21].

Phosphotriesterase (PTE) is a membrane-associated protein that is translated as a 29 amino acid-long target peptide [22]. PTE is classified as a metalloenzyme or metal-dependent enzyme which means that without the metal, the enzyme will not function [23]. The natural substrate for PTE is still as yet unknown; however, the ability of PTE to hydrolyze the insecticide paraoxon at a rate that approaches the diffusion-controlled limit, is remarkable [9,19,24]. Although PTE can hydrolyze paraoxon efficiently, the PTE hydrolysis rates for malathion, phosalone, and acephate are approximately 1000-fold slower compared to that of paraoxon [13]. Thus, many efforts have been done to improve the hydrolysis rate using molecular approaches such as mutation at the active site pocket, varying the ribosomal operon and expression host. The enzyme has been reported to have three binding pockets to accommodate the substrate [6,25,26]. The active site pocket of the PTE in a hydrophobic condition contains two metal ions (Zn^{2+} Mn^{2+}, Co^{2+}, and Cd^{2+}) bridged with side-chains of histidine residues that are clustered around the metal atoms [22,27]. A number of the crystal structures of PTE isolated from bacteria such as *Pseudomonas pseudoalcaligenes* (2.1 Å), *Pseudomonas diminuta* (2.0 Å), *Mycobacterium tuberculosis* (2.27 Å), *Agrobacterium radiobacter* (1.99 Å), *Geobacillus stearothermophilus* (2.09 Å), and *Geobacillus kaustophilus* HTA426 (2.05 Å) have been determined [4,7,16,28,29]. In this review, three crystal structures of PTEs from *Pseudomonas diminuta*, *Sphingobium fuliginis* strain ATCC 27551, and *Geobacillus stearothermophilus* (PLL) are used as examples in this section to shown the similarities and differences in their 3D structure Figure 2. All crystal structures reveal that there is some degree of similarity in their structures, such as each containing an α/β-barrel motif Figure 2. All of these PTEs are metal-dependent hydrolases and each naturally exists as a homodimer [5]. Although some researchers reported that PTEs had varying protein folds ((β/α) 8 or TIM-barrel, β-lactamase folds, and pita bread folds), all of these crystal structures are actually TIM-barrel folds [2,6,19]. In addition, all PTEs require divalent metals to directly ligate the substrate to simplify the hydrolysis process Figure 2(B1–B3) [5,6]. In this case, the metal ion in *Pseudomonas diminuta* and *Sphingobium fuliginis* strain ATCC 27551 PTE have Zn^{2+}, as the ion, whereas for the *Geobacillus stearothermophilus* PTE it is Co^{2+} [7,30]. In the absence of a substrate, the α zinc metal was reported as buried near to His55, His 57, and Asp 301, and the β zinc metal was more exposed to the solvent [6,7,30]. The divalent metal ions in the active site center are reported to contribute to the activity and stability of the enzyme. The zinc is reported to have the greatest effect on stability while, Co in the center of the enzyme, increases the enzyme activity up to 300% [21]. Moreover, the carboxylated lysine and four histidine residues, which coordinate the two divalent cations, are required in the active center for activity [28,31]. The metals ions in turn coordinate an activated hydroxyl group that functions as the nucleophile in the OP hydrolysis reaction [32].

The 3D structure shows that the PTE enzymes have three detached binding pockets (small, large and leaving groups) that allow the substrate (paraoxon) to attach to the phosphorus center Figure 2B [2,6,32]. Due to their high similarity (of up to 90%) in structure and protein homology [10,12].

PTE from *Pseudomonas* sp. and *Sphingobium fuliginis* have similar active site pockets. The small binding pockets for *Pseudomonas diminuta* and *Sphingobium fuliginis* are Gly-60, Leu-303/Thr-303, Ser-308, and Ile-106 [11,12,28]. Similarly, with the leaving group pocket, all hydrophobic residues were conserved (Phe-306, Phe-132, Trp131, and Tyr-309) Figure 2(B1,B2) [16,23]. Unlike the large pocket, His-254 and 257 are substituted by Gly-254 and Trp-257. Other residues namely Leu-271 and Met-317 are also well conserved [33]. The active site pocket for *Geobacillus stearothermophilus* was found to be different from that of *Pseudomonas* sp. and *Sphingobium fuliginis* where the small binding pocket contained F28, Y30, T70, C74, V268, W271, the large pocket (R230, I233, M236, V237, and W289), and the leaving group pocket (Y100 and E103), see Figure 2(B3) (17) [30]. Moreover, the metal ion at the center of *Geobacillus stearothermophilus* PTE is different from that of *Pseudomonas* sp. and *Sphingobium fuliginis*, which is Co^{2+} [30]. The two metal ions are thus designated as α and β. The α metal is more buried and is in direct coordination with His55, His57, and Asp301 (*Pseudomonas* sp. and *Sphingobium fuliginis*) [17]. The β-metal is more solvent-exposed and coordinated towards His201 and His230 [7,16].

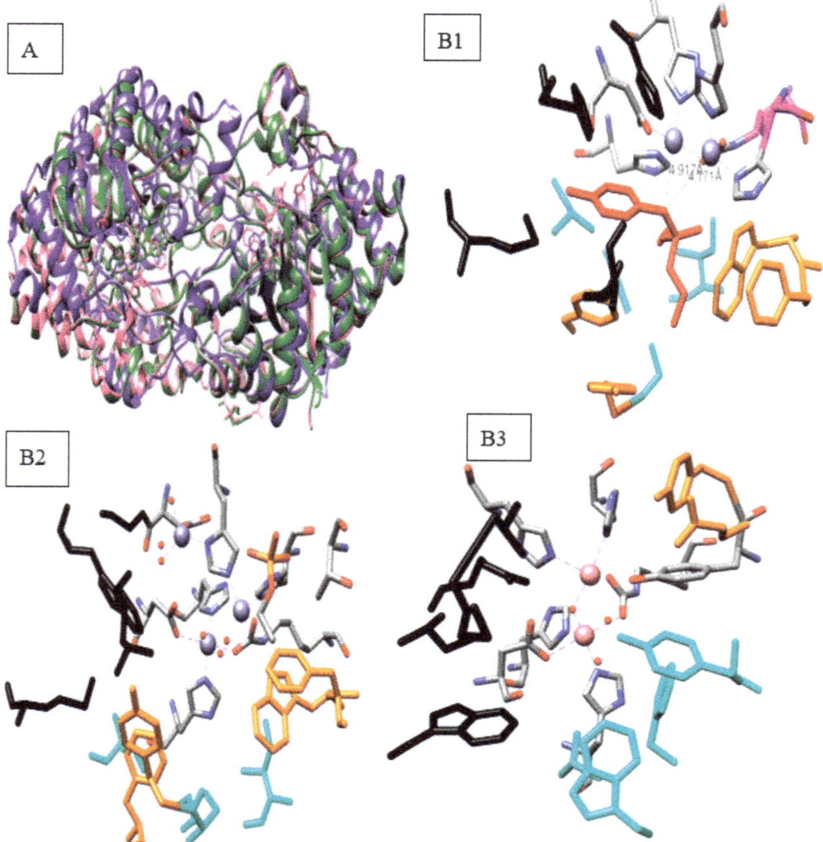

Figure 2. Superposition of 3D structure and substrate binding sites of phosphotriesterase. (**A**) Colored pink: *Pseudomonas dimunita* PTE; green: *Sphingobium fuliginis PTE*; blue: *Geobacillus stearothermophilus PTE*. (**B1**) *Pseudomonas dimunita* PTE; (**B2**) *Sphingobium fuliginis PTE*; (**B3**) *Geobacillus stearothermophilus PTE*. Residue colors: Gray: hexahedron residue for Zn2+; Cyan: Small binding pocket; Orange: Large pocket; Black: Leaving group pocket. All coordinates were taken from PDB 1DPM, 1P6B, and 3F4D. The cartoon diagram was generated using Chimera software.

2.2. Mechanism of Hydrolysis and Stereoselectivity of PTE

Hydrolysis of OP by PTE depends on three factors: the metal ion in the enzyme, the pH of the environment, and the substrate itself [22,34]. In detail, the structure of the PTE shows that the two metal ions coordinated in the enzyme interact with a carboxylated lysine (Lys169) and a hydroxide ion or water molecule [17,24,35]. The nucleophilic attack starts with these three components which are water bridged between two divalent cations, the bimetal ion, and substrate in the active PTE site [2,6]. The 3D crystal structure shows the presence of this hydroxide or water bridge between the metal [6,19]. The nucleophilic attack occurs on the phosphorus center (P–O bond) followed by a proton transfer [6,11,17]. Histidine 254 and Aspartic acid 233 extend the hydrogen bond network in the binuclear metal center and Asp301 helps to shuttle protons from the active site to the solvent [6,17]. The polarization of the P–O bond of the substrate (pH and substrate factors) also contributes to the nucleophilic attack [17,21,24]. The OH− then attacks the phosphorus center of the substrate, followed by a proton transfer event to Asp301 Figure 3 [17,19]. The P–O bond is thus broken, and the products are released from the active site. The proton is shuttled away from the active site with the assistance of His254 and Asp233 [6,17]. This mechanism is called the SN2 mechanism where one bond is broken and one bond is formed synchronously [5,36]. Moreover, the metal ion was found to depend on the pH where protonation occurred from the bridging hydroxyl group [11,17]. The catalytic activity is lost at a low pH because of the protonation [17]. Furthermore, polarizing the P–O bond of the substrate, makes it more susceptible for a nucleophilic attack to occur [6,21,24].

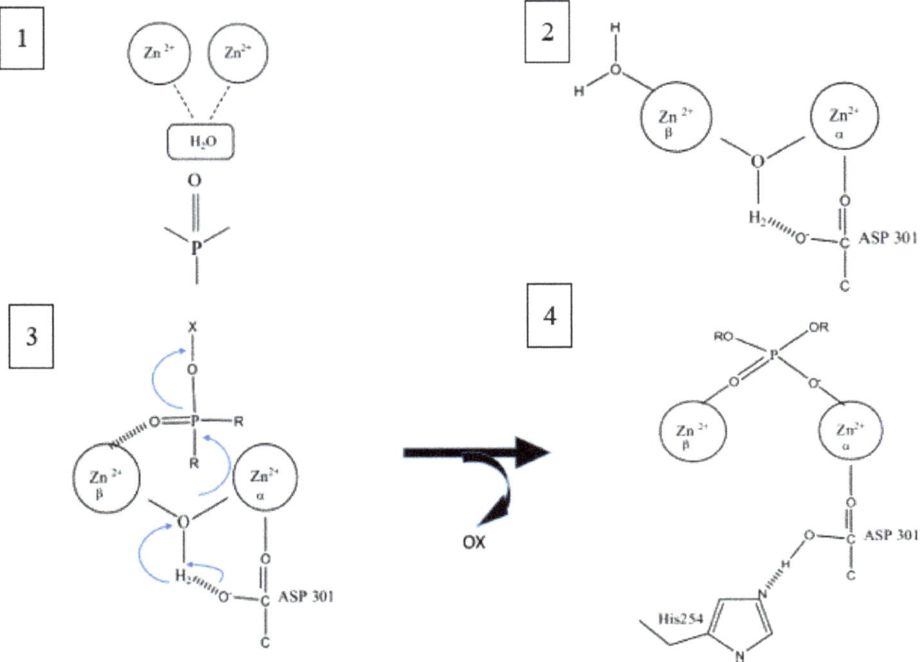

Figure 3. Proposed mechanism of PTE hydrolysis. 1: Three components for nucleophilic attack; 2: Active site of PTE without the substrate; 3: Mechanism hydrolysis of OP compound; 4: Proton transfer by Asp301.

The three distinct active site pockets (small, large, and leaving group) play an important role in the stereoselectivity of PTE. Wild type PTE prefers paraoxon as a substrate and can hydrolyze a variety of insecticides including phosphotriesters, thiophosphotriesters, and phosphorothioesters [2,37].

Each pocket plays its own role in stereoselectivity [19,25,37]. The leaving group pocket has been reported to play an important role in dictating the enzyme specificity [6,25]. The binding site pockets which are the small pockets and large pockets play a role in the determination of the specificity for the side ester groups on the substrate [6,27]. The PTE of the wild type is known to prefer S_P-enantiomers for most phosphate substrates [9,25]. The small pocket of the PTE has been reported as too small for malathion to bind and a mutation has been done to enlarge the pocket [21,33]. Reducing the size of the small pocket has been done by mutation at G60A and was found to increase the selectivity for the R_P-enantiomer compound from 9.4×10^{-1} M^{-1} s^{-1} to 2.1×10^4 M^{-1} s^{-1} by factor of 2×10^4 [2,6]. Similarly with the large pocket of the PTE, mutation has been done by switching the residue H254Q/H257F, the catalytic constant for S_p-enantiomer enhanced 2 times from the wild type [9,11]. Moreover, mutation at the substrate binding pocket (H257Y/L303T) also enhances the k_{cat}/K_m of YT mutant towards G-type nerve gases as compared to the wild type from 2.4×10^5 to 2×10^6; 1.5×10^4 to 5×10^5; and 2.3×10^5 to 8×10^5 M^{-1} s^{-1} for sarin (GB), soman (GD), and cyclosarin (GF), respectively [9,38]. The manipulation of large and small pockets of PTE allows the researcher to enhance, diminish, or reverse the hydrolysis of the substrate [6,7,15]. These approaches have enhanced the hydrolysis of the most toxic enantiomers of GB and GD by over two orders of magnitude and approximately 25-fold for the nerve agent VX [11,39]. In addition, mutation near the small pocket of PTE or Loop 7 using error prone PCR has also given reliable results where the mutant L7ep-3a enhanced a 600-fold (2.6×10^3 M^{-1} s^{-1}) faster action in the detoxification of VR and VX nerve agents [40]. The substitution of the metal ion has also been done to identify the influence of the metal ion towards selectivity and stability of the PTE [4,22] Zinc is the native metal ion for PTE but can be substituted with other metals such as Co^{2+}, Ni^{2+}, Cd^{2+}, or Mn^{2+} to retain the catalytic activity [17]. The value of k_{cat}/K_m has been shown to decrease when the Zn^{2+}/Zn^{2+} (6.8×10^6 M^{-1} s^{-1}) was substituted with Cd^{2+}/Cd^{2+} (1.32×10^6 M^{-1} s^{-1}) and Zn^{2+}/Cd^{2+} (2.8×10^6 M^{-1} s^{-1}) with paraoxon as the substrate, but not for parathion. The PTE with Zn^{2+}/Cd^{2+} showed an increment value of up to 5.0×10^5 M^{-1} s^{-1} for this substrate [17] Therefore, by modifying the active sites and their surrounding areas, the applicability of these enzyme can be expanded for industrial purpose.

2.3. *Industrial Application of Bacterial Phosphotriesterase*

2.3.1. Biosensors

The specificity of PTE and its mutant to hydrolyze a broad range of OP compounds even at low-level concentrations of these OPs is well documented [19,33]. This enzyme characteristic has been exploited to develop detectors to identify the presence of OP compounds in the environment. Even though many methods have been proposed for this purpose, the techniques are time-consuming and may not be easily portable for use. Moreover, the techniques require trained technicians to perform the analysis competently [41]. Thus, portable and specific enzymatic biosensors (especially for use in field OP monitoring), with high selectivity and sensitivity, would be very beneficial [42,43].

In enzymatic biosensing of OP compounds, the detection is based on either their inhibition of AChE or the hydrolysis process by the OP hydrolysis enzyme [42,43]. Detection using an inhibition-based sensor has been reported to have a high sensitivity and is capable of ultralow OP detection [42,44]. However, enzymes can easily be inhibited by many other factors which include heavy metals, temperature, pH, and detergents that normally found in the environment and may be subject to a slow detection rate (owing to prolonged incubation and regeneration periods) [41,45,46]. Therefore, direct measurement of the hydrolysis reaction is preferable compared to just measurement of inhibition. Furthermore, the detection limit of sensors based on the PTE reaction is better when compared with that of AChE activity detection [31,43].

PTE isolated from *Brevudimonas dimunita* and *Flavobacterium* sp. have been studied and used as potentiometric, amperometric and optical biosensors [45,47,48]. The PTE biosensor is reported to have high reusability (0.3% per time sensed), achieving up to 90% stability in 1000 s and 70%

longevity and selectivity (with the detection of six similar nerve agents) [41]. The development of this biosensor is based on the mechanism of hydrolysis of PTE in the OP compound [19,49,50]. Products from the hydrolysis process will change the pH and electrical charge, proton transfer to Asp 301 will occur [11,17,42]. Immobilized PTE has also been used as a bioreceptor to detect OP compounds. Immobilization supports such as graphene, chitosan, sulphate chitosan, bovine serum, and glutaraldehyde have been used to stabilize and enhance the enzyme properties [41,50,51]. Currently, a novel microelectronic device based on a CMOS-compatible ISFET chip and PTE enzyme, combined with a microfluidic system has been developed by Pozio and his coworkers to detect OP compounds [43,49].

2.3.2. Bioremediation and Detoxification

Currently, PTE and other OPH enzymes have emerged as one of the biocatalyst candidates for bioremediation purposes and the detoxification of agricultural pesticides as well as chemical warfare agents [3,34,39]. Since the OPs are not toxic to bacteria and some species of bacteria are also able to utilize OPs as a nutrient and carbon source, OPH bacteria like Pseudomonas, Aspergillus, Arthrobacter, and Chlorella have been used as organophosphate-degrading microorganisms [15,37,52,53]. The use of microbes offers effectiveness, low-cost, and an environmentally friendly method of removing these toxic compounds from the environment [2]. Similarly, PTE isolated from Pseudomonas, Altromonas, Geobacillus, and yeast have been also used as bioremediation to hydrolyze OP compounds [13,53,54].

PTE has also been considered as an important enzyme in the development of an enzyme-based cocktail for the decontamination of chemical warfare agents and pesticides in a certain country [55,56]. In 2004, an OPH enzyme was patented by Genencor International (a company manufacturing industrial enzymes and the Edgewood Chemical and Biological Center) for scale up and commercial purposes [8]. Detoxification of environments exposed to OP compounds is also a matter of crucial concern. There is a need to develop an enzyme that meets the requirements for a decontamination system that is easy to store and to transport. On top of that, the Defense Threat Reduction Agency (DTRA) has also their own guidelines for an enzyme candidate that they want to use as a decontamination agent [11]. The enzyme must be environmentally friendly, have activity/stability over broad pH and temperature ranges, and be stable in the presence of harsh solvents, metals, detergents, and/or denaturants (DTRA 2008) [8,56]. Contamination can occur in the air, water and soil [3,13,15]. As OP compounds can easily seep or dissolve into water, Istamboulie and his coworkers have tried to detoxify water using a PTE enzyme as the detoxification agent. The pilot batch was tested in a column where water mixes with the OP compound. The water was successfully detoxified by the PTE using an enzyme concentration of 500 IU [57]. These findings show that the role of this enzyme is very important in developing an ability to clean up the environment and has a bright future in industry.

2.3.3. Enzyme Therapy

Recently, OP compound intoxication cases have been seen to gradually increase in occurrence and represents a major public health problem [3]. An urgent need for the development of efficient and safe detoxification treatments for OP poisoning is a must in the management of cases involving the accidental use or misuse of these compounds [44]. A lot of effort has been made towards this goal. OPs are inhibitors of acetylcholinesterase (AChE) and this enzyme is a neurotransmitter in the peripheral nervous system [3,31]. Late treatment will result in the AChE aging and will ultimately result in death [10,44]. Symptoms such as the acute effects of respiratory failure, central and peripheral nervous damage, loss of muscle control and cardiac arrest are consequences of exposure to OP compounds [31,35,44]. Current treatment of over exposure to OP compound is by the use of oximes, atropine and benzodiazepines which are not necessarily successful [54].

PTE is also used in medical applications as an antidote or a therapeutic in preventing OP poisoning [2,56]. PTE has emerged as one of leading candidates for the development of prophylactic and postexposure treatments against OP compounds [2,8]. These enzymes break down the OP compounds to become nontoxic products before they enter the blood circulation and produce their toxic effects at the neuromuscular junctions [2,35]. Moreover, some tests have been done to look at the efficacy of PTE as a prophylactic in various animal models [2,10,40]. PTE was tested in mice which were pretreated with PTE. The treated mice were able to resist even higher doses of nerve agents [3,10,19].

PTE can also be considered to be a bioscavenger in OP poisoning treatments. This has been demonstrated from PTEs which were isolated from *Pseudomonas* sp., *Plesiomonas* sp., as well as the dimethoate-degrading enzyme from *Aspergillus niger*, and chlorpyrifos-degrading enzyme from an Enterobacter strain which showed an improvement in animal survivability in vitro [10,53,56,58]. The enzymes from *Pseudomonas diminuta*, *Flavobacterium* sp., and *Alteromonas* sp. can be considered the best characterized bacterial PTEs [10]. The current issue at hand is how to deliver the enzyme into our body and at the same time protect the enzyme from our own immune system. A number of experiments have been done using erythrocytes and liposomes as carriers for PTE inside mice [3,49]. However, the enzyme remained active for only 45 h at the most [8,14]. Similar results using encapsulation techniques showed the enzyme activity could last up to 45 h, whereafter the enzyme began to decrease in activity once beyond its half-life [1,35]. To prolong the enzyme half-life and to retain enzyme in an active state inside the immune system, a safe and efficient nanoparticle delivery system is needed and will be the focal point of research in the years to come.

3. Conclusions

In this review, it is clear that bacterium PTE possess an amazing property in which they are able to hydrolyze OP compounds efficiently. Microbial PTEs are easy to manipulate and can be overexpressed compared to the PTE of other organisms. These enzymes have wide hydrolysis activity levels with varying OP compounds which occurs at the hydrophobic active site. The 3D structures of PTEs show variations in protein folding and contain bimetal at the active site. Biotechnological applications using microbial PTEs possess huge potential impact on detoxification, detection and therapy activities against OP compound contamination. Large-scale commercialization of the production of these enzymes will make them easier to obtain, and therefore their application in the detection, treatment and removal of OP contamination which therefore can be carried out effectively. The use of PTEs as biosensors, or in detoxification and biomedical applications, shows that they have a wide spectrum of application. However, further work is crucial to engineer robust enzymes that meet the various requirements of the industry and address the characteristics of OP contamination events. A molecular strategy can be one of the tools used to enhance the stability and stereo selectivity of these enzymes, in addition to increasing their catalytic activity. It is expected that through these approaches, the development of PTE enzymes as one of the leading biocatalysts for OP compound contamination management will be achieved in the near future.

Author Contributions: W.L. wrote and improved the draft. M.S.M.A., V.F.K., S.A.M.N., K.K.O., N.A.M.K., N.A.H., and W.M.Z.W.Y. reviewed the draft of the article.

Funding: This research was funded by Ministry of Education Malaysia, Development Fund F0020.

Acknowledgments: Ministry of Education Malaysia.

Conflicts of Interest: The authors declare no conflict of interest.

References

1. Singh, B.K. Organophosphorus-degrading bacteria: Ecology and industrial applications. *Nat. Rev. Microbiol.* **2009**, *7*, 156–164. [CrossRef] [PubMed]
2. Iyengar, A.R.S.; Pande, A.H. Organophosphate-Hydrolyzing Enzymes as First-Line of Defence Against Nerve Agent-Poisoning: Perspectives and the Road Ahead. *Protein J.* **2016**, *35*, 424–439. [CrossRef] [PubMed]
3. Alejo-González, K.; Hanson-Viana, E.; Vazquez-Duhalt, R. Enzymatic detoxification of organophosphorus pesticides and related toxicants. *J. Pestic. Sci.* **2018**, *43*, 1–9. [CrossRef] [PubMed]

4. De Castro, A.A.; Caetano, M.S.; Silva, T.C.; Mancini, D.T.; Rocha, E.P.; da Cunha, E.F.F.; Ramalho, T.C. Molecular Docking, Metal Substitution and Hydrolysis Reaction of Chiral Substrates of Phosphotriesterase. *Comb. Chem. High Throughput Screen.* **2016**, *19*, 334–344. [CrossRef] [PubMed]
5. Benning, M.M.; Kuo, J.M.; Raushel, F.M.; Holden, H.M. Three-Dimensional Structure of Phosphotriesterase: An Enzyme Capable of Detoxifying Organophosphate Nerve Agents. *Biochemistry* **1994**, *33*, 15001–15007. [CrossRef] [PubMed]
6. Andrew, N.B.; Raushel, F.M. Catalytic Mechanism for phosphotriesterases. *Biochim. Biophys. Acta* **2014**, *1834*, 443–453.
7. Manco, G.; Porzio, E.; Suzumoto, Y. Enzymatic detoxification: A sustainable means of degrading toxic organophosphate pesticides and chemical warfare nerve agents. *J. Chem. Technol. Biotechnol.* **2018**, *93*, 2064–2082. [CrossRef]
8. Theriot, C.M.; Grunden, A.M. Hydrolysis of organophosphorus compounds by microbial enzymes. *Appl. Microbiol. Biotechnol.* **2011**, *89*, 35–43. [CrossRef]
9. Tsai, P.C.; Fox, N.; Bigley, A.N.; Harvey, S.P.; Barondeau, D.P.; Raushel, F.M. Enzymes for the homeland defense: Optimizing phosphotriesterase for the hydrolysis of organophosphate nerve agents. *Biochemistry* **2012**, *51*, 6463–6475. [CrossRef]
10. Bird, S.B.; Dawson, A.; Ollis, D. University of Massachusetts Medical Center, Worcester, Massachusetts, USA, 2 South Asian Clinical Toxicology Research Collaboration (SACTRC), Faculty of Medicine, University of Peradeniya, Peradeniya, Sri Lanka and School of Population and Health, Univers. *Front. Biosci.* **2010**, *S2*, 209–220. [CrossRef]
11. Bigley, A.N.; Mabanglo, M.F.; Harvey, S.P.; Raushel, F.M. Variants of Phosphotriesterase for the Enhanced Detoxification of the Chemical Warfare Agent VR. *Biochemistry* **2015**, *54*, 5502–5512. [CrossRef] [PubMed]
12. Chu, X.Y.; Wu, N.F.; Deng, M.J.; Tian, J.; Yao, B.; Fan, Y.L. Expression of organophosphorus hydrolase OPHC2 in Pichia pastoris: Purification and characterization. *Protein Expr. Purif.* **2006**, *49*, 9–14. [CrossRef] [PubMed]
13. Makkar, R.S. Enzyme-Mediated Bioremediation of Organophosphates Using Stable Yeast Biocatalysts. *J. Bioremed. Biodeg.* **2013**, *4*, 2. [CrossRef]
14. Merone, L.; Mandrich, L.; Rossi, M.; Manco, G. Enzymes with Phosphotriesterase and Lactonase Activities in Archaea Enzymes with Phosphotriesterase and Lactonase Activities in Archaea. *Curr. Chem. Biol.* **2008**, *2*, 237–248.
15. Kumar, S.; Kaushik, G.; Dar, M.A.; Nimesh, S.; LÓpez-chuken, U.J.; Villarreal-chiu, J.F. Microbial Degradation of Organophosphate Pesticides: A Review. *Pedosphere* **2018**, *28*, 190–208. [CrossRef]
16. Gotthard, G.; Hiblot, J.; Gonzalez, D.; Elias, M.; Chabriere, E. Structural and enzymatic characterization of the phosphotriesterase OPHC2 from Pseudomonas pseudoalcaligenes. *PLoS ONE* **2013**, *8*, e77995. [CrossRef]
17. Aubert, S.D.; Li, Y.; Raushel, F.M. Mechanism for the Hydrolysis of Organophosphates by the Bacterial Phosphotriesterase. *Biochemistry* **2004**, *43*, 5707–5715. [CrossRef]
18. Wang, J.B.; Chen, D.C.; Hu, D.X.; Su, X.F.; Tang, X.M. Cloning, Expression, Purification, and Characterization of an Organophosphate-Degrading Enzyme in Escherichia Coli. *Adv. Mater. Res.* **2012**, *610*, 203–209. [CrossRef]
19. Ghanem, E.; Raushel, F.M. Detoxification of organophosphate nerve agents by bacterial phosphotriesterase. *Toxicol. Appl. Pharmacol.* **2005**, *207*, 459–470. [CrossRef]
20. Dave, K.I.; Miller, C.E.; Wild, J.R. Characterization of organophosphorus hydrolases and the genetic manipulation of the phosphotriesterase from Pseudomonas diminuta. *Chem. Biol. Interact.* **1993**, *87*, 55–68. [CrossRef]
21. Del Giudice, I.; Coppolecchia, R.; Merone, L.; Porzio, E.; Carusone, T.M.; Mandrich, L.; Worek, F.; Manco, G. An efficient thermostable organophosphate hydrolase and its application in pesticide decontamination. *Biotechnol. Bioeng.* **2016**, *113*, 724–734. [CrossRef]
22. Carletti, E.; Jacquamet, L.; Loiodice, M.; Rochu, D.; Masson, P.; Nachon, F. Update on biochemical properties of recombinant Pseudomonas diminuta phosphotriesterase. *J. Enzyme Inhib. Med. Chem.* **2009**, *24*, 1045–1055. [CrossRef]
23. Roodveldt, C.; Tawfik, D.S. Directed evolution of phosphotriesterase from Pseudomonas diminuta for heterologous expression in Escherichia coli results in stabilization of the metal-free state. *Protein Eng. Des. Sel.* **2005**, *18*, 51–58. [CrossRef]

24. Buchbinder, J.; Stephenson, R.; Dresser, M.; Pitera, J.; Scanlan, T.; Fletterick, R. Biochemical characterization and crystallographic structure of an Escherichia coli protein from the phosphotriesterase gene family. *Biochemistry* **1998**, *37*, 5096–5106. [CrossRef]
25. Chen-Goodspeed, M.; Sogorb, M.A.; Wu, F.; Raushel, F.M. Enhancement, Relaxation, and Reversal of the Stereoselectivity for Phosphotriesterase by Rational Evolution of Active Site Residues† Misty. *Biochemistry* **2004**, *40*, 1332–1339. [CrossRef] [PubMed]
26. Manuscript, A.; Niswender, C.M.; Conn, P.J.; Sommer, M.A.; Wurtz, R.H. NIH Public Access. *Changes* **2010**, *38*, 319–335.
27. Shim, H.; Hong, S.B.; Raushel, F.M. Hydrolysis of phosphodiesters through transformation of the bacterial phosphotriesterase. *J. Biol. Chem.* **1998**, *273*, 17445–17450. [CrossRef] [PubMed]
28. Pedroso, M.M.; Ely, F.; Mitić, N.; Carpenter, M.C.; Gahan, L.R.; Wilcox, D.E.; Larrabee, J.L.; Ollis, D.L.; Schenk, G. Comparative investigation of the reaction mechanisms of the organophosphate-degrading phosphotriesterases from Agrobacterium radiobacter (OpdA) and Pseudomonas diminuta (OPH). *J. Biol. Inorg. Chem.* **2014**, *19*, 1263–1275. [CrossRef]
29. Jacquet, P.; Hiblot, J.; Daudé, D.; Bergonzi, C.; Gotthard, G.; Armstrong, N.; Chabrière, E.; Elias, M. Rational engineering of a native hyperthermostable lactonase into a broad spectrum phosphotriesterase. *Sci. Rep.* **2017**, *7*, 16745. [CrossRef]
30. Hawwa, R.; Aikens, J.; Turner, R.J.; Santarsiero, B.D.; Andrew, D.; Brook, O. Structural basis for thermostability revealed through the identification and characterization of a highly thermostable phosphotriesterase-like lactonase from Geobacillus stearothermophilus. *Arch. Biochem. Biophys.* **2017**, *488*, 109–120. [CrossRef]
31. Kern, R.J. Enzyme-Based Detoxification of Organophosphorus Neurotoxic Pesticides and Chemical Warfare Agents. Ph. D. Thesis, Texas A&M University, College Station, TX, USA, 2007.
32. Bigley, A.N.; Xiang, D.F.; Ren, Z.; Xue, H.; Hull, K.G.; Romo, D.; Raushel, F.M. Chemical Mechanism of the Phosphotriesterase from Sphingobium sp. Strain TCM1, an Enzyme Capable of Hydrolyzing Organophosphate Flame Retardants. *J. Am. Chem. Soc.* **2016**, *138*, 2921–2924. [CrossRef]
33. Jeong, Y.S.; Choi, J.M.; Kyeong, H.H.; Choi, J.Y.; Kim, E.J.; Kim, H.S. Rational design of organophosphorus hydrolase with high catalytic efficiency for detoxifying a V-type nerve agent. *Biochem. Biophys. Res. Commun.* **2014**, *449*, 263–267. [CrossRef]
34. Munnecke, D.M. Enzymatic hydrolysis of organophosphate insecticides, a possible pesticide disposal method. *Appl. Environ. Microbiol.* **1976**, *32*, 7–13.
35. Nachon, F.; Brazzolotto, X.; Trovaslet, M.; Masson, P. Progress in the development of enzyme-based nerve agent bioscavengers. *Chem. Biol. Interact.* **2013**, *206*, 536–544. [CrossRef]
36. Wong, K.Y.; Gao, J. The reaction mechanism of paraoxon hydrolysis by phosphotriesterase from combined QM/MM simulations. *Biochemistry* **2007**, *46*, 13352–13369. [CrossRef]
37. Horne, I.; Sutherland, T.D.; Oakeshott, J.G.; Russell, R.J. Cloning and expression of the phosphotriesterase gene hocA from Pseudomonas monteilii C11. *Microbiology* **2002**, *148*, 2687–2695. [CrossRef]
38. Tsai, P.-C.; Fan, Y.; Kim, J.; Yang, L.; Almo, S.C.; Gao, Y.Q.; Raushel, F.M. Structural Determinants for the Stereoselective Hydrolysis of Chiral Substrates by Phosphotriesterase. *Biochemistry* **2010**, *49*, 7988–7997. [CrossRef]
39. Reeves, T.E.; Wales, M.E.; Grimsley, J.K.; Li, P.; Cerasoli, D.M.; Wild, J.R. Balancing the stability and the catalytic specificities of OP hydrolases with enhanced V-agent activities. *Protein Eng. Des. Sel.* **2008**, *21*, 405–412. [CrossRef]
40. Bigley, A.N.; Xu, C.; Henderson, T.J.; Harvey, S.P.; Raushel, F.M. Enzymatic neutralization of the chemical warfare agent VX: Evolution of phosphotriesterase for phosphorothiolate hydrolysis. *J. Am. Chem. Soc.* **2013**, *135*, 10426–10432. [CrossRef]
41. Hondred, J.A.; Breger, J.C.; Alves, N.J.; Trammell, S.A.; Walper, S.A.; Medintz, I.L.; Claussen, J.C. Printed Graphene Electrochemical Biosensors Fabricated by Inkjet Maskless Lithography for Rapid and Sensitive Detection of Organophosphates. *ACS Appl. Mater. Interfaces* **2018**, *10*, 11125–11134. [CrossRef]
42. Anh, D.H.; Cheunrungsikul, K.; Wichitwechkarn, J.; Surareungchai, W. A colorimetric assay for determination of methyl parathion using recombinant methyl parathion hydrolase. *Biotechnol. J.* **2011**, *6*, 565–571. [CrossRef]
43. Andrianova, M.S.; Gubanova, O.V.; Komarova, N.V.; Kuznetsov, E.V.; Kuznetsov, A.E. Development of a Biosensor Based on Phosphotriesterase and n-Channel ISFET for Detection of Pesticides. *Electroanalysis* **2016**, *28*, 1311–1321. [CrossRef]

44. Nurulain, S.M. Different approaches to acute organophosphorus poison treatment. *J. Pak. Med. Assoc.* **2012**, *62*, 712–717.
45. Wang, J.; Krause, R.; Block, K.; Musameh, M.; Mulchandani, A.; Mulchandani, P.; Chen, W.; Schöning, M.J. Dual amperometric-potentiometric biosensor detection system for monitoring organophosphorus neurotoxins. *Anal. Chim. Acta* **2002**, *469*, 197–203. [CrossRef]
46. Latip, W.; Raja Abd Rahman, R.N.Z.; Leow, A.T.C.; Mohd Shariff, F.; Kamarudin, N.H.A.; Mohamad Ali, M.S. The effect of N-terminal domain removal towards the biochemical and structural features of a thermotolerant lipase from an antarctic Pseudomonas sp. Strain AMS3. *Int. J. Mol. Sci.* **2018**, *19*, 560. [CrossRef]
47. Wang, J.; Krause, R.; Block, K.; Musameh, M.; Mulchandani, A.; Schöning, M.J. Flow injection amperometric detection of OP nerve agents based on an organophosphorus-hydrolase biosensor detector. *Biosens. Bioelectron.* **2002**, *18*, 255–260. [CrossRef]
48. Lei, Y.; Mulchandani, P.; Wang, J.; Chen, W.; Mulchandani, A. Highly sensitive and selective amperometric microbial biosensor for direct determination of p-nitrophenyl-substituted organophosphate nerve agents. *Environ. Sci. Technol.* **2005**, *39*, 8853–8857. [CrossRef]
49. Porzio, E.; Bettazzi, F.; Mandrich, L.; Del Giudice, I.; Restaino, O.F.; Laschi, S.; Febbraio, F.; De Luca, V.; Borzacchiello, M.G.; Carusone, T.M.; et al. Innovative Biocatalysts as Tools to Detect and Inactivate Nerve Agents. *Sci. Rep.* **2018**, *8*, 13773. [CrossRef]
50. Efremenko, E.; Peregudov, A.; Kildeeva, N.; Perminov, P.; Varfolomeyev, S. New enzymatic immobilized biocatalysts for detoxification of organophosphorus compounds. *Biocatal. Biotransformation* **2005**, *23*, 103–108. [CrossRef]
51. Mulchandani, A.; Mulchandani, P.; Chen, W. Enzyme Biosensor for Determination of Organophosphates. *Field Anal. Chem. Technol. Chem. Technol.* **1998**, *2*, 363–369. [CrossRef]
52. Rosenberg, A.; Alexander, M. Microbial cleavage of various organophosphorus insecticides. *Appl. Environ. Microbiol.* **1979**, *37*, 886–891.
53. Singh, B.K.; Walker, A.; Morgan, J.A.W.; Wright, D.J. Biodegradation of chlorpyrifos by Enterobacter strain B-14 and its use in bioremediation of contaminated soils. *Appl. Environ. Microbiol.* **2004**, *70*, 4855–4863. [CrossRef]
54. De Castro, A.A.; Prandi, I.G.; Kuca, K.; Ramalho, T.C. Organophosphorus degrading enzymes: Molecular basis and perspectives for enzymatic bioremediation of agrochemicals. *Cienc. Agrotecnologia* **2017**, *41*, 471–482.
55. Vyas, N.K.; Nickitenko, A.; Rastogi, V.K.; Shah, S.S.; Quiocho, F.A. Structural insights into the dual activities of the nerve agent degrading organophosphate anhydrolase/prolidase. *Biochemistry* **2010**, *49*, 547–559. [CrossRef]
56. Wille, T.; Neumaier, K.; Koller, M.; Ehinger, C.; Aggarwal, N.; Ashani, Y.; Goldsmith, M.; Sussman, J.L.; Tawfik, D.S.; Thiermann, H.; et al. Single treatment of VX poisoned guinea pigs with the phosphotriesterase mutant C23AL: Intraosseous versus intravenous injection. *Toxicol. Lett.* **2016**, *258*, 198–206. [CrossRef]
57. Istamboulie, G.; Durbiano, R.; Fournier, D.; Marty, J.L.; Noguer, T. Biosensor-controlled degradation of chlorpyrifos and chlorfenvinfos using a phosphotriesterase-based detoxification column. *Chemosphere* **2010**, *78*, 1–6. [CrossRef]
58. Liu, Y.H.; Chung, Y.C.; Xiong, Y. Purification and Characterization of a Dimethoate-Degrading Enzyme of Aspergillus niger ZHY256, Isolated from Sewage. *Appl. Environ. Microbiol.* **2001**, *67*, 3746–3749. [CrossRef]

MDPI
St. Alban-Anlage 66
4052 Basel
Switzerland
Tel. +41 61 683 77 34
Fax +41 61 302 89 18
www.mdpi.com

Catalysts Editorial Office
E-mail: catalysts@mdpi.com
www.mdpi.com/journal/catalysts

www.ingramcontent.com/pod-product-compliance
Lightning Source LLC
LaVergne TN
LVHW070106170726
843515LV00007B/1619

* 9 7 8 3 0 3 6 5 6 2 6 5 0 *